Principles of Metallurgy

Principles of Metallurgy

Edited by **Ricky Peyret**

CWILLFORD PRESS

New York

Published by Willford Press,
118-35 Queens Blvd., Suite 400,
Forest Hills, NY 11375, USA
www.willfordpress.com

Principles of Metallurgy
Edited by Ricky Peyret

International Standard Book Number: 978-1-68285-120-3 (Hardback)

Printed in the United States of America.

Contents

Ultrafine-Grained Austenitic Stainless Steels X4CrNi18-12 and X8CrMnNi19-6-3 Produced by Accumulative Roll Bonding
Mathis Ruppert, Lisa Patricia Freund, Thomas Wenzl, Heinz Werner Höppel and Mathias Göken

Permissions

List of Contributors

Preface

Metallurgy has emerged as a rapidly growing discipline for production of metals and their respective alloys. It focuses on analyzing the physical and chemical properties of metals and their applications for combining metallic compounds. Metallurgy is widely practiced across industries for extraction of metallic compounds from their ores and manufacturing of alloys. This book compiles researches by renowned academicians and experts from around the world that discuss both theoretical developments as well as latest technological advancements in metal and alloy manufacturing. The significant topics included in this book are physico-chemical properties of metals, processing of ores, structure and strength of metallic compounds and alloys, etc. It would prove to be immensely beneficial for students, researchers and engineers involved in this field.

After months of intensive research and writing, this book is the end result of all who devoted their time and efforts in the initiation and progress of this book. It will surely be a source of reference in enhancing the required knowledge of the new developments in the area. During the course of developing this book, certain measures such as accuracy, authenticity and research focused analytical studies were given preference in order to produce a comprehensive book in the area of study.

This book would not have been possible without the efforts of the authors and the publisher. I extend my sincere thanks to them. Secondly, I express my gratitude to my family and well-wishers. And most importantly, I thank my students for constantly expressing their willingness and curiosity in enhancing their knowledge in the field, which encourages me to take up further research projects for the advancement of the area.

Editor

Titanium Matrix Composite Ti/TiN Produced by Diode Laser Gas Nitriding

Aleksander Lisiecki

Welding Department, Silesian University of Technology, Konarskiego 18A, 44-100 Gliwice, Poland;
E-Mail: aleksander.lisiecki@polsl.pl

Academic Editor: M. T. Whittaker

Abstract: A high power direct diode laser, emitting in the range of near infrared radiation at wavelength 808–940 nm, was applied to produce a titanium matrix composite on a surface layer of titanium alloy Ti6Al4V by laser surface gas nitriding. The nitrided surface layers were produced as single stringer beads at different heat inputs, different scanning speeds, and different powers of laser beam. The influence of laser nitriding parameters on the quality, shape, and morphology of the surface layers was investigated. It was found that the nitrided surface layers consist of titanium nitride precipitations mainly in the form of dendrites embedded in the titanium alloy matrix. The titanium nitrides are produced as a result of the reaction between molten Ti and gaseous nitrogen. Solidification and subsequent growth of the TiN dendrites takes place to a large extent at the interface of the molten Ti and the nitrogen gas atmosphere. The direction of TiN dendrites growth is perpendicular to the surface of molten Ti. The roughness of the surface layers depends strongly on the heat input of laser nitriding and can be precisely controlled. In spite of high microhardness up to 2400 HV0.2, the surface layers are crack free.

Keywords: titanium; titanium nitrides; composite; laser gas nitriding; laser surface treatment

1. Introduction

Titanium alloys are attractive structural materials, compared to other light alloys and modern high strength steels as investigated by A. Grajcar and A. Kurc [1–3]. However, due to poor tribological properties of titanium and titanium alloys, their service life may not be satisfactory under certain conditions, as reported by A. Biswas and P. Schaaf [4–8]. The low surface hardness and wear resistance limit the application of titanium alloys under conditions of friction and contact loads [9–11].

Tribological properties of titanium alloys may be improved or modified by coatings on the surface, changing the microstructure of surface layers, or thermochemical treatments such as nitriding, carburizing, oxidizing or nitro-oxidizing [12–14]. Nitriding of surface layers of titanium alloys is one of the most widely used thermochemical treatments in industry [15,16]. The methods of surface nitriding of titanium alloys such as plasma, physical vapor deposition (PVD), chemical vapor deposition (CVD), salt-bath nitriding and ion implantation have been developed and applied for many years [17–19].

One of the most promising and effective methods for improving the surface characteristics and properties of titanium alloys is laser surface treatment in a nitrogen atmosphere or rich in nitrogen-laser gas nitriding (LGN) [20,21].

This is the result of the large chemical affinity of titanium for nitrogen at elevated temperatures, particularly in a liquid state. This feature leads to the *in situ* synthesis of metal matrix composite (MMC) in the surface layer with a high content of titanium nitrides in the titanium metallic matrix (precisely Titanium Matrix Composite-TMC). Composite surface layers of titanium alloys modified in such a way have a significantly higher wear resistance compared to conventional titanium alloys, as well as to other wear-resistant surface layers reinforced by hard carbides or ceramic nano particles in metallic matrix produced by laser alloying or cladding, as those investigated by D. Janicki [22,23].

Despite the fact that the first trials of laser nitriding of titanium were carried out by Katayama [1] in the eighties of the last century, the process is still being developed. Initially trials of titanium LGN were carried out only by gaseous CO_2, mainly continuous wave (CW) lasers [1,24]. The fundamental problem during laser nitriding of titanium in the liquid state reported by most researchers was high surface roughness and cracking of the nitrided titanium, as well as short fatigue life [1,7,24]. Other researchers have noted that both the type and also the mode of laser beam affect significantly the nitriding process and have a strong influence as well on the characteristics and properties of the nitrided surface. L. Xue and M. Islam [24] found that in terms of surface finish and cracking severity, a solid state Nd:YAG (neodymium-doped yttrium aluminum garnet; $Nd:Y_3Al_5O_{12}$) laser in pulse mode (emitted at 1.06 μm) was better than a CO_2 laser in continuous wave mode (emitted at 10.6 μm).

Edson Costa Santos *et al.* [11] showed the difference between laser gas nitriding by Nd:YAG laser in CW, Q-sw and pulsed modes. Naofumi Ohtsu *et al.* [20] investigated the surface hardness of titanium modified by laser irradiation at different wavelengths in a nitrogen atmosphere. They found that the effect of irradiation is clearly different for different wavelengths, and reported that craters formed by the 532-nm laser were deeper than those formed by the 1064-nm laser for the same remaining parameters.

E. Carpene *et al.* [25] investigated the laser nitriding of different metals by various lasers with different wavelengths in a range from 0.3–3.0 μm. They indicated that for some metals the wavelength strongly effects the irradiation and efficiency of the nitriding process. A.L. Thomann *et al.* [15] investigated the surface nitriding of titanium and aluminum by laser-induced plasma. They used a pulsed CO_2 laser (emitting at 1.06 μm) and pulsed excimer XeCl (emitting at 308 nm), and found that the synthesized layers exhibit differences that depend on the type of laser used.

Due to the dynamic development of laser devices and laser technologies, new types and new generations of lasers are currently available [21,26]. One of the new generations of lasers introduced for material processing is the high power diode laser (HPDL) characterized by a rectangular laser beam spot and a uniform energy distribution across the beam spot [5,6,16,22]. The number of publications on titanium nitriding by HPDL lasers is still very limited but every author points to the high potential and benefits of this type of laser in surface treatment, especially in titanium gas nitriding [4–6,16,19,20].

Results of investigations on the nitriding of titanium alloys by HPDL lasers show the advantages of the HPDL lasers over the Nd:YAG and CO_2 lasers with a circular laser beam. M. Gołębiewski et al. [19] investigated different processes of titanium nitride production and revealed that the lowest residual stresses were measured for the surfaced nitrided by HPDL with a rectangular beam spot. The residual stresses in the layers produced by HPDL are compressive in nature and about ten times lower (−300 to −800 MPa) compared to the stresses in layers produced by pulsed laser deposition (−500 to −8000 MPa). A. Biswas et al. [6] investigated the laser gas nitriding of Ti6Al4V alloy applying a high power diode laser. They also found that the residual stresses are mainly compressive in nature and strongly depend on process parameters; moreover the lowest compressive residual stresses started from 50 MPa.

To summarize, the advantages of the specific type of diode lasers beam for surface processing of metals are as follows:

- high absorption in the very near infrared band of laser radiation, especially compared to the gaseous CO_2 lasers (even several times higher) makes the diode laser beam a very efficient heat source,
- uniform energy distribution across the laser beam spot (Top-hat, multimode beam profile, shown in Figures 1a and 2b),
- rectangular shape of the laser beam spot,
- uniform heating of the metal surface across the processing track as a result of the above two advantages.

The rectangular shape of a laser beam spot is very profitable in the case of surface treatment, when the heating intensity, temperature range, and a thermal cycle on the treated surface are crucial parameters. When the rectangular laser beam is moved linearly on the treated surface and the laser beam is emitted in CW mode, the irradiation time (i.e., time of beam interaction on the metal surface τ) is constant across the processing track, unlike in the case of the circular laser beam of commonly used gaseous CO_2 or solid state YAG lasers, as shown in Figures 1 and 2b. The time of beam interaction on the treated metal surface "τ" is defined as the ratio of the beam width (or length) "w" to the scanning speed "v":

$$\tau = w/v \text{ (s)} \tag{1}$$

where:

w, is laser beam width (or length, mm) in the case of a rectangular or square beam and a diameter in a case of a circular beam (just for calculation of the maximum interaction time τ_{max}, as illustrated in Figure 1a),

v, is scanning speed (mm/s).

1) $y_0 = 0$ $w(y_0) = 2r$ $\tau(w) = \tau_{max}$
   $\tau(w) = \dfrac{w(y)}{v}$
2) $y = +/- r$ $w(r) = 0$ $\tau(r) = 0$

(a)

$w = const$ $\tau = const$ $I(y) = const$

(b)

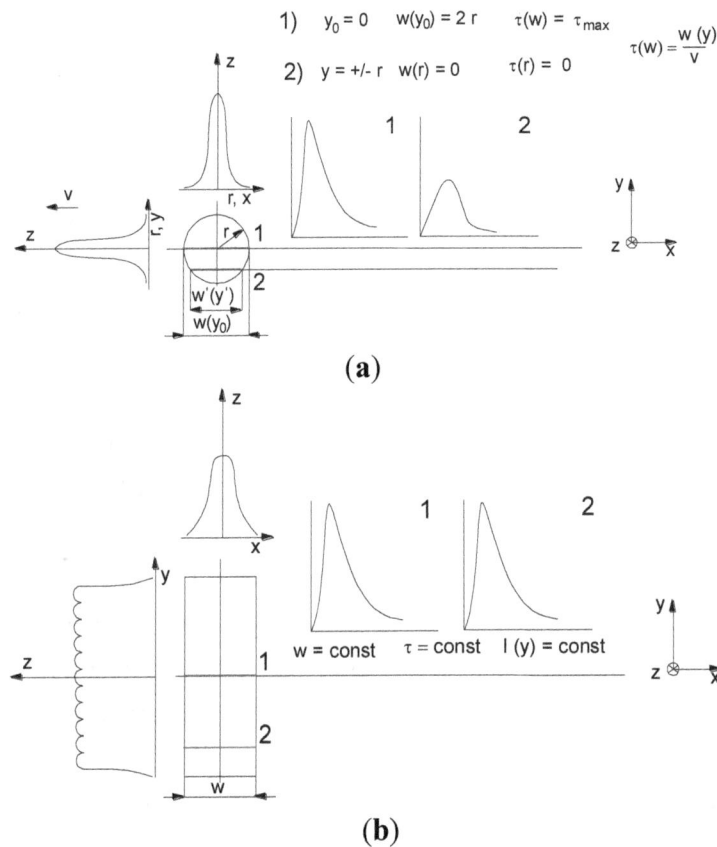

Figure 1. The differences in surface heating by (**a**) the circular laser beam with a Gaussian single mode of energy distribution (TEM00) and (**b**) the rectangular laser beam with uniform energy distribution (Top-hat, multimode profile, shown also in Figure 2).

2. Experimental Section

The experiments of laser gas nitriding were carried out by means of a prototype, numerically controlled stand, equipped with the continuous wave high power direct diode laser HPDDL Rofin DL020 with unique characteristics, emitting a laser radiation at a dominant wavelength 808 nm and with a rectangular beam spot. Additionally, the rectangular laser beam spot with a width of 1.8 mm and a length of 6.8 mm is characterized by multimode, uniform intensity of laser radiation across the beam spot, as show in Figure 2b and given in Table 1.

The specimens of titanium alloy Ti6Al4V were cut into coupons 50.0 × 100.0 mm from a hot-rolled sheet with a thickness of 3.0 mm. The structure of the base metal (BM) of titanium alloy Ti6Al4V is shown in Figure 3. Surfaces of the specimens were ground by 180-grade SiC paper to remove oxides and other contaminations from the surface. Prior to nitriding, the surfaces were degreased by acetone. The specimens of titanium alloy were placed in a gas chamber filled up with gaseous nitrogen of 99.999% purity (Figure 2a).

For precise control of the gas atmosphere in the gas chamber, an electronic system of gas delivery was used and the nitrogen flow rate was kept at 10.0 L/min at the pressure 1.0 atm. Flow of the nitrogen through the gas chamber was switched on 90 s prior to the laser nitriding process in order to totally remove the air from the gas chamber (Figure 2a).

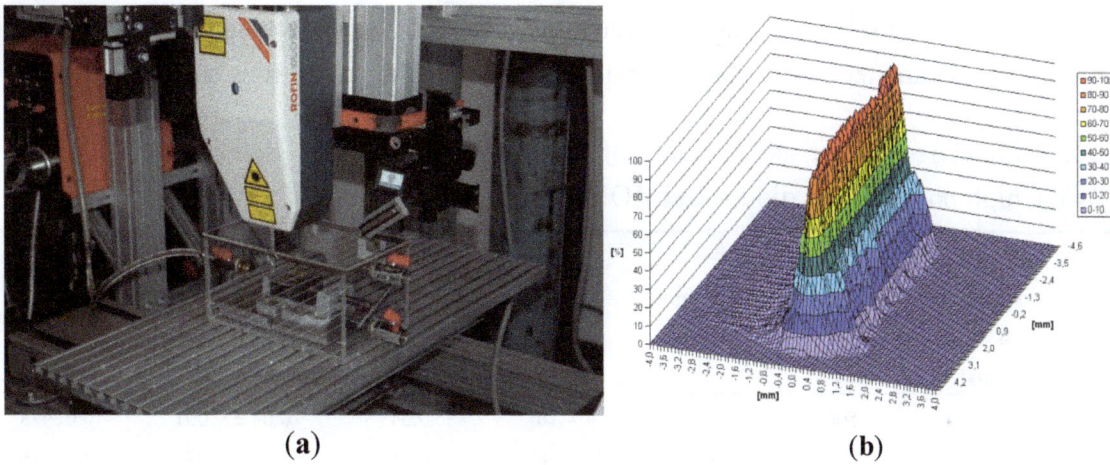

(a) (b)

Figure 2. (a) A view of the experimental setup equipped with the continuous wave high power direct diode laser (HPDDL) ROFIN DL 020; (b) 3D laser beam intensity profile (energy distribution) determined at the focal plane (beam spot) for the applied laser.

Table 1. Technical data of the continuous wave (CW) high power direct diode laser ROFIN DL 020.

Parameter	Value
Wavelength of the laser radiation (nm)	$808 \div 940$ * (± 5)
Maximum output power of the laser beam (kW)	2.2
Range of laser power (kW)	$0.1 \div 2.2$
Focal length (mm)	82/32
Laser beam spot size (mm)	1.8×6.8 or 1.8×3.8 **
Range of laser power intensity (kW/cm^2)	$0.8 \div 32.5$

* the dominant wavelength is 808 nm; ** size of the beam spot when an additional lens is applied with a focal length of 32 mm.

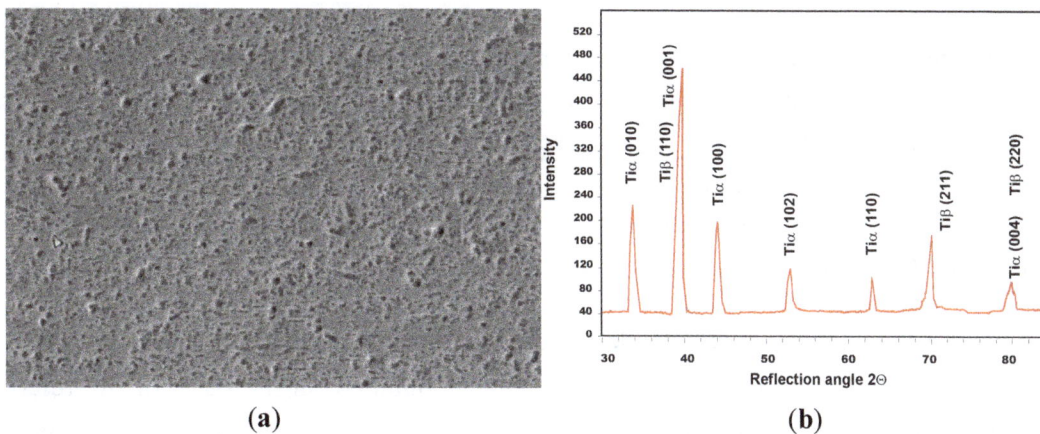

(a) (b)

Figure 3. (a) Scanning electron micrograph (SEM) and (b) XRD spectrum of the as-received titanium alloy Ti6Al4V specimen 3.0 mm thick ($\alpha + \beta$ phases).

The nitrided surface layers on titanium alloy specimens were produced as single stringer beads. The rectangular laser beam spot was focused on the top surface of specimens and set perpendicularly to the direction of scanning. Laser gas nitriding (LGN) tests of the titanium alloy Ti6Al4V surface were conducted over a wide range of processing parameters, at different laser power and different scanning speed. The processing parameters, as well as the detailed technological conditions are given in Table 2.

Table 2. Parameters of laser gas nitriding of titanium alloy Ti6Al4V specimens 3.0 mm thick by the high power direct diode laser ROFIN DL 020.

Sample No.	Scanning speed (mm/min) (mm/s)	Output laser power (W)	Heat input (J/mm)	Power * density (W/cm²)	Time of laser beam interaction (s)	Thickness of TiN layer (mm)	Remarks
P1	200 (3.33)	500	150	4×10^3	0.54	0.04 ± 0.001	NC, US
P2	600 (10.0)	1800	180	1.5×10^4	0.18	1.08 ± 0.037	NC, US
P3	400 (6.66)	1200	180	0.98×10^4	0.27	0.98 ± 0.028	NC
P4	200 (3.33)	600	180	0.49×10^4	0.54	0.90 ± 0.003	NC, UW
P5	200 (3.33)	700	210	0.57×10^4	0.54	1.08 ± 0.007	NC
P6	1000 (16.66)	2000	120	1.6×10^4	0.108	0.26 ± 0.009	NC, US
P7	900 (15.0)	1800	120	1.5×10^4	0.12	0.24 ± 0.008	NC, UW
P8	600 (10.0)	1200	120	0.98×10^4	0.18	0.28 ± 0.007	NC
P9	300 (5.0)	600	120	0.49×10^4	0.36	0.05 ± 0.001	NC, US
P10	800 (13.33)	2000	150	1.6×10^4	0.135	0.79 ± 0.02	NC
P11	600 (10.0)	1500	150	1.2×10^4	0.18	0.67 ± 0.021	NC
P12	1000 (16.66)	1500	90	1.2×10^4	0.108	0.14 ± 0.003	NC, US, HR
P13	800 (13.33)	1200	90	0.98×10^4	0.135	0.17 ± 0.002	NC
P14	600 (10.0)	900	90	0.74×10^4	0.18	0.13 ± 0.002	NC, UW
P15	400 (6.66)	600	90	0.49×10^4	0.27	0.02 ± 0.001	NC, UW, SF
P16	400 (6.66)	1800	270	1.5×10^4	0.27	1.74 ± 0.04	NC
P17	200 (3,33)	900	270	0.74×10^4	0.54	0.67 ± 0.018	NC, US
P18	2000 (33.33)	2000	60	1.6×10^4	0.054	0.19 ± 0.005	NC, US
P19	1800 (30.0)	1800	60	1.5×10^4	0.06	0.09 ± 0.001	NC, US
P20	1500 (25.0)	1500	60	1.2×10^4	0.072	0.14 ± 0.006	NC
P21	1200 (20.0)	1200	60	0.98×10^4	0.09	0.098 ± 0.002	NC, US
P22	900 (15.0)	900	60	0.74×10^4	0.12	0.12 ± 0.004	NC, US
P23	1200 (20.0)	1800	90	1.5×10^4	0.09	0.11 ± 0.005	NC, US

Remarks: laser beam spot size 1.8 × 6.8 mm, focal length 82.0 mm, nitrogen gas flow rate 10.0 L/min; NC, no crack; US, uneven surface; HR, high roughness of the bead face; UW, uneven width of the bead; SF, smooth surface of the bead face; * in laser physics, the power density of laser beam is defined as a beam intensity.

In order to compare the surface layers nitrided at the same heat input (J/mm) but at different beam interaction time τ (s), the experiment (trial of nitriding) was planned on the basis of the relationship between laser power and scanning speed illustrated in Figures 4 and 5. The time of laser beam interaction on the treated surface of titanium alloy was calculated for the beam spot width of 1.8 mm according to Equation (1). In welding and surface treatment technologies, when the moving heat source is applied, the term of heat input "E_v" (J/mm) is commonly used and it is defined as the power of a heat source "P" (W) divided by the processing speed "v" (mm/s) (e.g., welding or scanning speed):

$$E_v = P/v \text{ (J/mm)} \tag{2}$$

Therefore the linear relationships between laser power and scanning speed for different heat inputs were determined as illustrated in Figure 5. These relationships allowed the selection of processing parameters over a wide range of laser power and scanning speed for precisely set and constant heat input.

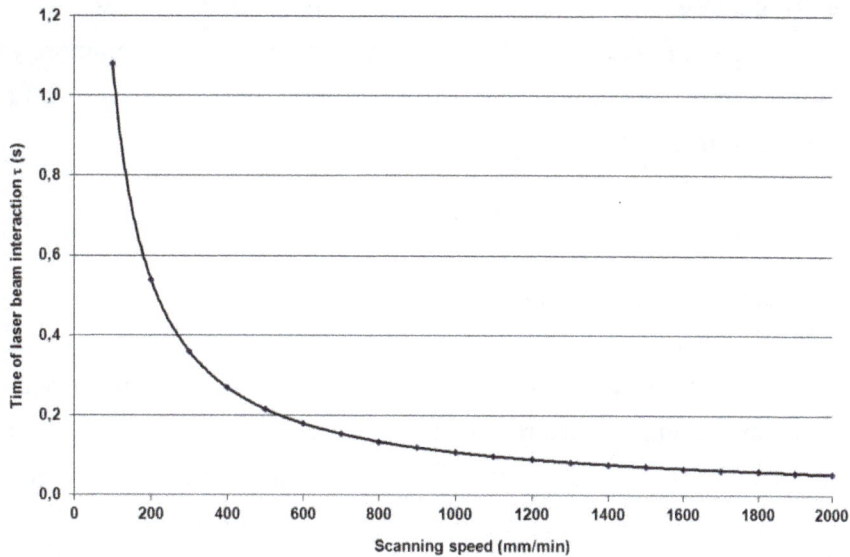

Figure 4. Relationship between the interaction time τ of laser beam and scanning speed v determined for the rectangular beam spot of a width of 1.8 mm.

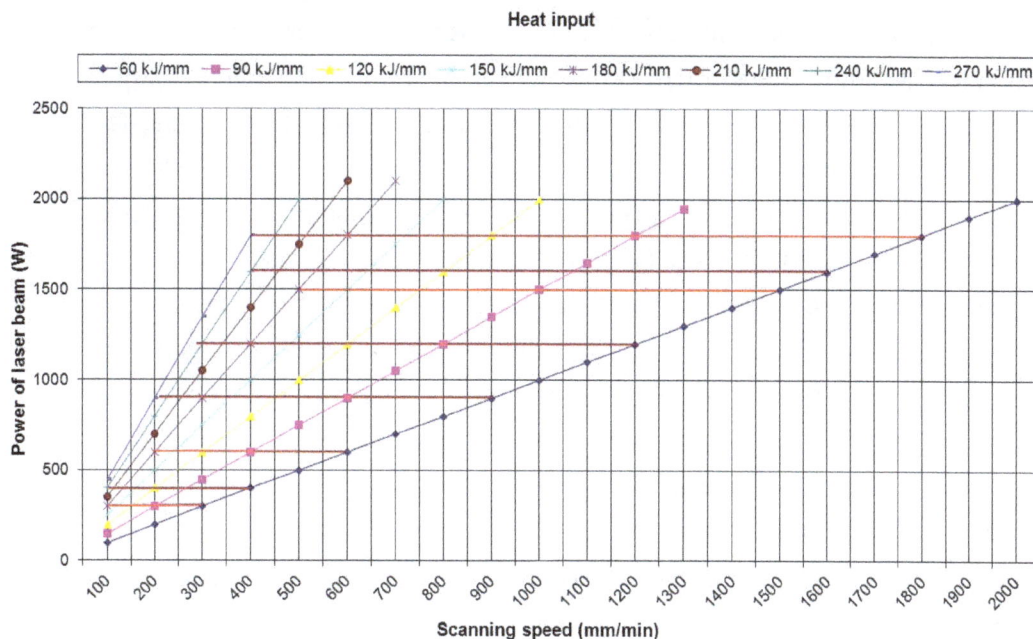

Figure 5. Graphic illustration of the relationship between laser power and scanning speed for a constant heat input.

The surface layers produced during laser gas nitriding of the titanium alloy were examined by visual testing, roughness, and surface topography analysis, next the samples were cut for cross-section examinations. Roughness measurements were carried out by means of a portable surface roughness tester SJ-210 Surftest (Mitutoyo Corporation, Kanagawa, Japan) according to ISO 4288 standard.

Whereas the surface topography was measured and analyzed by an optical, non-contact Profilograph Micro Prof 100 FRT (FRT GmbH, Bergisch Gladbach, Germany). The microstructure and phase composition of base metal and surface layers were examined and analyzed by optical and scanning electron microscopy SEM (Carl Zeiss, Oberkochen, Germany), EBSD (EDAX Inc., Mahwah, NJ, USA) and X-Ray diffraction (Panalitycal, Almelo, The Netherlands). Microhardness was measured and determined on the cross-section of surfaced layers. The influence of the parameters of the laser gas nitriding on surface topography, microhardness, structure and the shape of surface layers was analyzed. The results are given in Tables 2 and 3, and also in Figures 3–13.

3. Results and Discussion

The surface layers of titanium alloy Ti6Al4V specimens 3.0 mm thick after laser gas nitriding (LGN) in an atmosphere of pure nitrogen are matt, and all present the golden shine characteristic of titanium nitrides (Figures 6 and 7). The topography and roughness of the surface layers depend strongly on the process parameters of laser gas nitriding, *i.e.*, laser power and scanning speed (Figures 6 and 7, Table 3). The results of laser gas nitriding conducted at different process parameters clearly indicate that the heat input, as a process parameter, is not sufficient for identifying and predicting the properties of nitrided surface layers (Figure 6). The nitrided surface layers produced at the same heat input of 90 J/mm but at different laser power output, different scanning speed, thus different time of laser beam interaction, as well as different power density, have very different surface topography, roughness, and thus different morphology (Figure 6, Table 2). In the range of high scanning speed over 1200 mm/s and simultaneously high laser power over 1200 W (relatively low heat input, in general below 90 J/mm) the nitrided surface is very even, flat and smooth (Figures 7a and 8a, Tables 2 and 3). Roughness parameter Ra of the surface layer nitrided at 1500 W and 1500 mm/min is about 25 μm (Table 3). Reducing the scanning speed of the LGN process below 1000 mm/min while at the same time the heat input ranging from 120–180 J/mm resulted in significant increase of the surface roughness and irregular surface topography (Figures 7b and 8b). The surface is strongly humped and the furrows are placed evenly and perpendicularly to the direction of laser beam scanning (bead direction) (Figure 7b). In this case the roughness of the surface is over the measurement range.

Further scanning speed reduction below 400 mm/min or laser power increase over 1200 W corresponding to a heat input over 210 J/mm resulted in subsequent surface topography change and simultaneous decrease of the surface roughness. The surface of the sample produced at 1800 W and 400 mm/min is smooth and flat (Figures 7c and 8c, Table 3). In this case the roughness parameter Ra of the nitrided surface layer ranges from about 8.0–15.0 μm (Table 3). The reason for this phenomenon is a different mechanism of liquid metal (weld pool) solidification, in particular the mechanism of titanium nitrides TiN nucleation and subsequent growth of the nitrides, depends on the thermal conditions, especially heating and cooling rate, temperature range, as well as volume and depth of the liquid metal (weld pool), surface tension and convection of the liquid metal. Detailed and comprehensive explanation of the phenomenon in the case of HPDDL gas nitriding requires broader research and study of the mechanisms of titanium nitrides nucleation, growth, and crystallization of the liquid metal of the matrix, as well as the kinetics of gaseous nitrogen absorption by the liquid metal at different thermal conditions during laser gas nitriding in the liquid state.

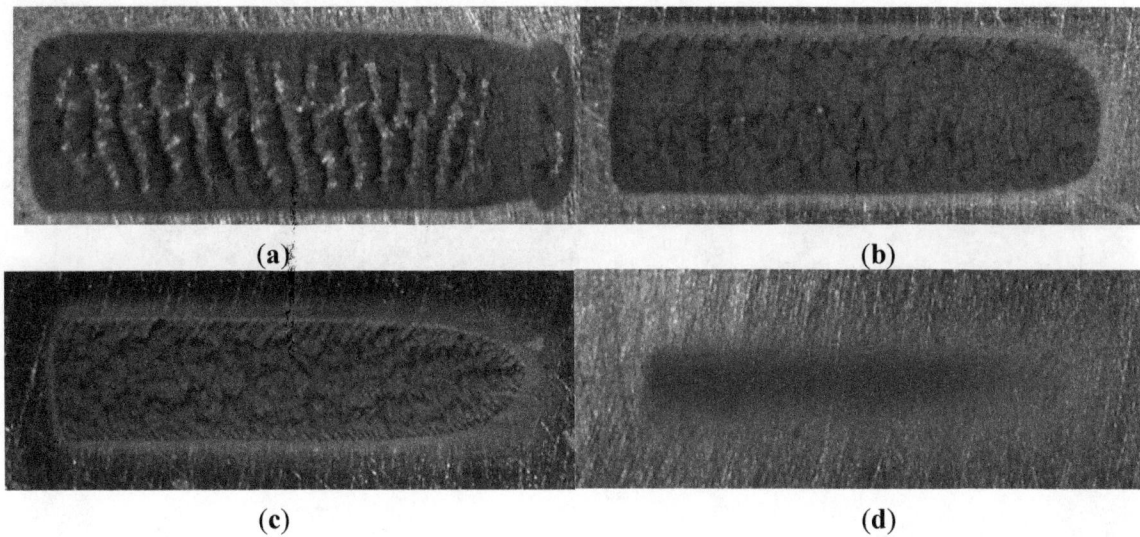

Figure 6. The surface morphology of single beads produced on titanium alloy Ti6Al4V by HPDDL laser gas nitriding at constant heat input of 90 J/mm, but different power density and different scanning speed, thus different time of laser beam interaction (Table 2, Figures 4 and 5); (**a**) No. P12 (laser power 1.5 kW, scanning speed 1.0 m/min); (**b**) No. P13 (laser power 1.2 kW, scanning speed 0.8 m/min); (**c**) No. P14 (laser power 0.9 kW, scanning speed 0.6 m/min); (**d**) No. P15 (laser power 0.6 kW, scanning speed 0.4 m/min).

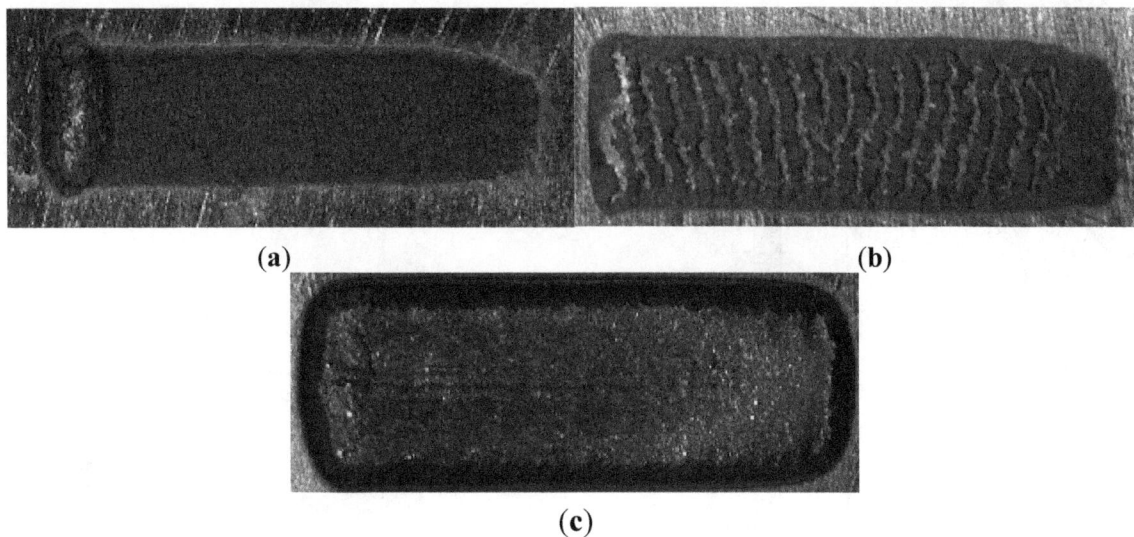

Figure 7. The surface morphology of single beads produced on titanium alloy Ti6Al4V by HPDDL laser gas nitriding at different heat inputs, (Table 2); (**a**) sample No. P20 (laser power 1.5 kW, scanning speed 1.5 m/min, heat input 60 J/mm); (**b**) sample No. P4 (laser power 0.6 kW, scanning speed 0.2 m/min, heat input 180 J/mm); (**c**) sample No. P16. (laser power 1.8 kW, scanning speed 0.4 m/min, heat input 270 J/mm).

(a)

(b)

(c)

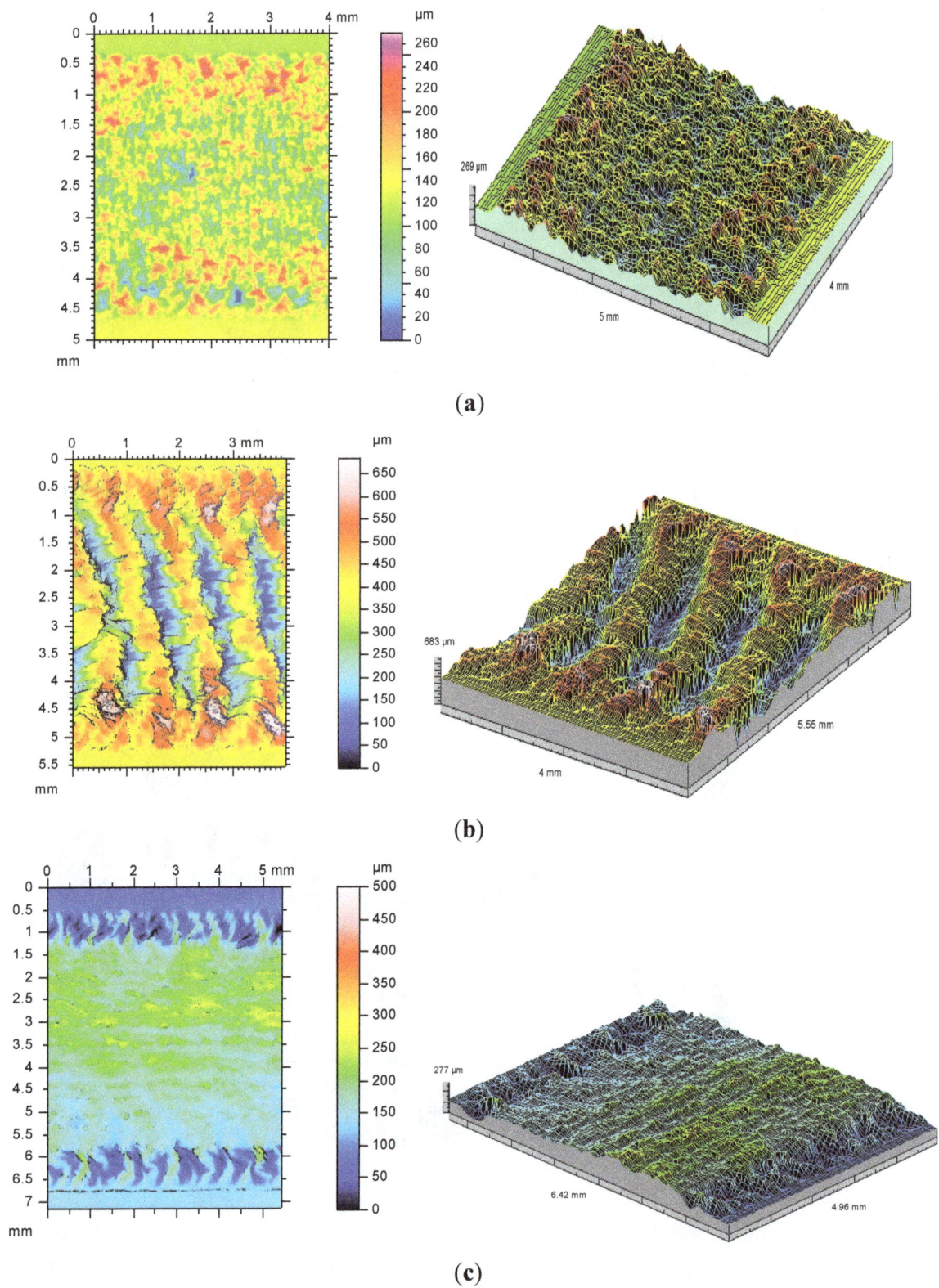

Figure 8. 2D and 3D topography of the surface layers produced on titanium alloy Ti6Al4V by HPDDL laser gas nitriding (Tables 2 and 3); (**a**) sample No. P20; (**b**) sample No. P12; (**c**) sample No. P16.

Table 3. Results of roughness measurements on the surface of titanium alloy Ti6Al4V specimen and on the nitrided surface layers (Table 2).

Sample type	Roughness parameters, µm					
	Rz	Rt	Ry	Sm	Ra	Rq
Ti6Al4V as received	4 ± 0.1	7 ± 0.18	6 ± 0.1	162 ± 3.9	0.4 ± 0.01	0.8 ± 0.01
Sample No. P5	76 ± 1.3	98 ± 1.9	97 ± 2.1	491 ± 5.5	15.4 ± 0.4	18.8 ± 0.5
Sample No. P8	53 ± 0.7	67 ± 1.6	62 ± 1.4	259 ± 4.3	9.4 ± 0.2	11.8 ± 0.3
Sample No. P16	34 ± 0.9	42 ± 0.9	41 ± 0.8	229 ± 3.9	8.2 ± 0.19	10 ± 0.25
Sample No. P20	136 ± 3.1	168 ± 3.7	157 ± 3.6	245 ± 4.1	25 ± 0.6	30.6 ± 0.9

Remarks: Rz, average distance between the highest peak and lowest valley in each sampling length; Rt, Maximum Height of the Profile; Ry, Maximum Height of the Profile; Sm, Mean Spacing of Profile Irregularities; Ra, arithmetic average of absolute values; Rq, Root Mean Square (RMS) Roughness.

The shape, penetration depth, and width of the surface layers depend strongly on the parameters of laser gas nitriding with the HPDDL laser (Figures 9 and 10, Table 2). The minimum thickness of the nitrided surface layer including the heat affected zone (HAZ), produced at the laser power 900 W, scanning speed 900 mm/min and thus heat input 60 J/mm is about 0.7 mm with the hard surface layer of TiN precipitations at about 0.05 mm (Table 2). While the maximum thickness of the TiN layer over 1.7 mm was produced at laser power 1800 W, scanning speed 400 mm/min and heat input 270 J/mm (Figures 7c and 9a). The width of the surface layers in general correlated with the size of the HPDDL laser beam spot. The width of the diode laser beam spot is 6.8 mm, set perpendicularly to the scanning direction, therefore the width of the surface layers is in the range from 4.0 mm up to over 7.0 mm (Figure 10).

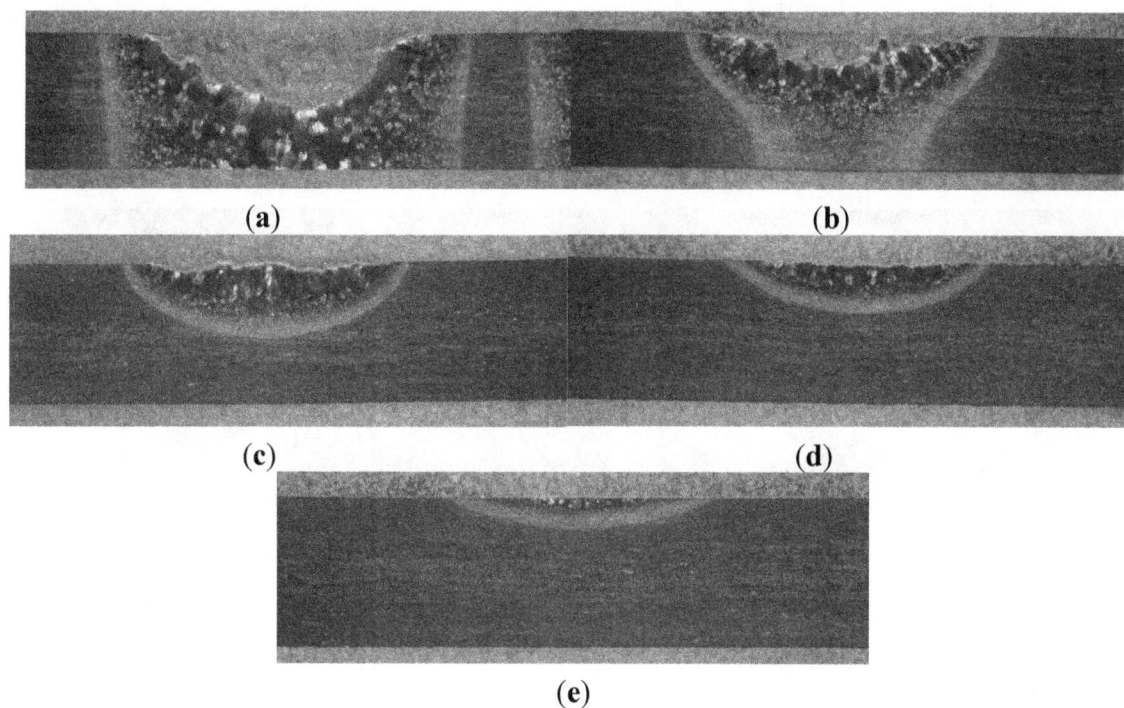

(a) **(b)** **(c)** **(d)** **(e)**

Figure 9. A view of cross-sections of surface layers after laser gas nitriding of titanium alloy Ti6Al4V by the HPDDL laser (Table 2); **(a)** sample No. P16; **(b)** sample No. P4; **(c)** sample No. P12; **(d)** sample No. P20; **(e)** sample No. P22.

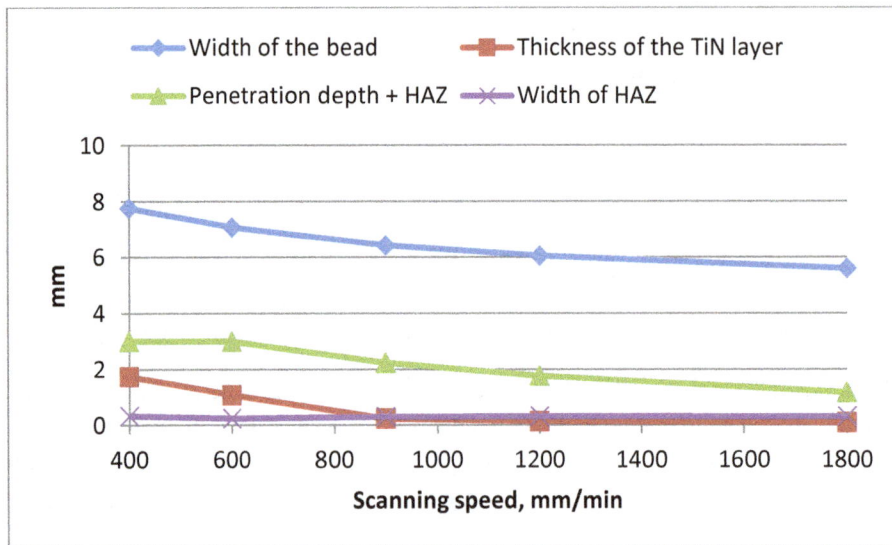

Figure 10. Influence of the scanning speed on the geometry of nitrided surface layers produced at constant output laser power 1800 W (Table 2).

Figure 3a shows the microstructure of as-received 3.0 mm thick plate of the titanium alloy Ti6Al4V (base metal BM) comprised of two phases Tiα and Tiβ (shown in Figure 3b). Figure 11a shows scanning electron micrographs (SEM) of the surface layer No. 16, with the thicker layer of titanium nitrides up to 1.74 mm, produced at laser power 1800 W and scanning speed 400 mm/min, thus a heat input of 270 J/mm. In contrast Figure 11b shows SEM of the surface layer No. 20 with a thickness of the TiN layer of just about 0.14 mm, produced at laser power 1500 W and scanning speed 1500 mm/min, thus a heat input of just 60 J/mm (Table 2). In both cases a thin homogenous and consistent layer, about 10.0 μm thick, tightly covers the surface layers (Figure 11). From the homogenous layer on the top surface, dendritic precipitations grow into the bulk of the surface layer and all of the dendrites, in the near surface region are oriented perpendicularly to the top surface (Figure 11).

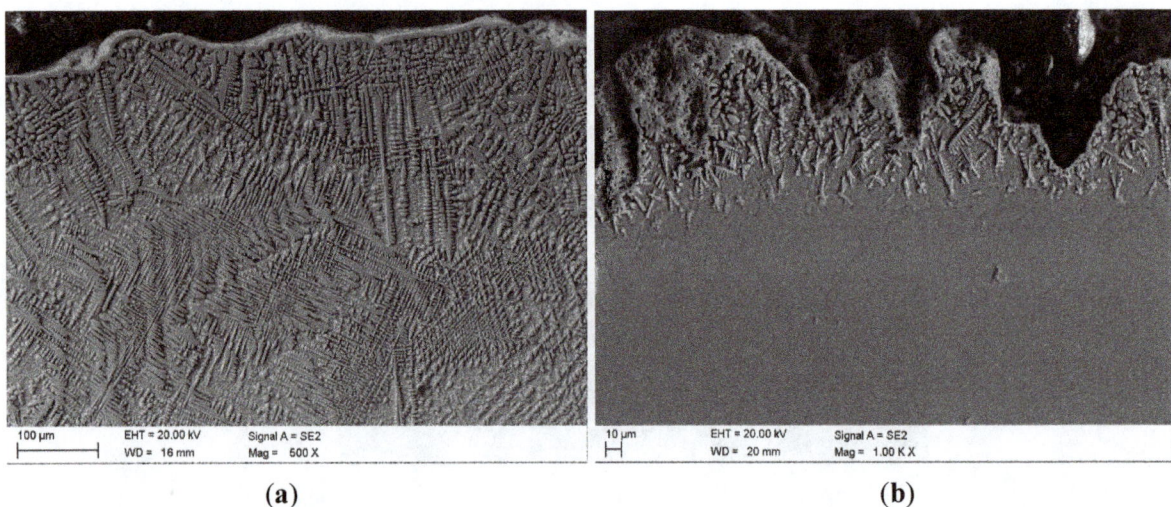

(a)   (b)

Figure 11. (**a**) Scanning electron micrograph of the nitrided surface layer of sample No. P16 (Table 2) and (**b**) sample No. 20.

The electron backscatter diffraction (EBSD) analysis conducted directly below the top surface showed that the near surface region consists mainly of the titanium nitrides TiN embedded in the metal matrix of Tiα (Figure 12).

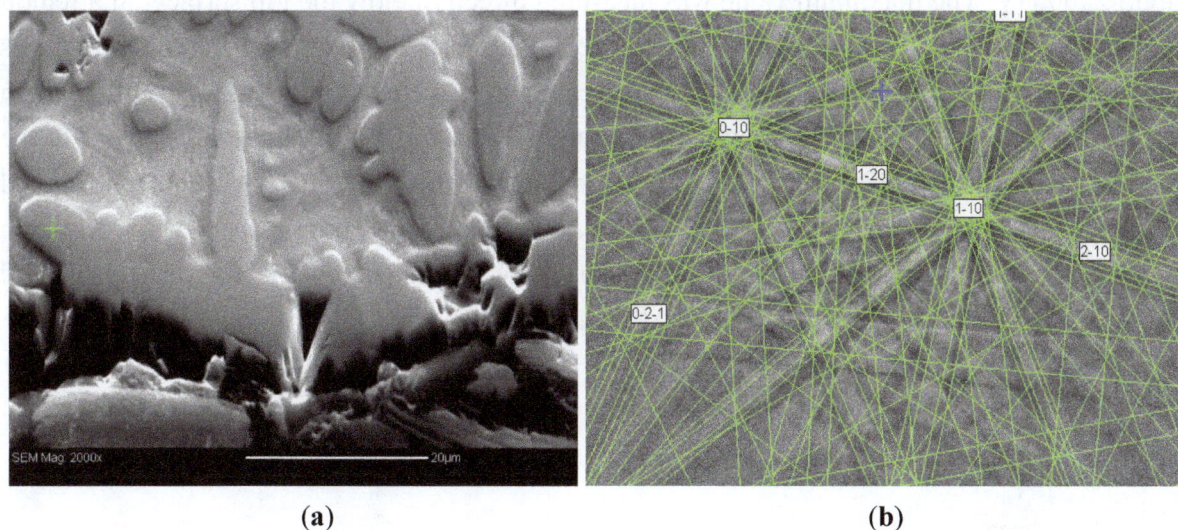

(a)　　　　　　　　　　　　　　　　　　　(b)

Figure 12. (a) Scanning electron micrograph of the nitrided surface layer of sample No. P16 in the region just below the top surface (Table 2) and (b) results of EBSD analysis of the precipitations in the metal matrix. Phase name: TiN, composition: N (50 at.%), Ti (50 at.%).

Microstructure study shows a good metallurgical bonding of the interface between nitrided layers rich in dendritic titanium nitrides and the substrate of titanium alloy Ti6Al4V, free of any internal imperfections or discontinuity such as transgranular and intergranular cracks, micro cracks, or porosity (Figures 11 and 12).

The morphology and orientation of the titanium dendrites in the near surface region clearly indicate that the nucleation of dendrites takes place directly below the top surface of weld pool, near the liquid/gas boundary, as a result of nitrogen absorption from the gaseous atmosphere and subsequent increase of the nitrogen content in the liquid metal of a weld pool. The growth of titanium nitride dendrites (crystallization) proceeds into the liquid metal (weld pool), perpendicularly to the top surface.

When the laser beam as a heat source is being moved relative to the heated surface of titanium alloy, the heat transfer and thermal conditions are quite different compared to the steady state conditions during stationary heating [21]. Thus given the fact that the scanning speeds ranged from several to several tens mm/s and simultaneously the volume of a weld pool was not greater than 5–6 mm^3, the time for nucleation and solidification of TiN dendrites was limited.

For example the volume of the weld pool of the surface layer produced at laser power 1800 W and scanning speed 400 mm/min (6.66 mm/s, Sample No. 16, Table 2) is about 5.5 mm^3. It was estimated on the basis of the penetration depth profile (Figure 9a), the shape and the dimensions of the top surface of the solidified weld pool which is 6 mm wide and 0.7–1.5 mm long in the middle, as shown on the left side of Figure 7c. Having regard to the scanning speed 6.66 mm/s, the growth rate of the 250–300 μm long dendrites is about 1.3–2.8 mm·s^{-1}.

The microhardness measured on cross-sections of the nitrided surface layers is generally characterized as a gradient relationship as shown in Figure 13. The maximum value of microhardness reaches about 2400 HV0.2 in the case of the surface layer No. 16, with the highest penetration depth and greatest thickness of the hard layer of the titanium nitrides (Table 2). According to the various literature sources, the maximum microhardness of titanium nitride TiN may range from 1800–2100 HV or even up to 2400 HV. The maximum value was measured directly below the top surface, in a region where the density and percentage of titanium nitride precipitations is highest. Next the microhardness decreases gradually with the distance from the top surface down to the level of about 300 HV0.2 which is the microhardness of the Ti6Al4V titanium alloy (Figure 13). Additionally Figure 13 shows that the maximum values of microhardness measured on cross-sections of the individual surface layers depend directly on the penetration depth. This phenomenon may be related to the thickness of the titanium nitrides layer, as well as on the phase composition, especially on the percentage and density of hard titanium nitrides TiN_x in the metal matrix (Figures 11 and 12).

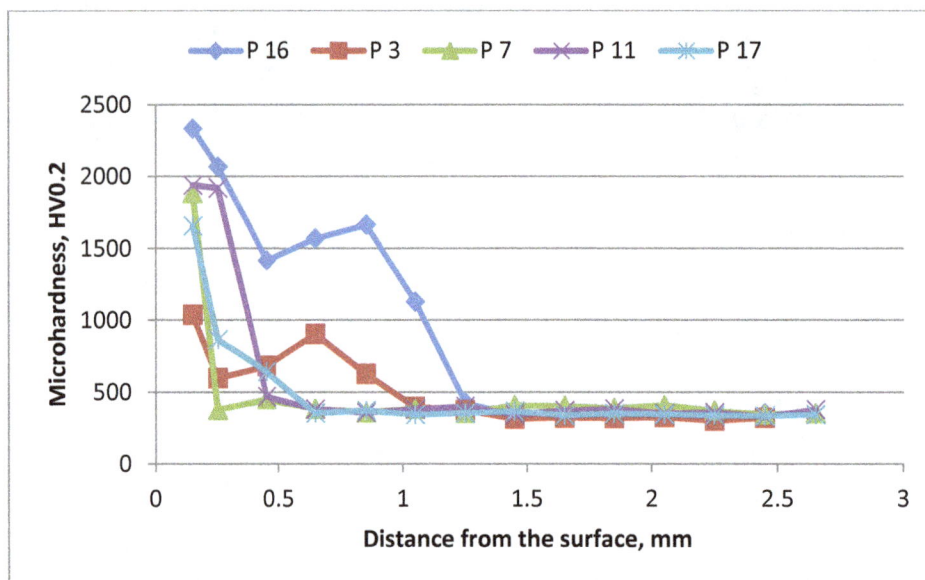

Figure 13. Microhardness distribution on the cross-section of surface layers after laser gas nitriding of titanium alloy Ti6Al4V specimen 3.0 mm thick by HPDDL laser (Table 2).

4. Conclusions

A high power direct diode laser with a rectangular laser beam spot, uniform energy distribution across the laser beam and emitting in the range of 808–940 nm was successfully applied for *in situ* production of composite surface layers, consisting of dendritic titanium nitrides precipitations TiN, embedded in the metal matrix of Tiα (Titanium Matrix Composite). Besides, a thin homogenous and consistent layer of titanium nitrides, about 10.0 µm thick, tightly covers the surface layers.

The composite surface layers produced by HPDDL laser nitriding of Ti6Al4V titanium alloy specimens in a pure nitrogen atmosphere, in the investigated range of processing parameters, were of high quality and free of any surface and internal defects such as cracks or porosity even in the case of very thick layers with very high hardness, often reported by many researchers who have used other types of lasers.

The topography and roughness of laser gas nitrided surface layers may be changed over a wide range from a very flat and even surface, through to a very rough, and up to uneven and humped, depending on the processing parameters. Similarly the thickness of laser gas nitrided surface layers may be changed over a wide range from about 0.1–0.15 mm up to very thick about 1.74 mm.

The heat input as a process parameter is not sufficient to identify and predict the properties of nitrided surface layers produced by the HPDDL laser. The nitrided surface layers produced at the same heat input but at different output laser power, different scanning speed, thus different time of laser beam interaction and also different power density, have significantly different surface topography, roughness, and thus different morphology. Therefore the properties of the nitrided surface layers are directly dependent on the power density of the HPDDL laser beam and the time of laser beam interaction on the treated surface of the titanium alloy.

Acknowledgments

I am grateful to A. Grabowski, K. Gołombek and A. Iwaniak for substantive and experimental support. This work was partially supported by the Center of Innovation and Technology Transfer of Silesian University of Technology under the project 05/TWIP/BP/2013.

Conflicts of Interest

The author declares no conflict of interest.

References

1. Katayama, S.; Matsunawa, A.; Morimoto, A.; Ishimoto, S.; Arata, Y. Surface hardening of titanium by laser nitriding. In Proceedings of the ICALEO'83, Los Angeles, CA, USA, 14–17 November 1983; pp. 127–134.

2. Kurc-Lisiecka, A.; Ozgowicz, W.; Ratuszek, W.; Kowalska, J. Analysis of deformation texture in AISI 304 steel sheets. *Sol. St. Phenom.* **2013**, *203–204*, 105–110.

3. Opiela, M.; Grajcar, A. Hot deformation behavior and softening kinetics of Ti-V-B microalloyed steels. *Arch. Civ. Mech. Eng.* **2012**, *12*, 327–333.

4. Biswas, A.; Li, L.; Chatterjee, U.K.; Manna, I.; Majumdar, J.D. Diode laser assisted surface nitriding of Ti-6Al-4V: Properties of the nitrided surface. *Metall. Mater. Trans. A* **2009**, *40A*, 3031–3037.

5. Biswas, A.; Li, L.; Maity, T.K.; Chatterjee, U.K.; Mordike, B.L.; Manna, I.; Majumdar, J.D. Laser surface treatment of Ti-6Al-4V for bio-implant application. *Lasers Eng.* **2007**, *17*, 59–73.

6. Biswas, A.; Li, L.; Chatterjee, U.K.; Manna, I.; Pabi, S.K.; Majumdar, J.D. Mechanical and electrochemical properties of laser surface nitrided Ti–6Al–4V. *Scr. Mater.* **2008**, *59*, 239–242.

7. Man, H.C.; Bai, M.; Cheng, F.T. Laser diffusion nitriding of Ti–6Al–4V for improving hardness and wear resistance. *Appl. Surf. Sci.* **2011**, *258*, 436–441.

8. Majumdar, J.D. Laser gas alloying of Ti-6Al-4V. *Phys. Proc.* **2011**, *12*, 472–477.

9. Burdzik, R.; Konieczny, Ł.; Stanik, Z.; Folega, P.; Smalcerz, A.; Lisiecki, A. Analysis of impact of chosen parameters on the wear of camshaft. *Arch. Metall. Mater.* **2014**, *59*, 961–967.

10. Kusiński, J.; Kac, S.; Kopia, A.; Radziszewska, A.; Rozmus-Górnikowska, M.; Major, B.; Major, L.; Marczak, J.; Lisiecki, A. Laser modification of the materials surface layer—A review paper. *Bull. Pol. Acad. Sci.* **2012**, *60*, 711–728.

11. Santos, E.; Morita, M.; Shiomi, M.; Osakada, K.; Takahashi, M. Laser gas nitriding of pure titanium using CW and pulsed Nd:YAG lasers. *Surf. Coat. Technol.* **2006**, *201*, 1635–1642.

12. Perez, M.G.; Harlan, N.R.; Zapirain, F.; Zubiri, F. Laser nitriding of an intermetallic TiAl alloy with a diode laser. *Surf. Coat. Technol.* **2006**, *200*, 5152–5159.

13. Rességuier, T.; Loison, D.; Dragon, A.; Lescoute, E. Laser driven compression to investigate shock-induced melting of metals. *Metals* **2014**, *4*, 490–502.

14. Węgrzyn, T. The Classification of Metal Weld Deposits in Terms of the Amount of Nitrogen. Available online: http://www.isope.org/publications/proceedings/ISOPE/ISOPE%202000/pdffiles/papers/Vol4/022.pdf (accessed on 29 December 2014).

15. Thomann, A.L.; Sicard, E.; Boulmer-Leborgne, C.; Vivien, C.; Hermann, J.; Andreazza-Vignolle, C.; Andreazza, P.; Meneau, C. Surface nitriding of titanium and aluminum by laser-induced plasma. *Surf. Coat. Technol.* **1997**, *97*, 448–452.

16. Lisiecki, A. Mechanism of laser surface modification of the Ti-6Al-4V alloy in nitrogen atmosphere using a high power diode laser. *Adv. Mater. Res.* **2014**, *1036*, 411–416.

17. Yilbas, B.S.; Karatas, C.; Uslan, I.; Keles, O.; Usta, I.Y.; Ahsan, M. CO_2 laser gas assisted nitriding of Ti–6Al–4V alloy. *Appl. Surf. Sci.* **2006**, *252*, 8557–8865.

18. Garcia, I.; Damborenea, J. Corrosion properties of TiN prepared by laser gas alloying of Ti and Ti5Al3V. *Corros. Sci.* **1998**, *40*, 1411–1419.

19. Gołębiewski, M.; Kruzel, G.; Major, R.; Mróz, W.; Wierzchoń, T.; Ebner, R.; Major, B. Morphology of titanium nitride produced using glow discharge nitriding, laser remelting and pulsed laser deposition. *Mater. Chem. Phys.* **2003**, *81*, 315–318.

20. Ohtsu, N.; Kodama, K.; Kitagawa, K.; Wagatsuma, K. Comparison of surface films formed on titanium by pulsed Nd:YAG laser irradiation at different powers and wavelengths in nitrogen atmosphere. *Appl. Surf. Sci.* **2010**, *256*, 4522–4526.

21. Lisiecki, A. Experimental and numerical study of heat conditions during diode laser gas nitriding of titanium alloy. *Adv. Mater. Res.* **2014**, *1036*, 320–325.

22. Janicki, D. High power diode laser cladding of wear resistant metal matrix composite coatings. *Sol. State Phenom.* **2013**, *199*, 587–592.

23. Casati, R.; Vedani, M. Metal matrix composites reinforced by nano-particles—a review. *Metals* **2014**, *4*, 65–83.

24. Xue, L.; Islam, M.; Koul, A.K.; Bibby, M.; Wallace, W. Laser gas nitriding of Ti-6Al-4V part 1: Optimization of the process. *Adv. Perf. Mater.* **1997**, *4*, 25–47.

25. Carpene, E.; Schaaf, P.; Han, M.; Lieb, K.P.; Shinn, M. Reactive surface processing by irradiating with excimer laser, Nd:YAG laser, free electron laser and Ti:sapphire laser in nitrogen atmosphere. *Appl. Surf. Sci.* **2002**, *186*, 195–199.

26. Bodzenta, J.; Kaźmierczak-Bałata, A.; Wokulska, K.; Kucytowski, J.; Szperlich, P.; Łukasiewicz, T.; Hofman, B. Analysis of influence of Yb concentration on thermal, elastic, optical and lattice parameters in YAG single crystal. *J. Alloy Compd.* **2009**, *473*, 245–249.

Vacuum Die Casting Process and Simulation for Manufacturing 0.8 mm-Thick Aluminum Plate with Four Maze Shapes

Chul Kyu Jin [1], Chang Hyun Jang [1] and Chung Gil Kang [2],*

[1] Graduate School of Mechanical and Precision Engineering, Pusan National University, San 30 Chang Jun-dong, Geum Jung-Gu, Busan 609-735, Korea; E-Mails: ckjeans82@pusan.ac.kr (C.K.J.); jangch@hyundai-wisco.com (C.H.J.)

[2] School of Mechanical Engineering, Pusan National University, San 30 Chang Jun-dong, Geum Jung-Gu, Busan 609-735, Korea

* Author to whom correspondence should be addressed; E-Mail: cgkang@pusan.ac.kr

Academic Editor: Anders E. W. Jarfors

Abstract: Using vacuum die casting, 0.8 mm-thick plates in complicated shapes are manufactured with the highly castable aluminum alloy Silafont-36 (AlSi9MgMn). The sizes and shapes of the cavities, made of thin plates, feature four different mazes. To investigate formability and mechanical properties by shot condition, a total of six parameters (melt temperatures of 730 °C and 710 °C; plunger speeds of 3.0 m/s and 2.5 m/s; vacuum pressure of 250 mbar and no vacuum) are varied in experiments, and corresponding simulations are performed. Simulation results obtained through MAGMA software show similar tendencies to those of the experiments. When the melt pouring temperature is set to 730 °C rather than 710 °C, formability and mechanical properties are superior, and when the plunger speed is set to 3.0 m/s rather than to 2.5 m/s, a fine, even structure is obtained with better mechanical properties. The non-vacuumed sample is half unfilled. The tensile strength and elongation of the sample fabricated under a melt temperature of 730 °C, plunger speed of 3.0 m/s, and vacuum pressure of 250 mbar are 265 MPa and 8.5%, respectively.

Keywords: vacuum die-casting; thin plate; aluminum alloy; casting simulation; casting defects

1. Introduction

Many industries, including the automobile, aircraft, and electronics sectors, have actively researched replacing ferrous alloys with light alloy materials, such as aluminum and magnesium. Replacing a ferrous alloy with a light alloy degrades mechanical properties and increases costs, which are large problems in these industries. The density of aluminum (2.7 g/cm^3) is one-third that of iron (7.9 g/cm^3) and, as its mechanical properties and workability are excellent, aluminum is used for casting and forging. Even with these advantages, because an aluminum alloy has substantially inferior mechanical properties compared with those of a ferrous alloy, alloy composition and process methods should be improved for better mechanical properties [1–6].

Aluminum process methods are classified into casting, forging, and plastic deformation. Die-casting is a method that produces a product by pouring melt into a mold followed by punch-pressing, which allows a complicated shape to be fabricated. However, because die-casting injects melt at a high velocity, gases and air remaining in the melt may cause internal defects; therefore, the product's mechanical properties are degraded. An enhanced die-casting method, vacuum die-casting, has been developed by adding a vacuum device. Because vacuum die casting creates a vacuum inside the mold cavity during casting, gases or air in the melt is removed, decreasing the volume of gas pockets and improving the mechanical properties and smoothness of the resulting surface [7,8]. Using an aluminum or magnesium alloy made by vacuum die-casting, aircraft and automotive parts in bulk shapes have been manufactured. Recently, studies on manufacturing 0.8-mm-thick, thin-plated products have been introduced [9–12].

When manufacturing a thin-plated part, the fluidity of a material is extremely important, so a casting alloy with 10% silicon content needs to be used. Moreover, it is necessary to analyze the behavior of melt when it is poured into a mold and fills the cavity, and to predict regions in which the flow of the melt is not smooth or unstable and casting defects occur. For this process, filling and solidification analyses need to be conducted by using computer-aided engineering (CAE).

In this study, we propose a method for manufacturing a thin aluminum plate with a complicated shape by using vacuum die-casting. The shape of the mold cavity with four mazes is complex and the thickness of the mold is 0.8 mm. Using CAE, the behavior of the melt and any defects are predicted, which are then compared against the actual results of an experiment. Then, the formability of a thin plate in accordance with shot conditions and mechanical properties of the manufactured plates are comparatively analyzed. For the experiment, a 660-ton cold chamber die-casting machine was used and the melt material was a Silafont-36 (AlSi9MgMn) alloy.

2. Experimental Section

2.1. Mold Design

Figure 1a presents an image of the cavity in the vacuum die casting mold. The mold structure for vacuum die casting consists of a biscuit, a runner, gates, a product, and overflows. The melt is poured into the biscuit and filled to the product after passing through the runner and gates. Because the behavior of the melt filled into the product is highly affected by the shape of the gates, designing a gate appropriate for the shape of the product is essential. A gate has an appropriate shape for a thin plate. Melt injected through a gate can fill the cavity in sequence up to the overflows [10,11,13–15].

The overflows are the parts where the melt is filled last and should be located at the position where material filling is incomplete; these are located on the left and right of the product and on the other side of the gate, which is the last region to be filled by the melt. The vacuum block is located at the end of the overflow on the other side of the gate. The valve opens during filling and it is closed by the pressure of the melt. The shape of the product is a thin plate with four mazes. In size the product is 220 mm high, 220 mm wide, and 0.8 mm thick. Figure 1b shows the cross section of a gate. The cross-sectional area of the gate connected with the product is 147.2 mm. Figure 2 presents plane and sectional images of the four mazes. The four mazes are shaped as squares with 0.3 mm-deep grooves and different paths in each maze. Mazes 1 and 4 are 41 mm wide and 58 mm high with the same 0.9-mm-wide grooves, but the maze shapes are different (Maze 1 is serpentine and Maze 4 is straight). Maze 2 is 45 mm wide and 58 mm high with 1.0-mm-wide grooves; its shape is serpentine, the same as that of Maze 1. Maze 4 is 45 mm wide and 58 mm high with 1.2-mm-wide grooves; its shape is serpentine, the same as those of Mazes 1 and 3. The thickness of the four mazes is 1.1 mm, including the depth of grooves, which is 0.3 mm.

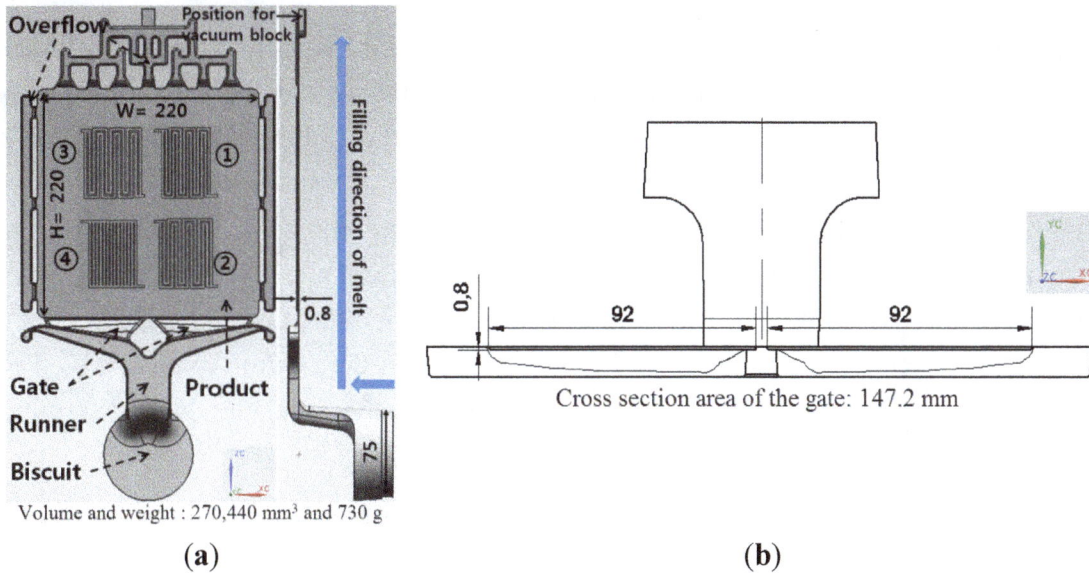

(a)   **(b)**

Figure 1. Geometry of die cavity with four mazes (unit: mm). **(a)** Geometry; **(b)** Gate.

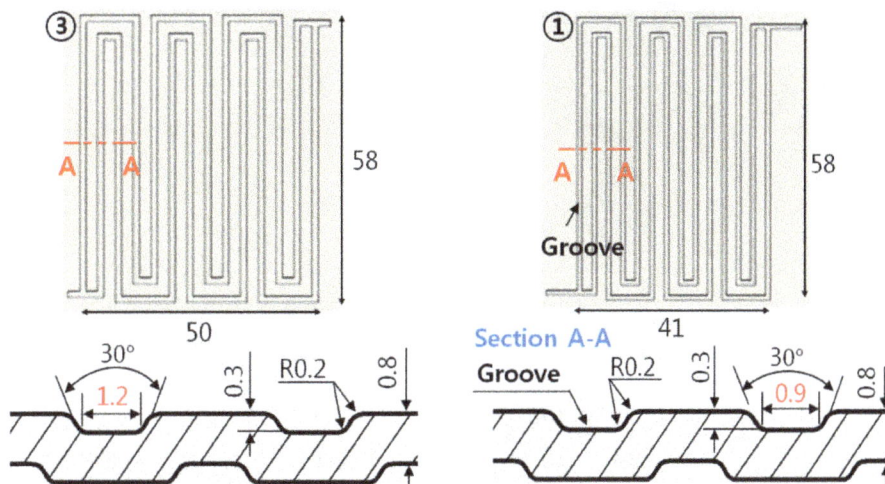

Figure 2. *Cont.*

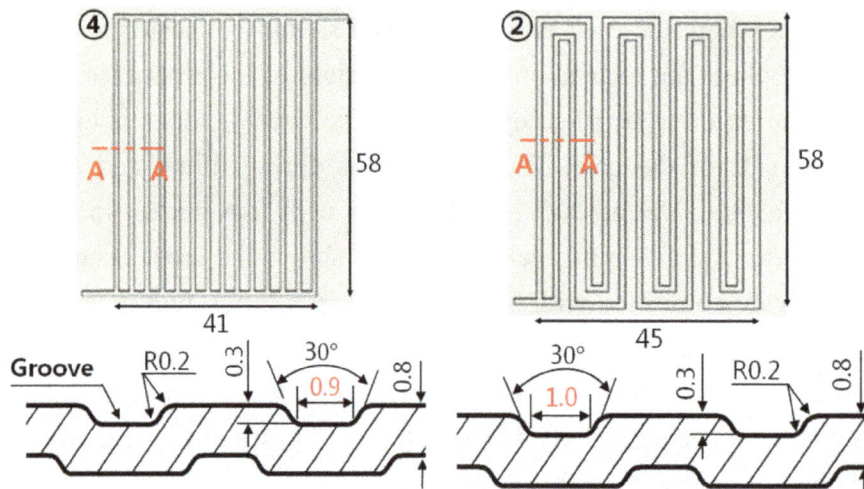

Figure 2. Detail view of four mazes (unit: mm): ① maze 1; ② maze 2; ③ maze 3 and ④ maze 4.

2.2. Conditions of the Experiment and Simulation

The casting simulation was performed using the MAGMA 5.1 software (MAGMA Giessereitechnologie GmbH, Aachen, Germany), set according to the melt temperature, plunger speed, and vacuum pressure. MAGMA's mesh-partitioning method is the finite volume method (FVM). An accurate simulation result requires at least three meshes per axis. Because too many meshes would require a substantial amount of time for calculation, an appropriate number of meshes needs to be determined. Accordingly, to generate three meshes in the direction of the plate's thickness, meshes were divided as much as possible in the y-axis, which is the direction of the thickness. The results were 594,275 meshes on the metal cell (cast) and 14,575,210 meshes on the control volume, which is the material group (mold) as shown in Figure 3. All other conditions in the simulation were determined to be the same as those of the experiment.

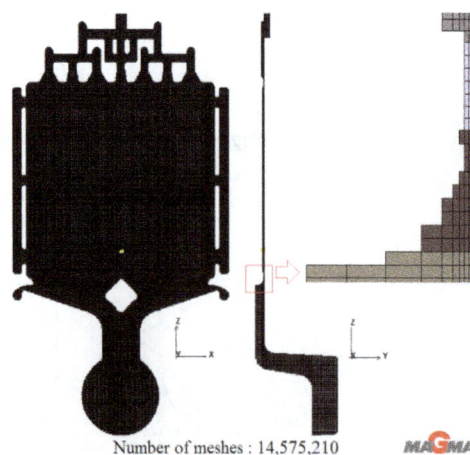

Number of meshes : 14,575,210

Figure 3. Mesh of die cavity with four mazes (unit: mm).

A 660-ton Buhler cold chamber die casting process was used for the shot experiment. In this process, the melt is poured into the sleeve, and the plunger inside the sleeve pushes the melt into the mold. The length of the shot sleeve is 470 mm and the diameter of the plunger is 75 mm. Figure 4 shows the vacuum die casting process. During the first process of vacuum die-casting, a vacuum pump creates a vacuum inside the mold cavity. The melt is injected through the pouring hole in the shot sleeve, and the plunger pushes it into the mold. The plunger moves at a low speed (V_1) up to a point 400 mm inside the sleeve and at a high speed (V_2) between the points 400 mm and 470 mm into the sleeve. As soon as the plunger reaches the point of 470 mm, it pressurizes the melt at 500 bar; this is when the melt solidifies in the cavity, forming into a product. Filling time is about 327 ms during which melt moves from the chamber to the overflow. It only takes about 12 ms to completely fill the die cavity. Table 1 presents the parameters of the vacuum die casting process, and Table 2 lists the shot conditions (experimental). We examined the effects of melt temperature, plunger speed, and vacuum pressure on formability and mechanical properties. The temperatures of the pouring melt were 710 °C and 730 °C, and the vacuum pressures were 250 mbar and no vacuum. The plunger speeds were 0.2 m/s for the low speed (V_1) and 2.5 and 3.0 m/s for the high speeds (V_2). As Table 2 shows, there were a total of six shot conditions, and each condition was tested five times. To eliminate variation of fluid characteristics of the melt attributable to the mold's changing temperature, five preparatory shots were made prior to the actual shots, maintaining the temperature of the mold at 250 °C. The melt was the die-casting alloy Silafont-36 (AlSi9MgMn). A Silafont-36 material has 10% Si content and has an excellent castability along with superior weldability and workability. Further, the Fe content of this die-casting alloy is less than 0.13% in Mn content, preventing the mold from sticking and improving mechanical properties (e.g., high ductility) owing to the Mg content [16,17]. The alloy composition of Silafont-36 is presented in Table 3.

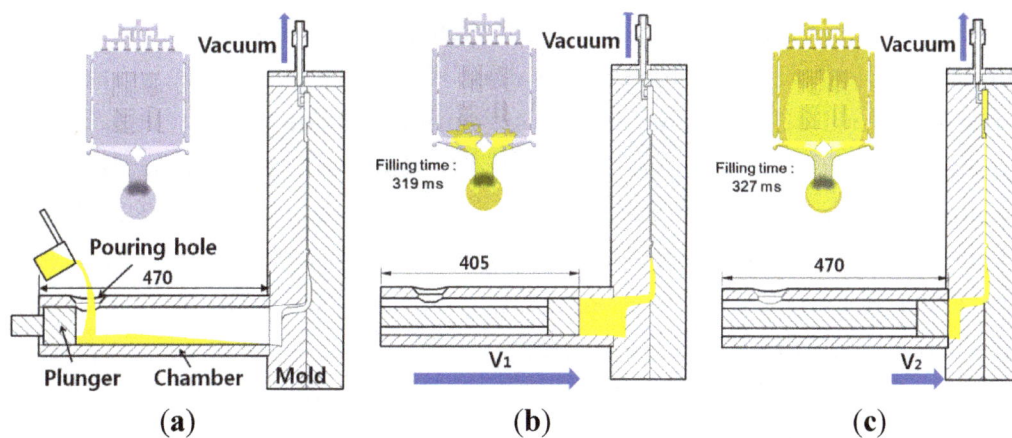

Figure 4. Vacuum die casting process: (**a**) Input melt in chamber; (**b**) Slow shot and (**c**) Fast shot.

Table 1. Process parameters of vacuum cold chamber die-casting.

Parameters		Values
Melt	Material	Silafont-36
	Pouring Temperature (T_m)	710, 730 °C
Die	Material	SKD 61
	Temperature	250 °C
Length of shot sleeve		470 mm
Length at low (V_1) and high (V_2) speed		400 and 70 mm
Working pressure		500 bar
Heat transfer coefficients	between material and mold	3000 W/m²·K
	between mold and mold	1000 W/m²·K

Table 2. Experimental conditions of vacuum cold chamber die-casting.

No	T_m (°C)	V_1 (m/s)	V_2 (m/s)	Vacuum (mbar)
1	700	0.2	2.5	250
2	700	0.2	3.0	no vacuum
3	700	0.2	3.0	250
4	730	0.2	2.5	250
5	730	0.2	3.0	no vacuum
6	730	0.2	3.0	250

Table 3. Chemical composition of Silafont-36 alloy (wt. %).

Si	Mn	Mg	Ga	Ti	Fe
9.91	0.662	0.372	0.011	0.07	0.07

Zn	Cr	Cu	Ca	Al
0.002	0.001	0.001	0.0005	Bal

3. Results and Discussion

3.1. Formability

Figure 5 presents samples, fabricated of thin plates, under the six shot conditions listed in Table 2. Each of the shot conditions was tested five times, producing to a total of 30 samples. Figure 5 shows a selection of the best samples from each condition. As Figure 5 indicates, the product area in each sample was incompletely filled. The six samples were all unfilled in the mazes and, in particular, Mazes 1 and 3, positioned at the end of the product area, were unfilled. While the melt passed through the bumps in the front-positioned Mazes 2 and 4, its fluidity substantially decreased and turbulence occurred. We conclude that the unstable turbulence caused a sudden solidification through Mazes 1 and 3. Of the six conditions, the best formability was found under Condition 6 (T = 730 °C, V_2 = 3.0 m/s, P_{vacuum} = 250 mbar), and the sample fabricated under this condition was unfilled only at the end of the product area and where the overflows were connected.

Better formability was found when the melt pouring temperature was 730 °C rather than 710 °C. Figure 6 presents simulation results (for varying filling temperature) for Conditions 3 and 6 with pouring temperatures of 710 °C and 730 °C, respectively, when the plunger speed was 3.0 m/s and the vacuum pressure was 250 mbar. The simulation result presents the temperature distributions immediately after the melt filled 88% and 100% of the cavity. In Figure 6a, when the pouring temperature of the melt was set to 730 °C, the temperature remained at approximately 725 °C from the gate to the entrances of Mazes 2 and 4. The temperature at the center of the product was roughly 722 °C, and from the ends of Mazes 1 and 3 to the entrances of the overflows the melt solidified rapidly and temperatures remained at approximately 715 °C. The parts at which the temperature rapidly dropped were more likely unfilled, as the fabricated samples show in Figure 5. Figure 6b presents a simulation result when the melt temperature was set to 710 °C. The front half of the product (from the gate to the entrances of Mazes 1 and 3) is at roughly 702 °C, and the other half (from the entrances of Mazes 1 and 3 to the overflows) is at 698 °C. This result suggests that the back half of a product is more likely to remain unfilled. In addition, because there are parts in Mazes and at the entrance of the overflow at which rapid solidification occurs at 700 °C and 695 °C, respectively, it is possible that the mazes could be partially unfilled. The image of the sample (Condition 3) presented in Figure 5, shows a similar result to that of the simulation. Approximately 2 mm at the ends of Mazes 2 and 4 was unfilled. Mazes 1 and 3 contain many unfilled regions, sized 2 mm and over 5 mm. The back half of the products had flow marks, indicating that the melt flow was very unstable.

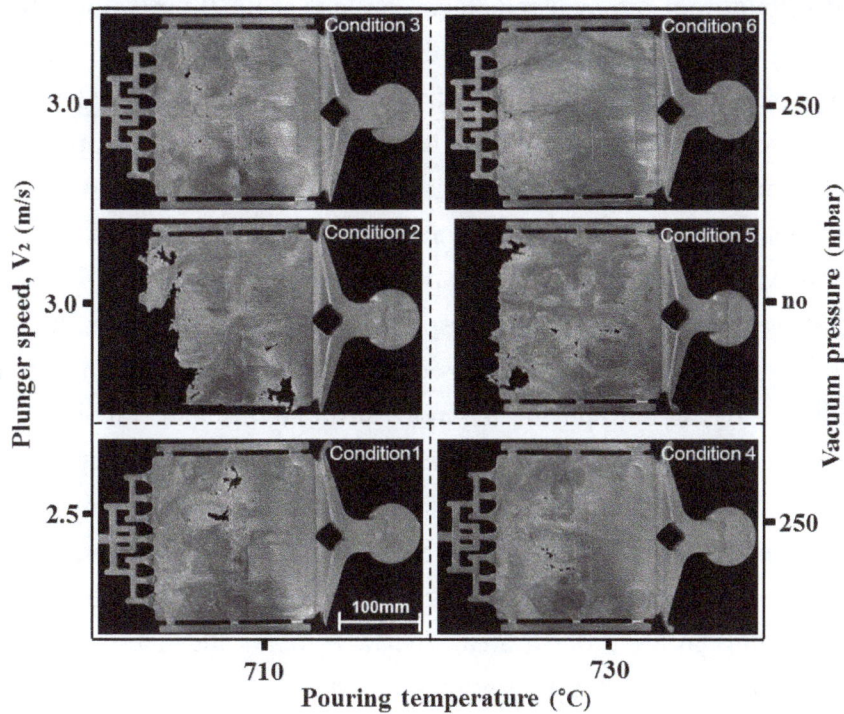

Figure 5. Samples fabricated by vacuum die casting with different parameters: Condition 1 (T = 710 °C, V_2 = 2.5 m/s, P_{vacuum} = 250 mbar); Condition 2 (T = 710 °C, V_2 = 3.0 m/s, no vacuum); Condition 3 (T = 710 °C, V_2 = 3.0 m/s, P_{vacuum} = 250 mbar); Condition 4 (T = 730 °C, V_2 = 2.5 m/s, P_{vacuum} = 250 mbar); Condition 5 (T = 730 °C, V_2 = 3.0 m/s, no vacuum); Condition 6 (T = 730 °C, V_2 = 3.0 m/s, P_{vacuum} = 250 mbar).

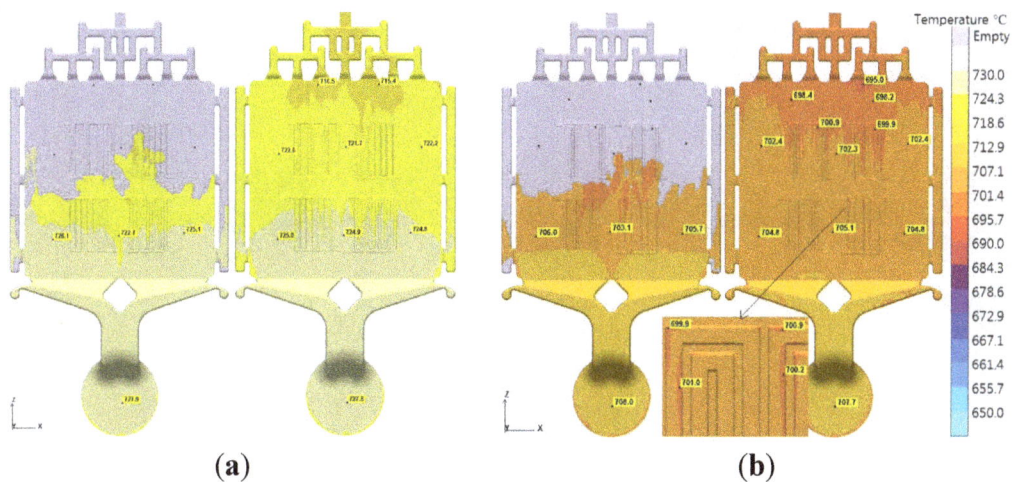

Figure 6. Temperature distribution in die cavity with different pouring temperatures during vacuum die-casting: (**a**) 730 °C and (**b**) 710 °C.

For the plunger speed parameter, *i.e.*, the speed of the plunger's travel, better formability was found at 3.0 m/s than at 2.5 m/s. Figure 7 presents simulation results for Conditions 6 and 4, with plunger speeds of 3.0 m/s and 2.5 m/s, respectively, when the melt temperature was 730 °C and the vacuum pressure was 250 mbar. The simulation results show velocity distributions and gas content immediately after the melt filled the cavity. In Figure 7a, showing Condition 6 with a plunger speed of 3.0 m/s, the velocity distribution of the melt remained constant at 82–90 m/s throughout the product area and was 83 m/s on the left and right edges and in the gap between the mazes. In Mazes 2 and 1 the velocity of the melt was 32 m/s and 36 m/s, respectively and in Mazes 4 and 3 it was 56 m/s and 23 m/s, respectively. This result indicates that the melt substantially slowed as it passed the bumps of the mazes. Under Condition 4 (Figure 7b), when the plunger speed was set to 2.5 m/s, the entire product area indicated approximately 67–81 m/s. Moreover, in the mazes the velocity decreased more than under Condition 6. In Mazes 2 and 1 the velocity of the melt was 24 m/s and 21 m/s, respectively, and in Mazes 4 and 3 it was 50 m/s and 30 m/s, respectively. It is predicted that the melt solidifies while passing through the mazes. The result in terms of air entrapment, indicating gas content in the cavity, shows that the higher the air entrapment, the more unstable the fluid flow in the cavity and the higher the chance of air pockets in the product. As the air entrapment results suggest under these two conditions, air entrapment was high from the ends of Mazes 1 and 3 to the entrances of overflows. In Condition 4, air entrapment was high in the mazes; Mazes 1 and 3, in particular, showed substantial air entrapments. As the fabricated sample images in Figure 5 suggest, Mazes 1 and 3 under Condition 4 had unfilled parts and hot tears in the product area. The flow marks indicate that the melt flow was extremely unstable while passing through Mazes 2 and 4.

Under Conditions 2 and 5, in which no vacuum was used, unfilled parts were found throughout the product and the overflows in the ends were completely unfilled. This result suggests that vacuum is an important factor, substantially affecting formability (castability) in the fabrication of a thin plate in the die-casting process.

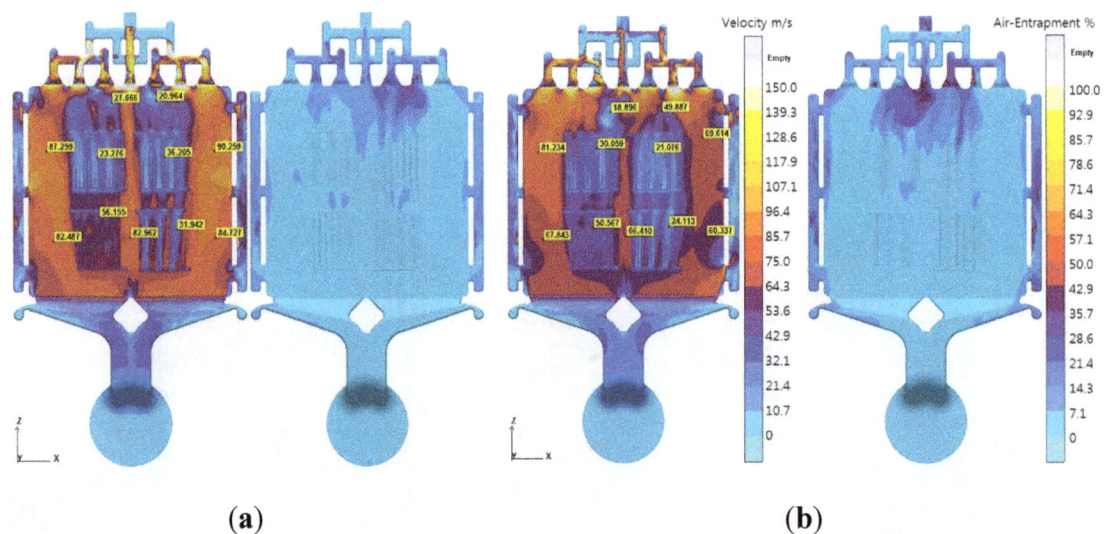

(a)   **(b)**

Figure 7. Velocity distribution of the melt and air content in the die cavity using different plunger speeds in vacuum die-casting: (**a**) 3.0 m/s and (**b**) 2.5 m/s.

3.2. Microstructures and Mechanical Properties

Each fabricated sample was examined for its microstructure and mechanical properties. To observe the microstructure of a sample, a 10×10 mm cut was made from the center of a specimen, mounted and polished, and observed using an optical microscope. To assess mechanical properties, tensile and Vickers hardness tests were performed. For the tensile test, rectangular tension test specimens (subsize specimens) were prepared according to ASTM E8M with gage length and width of 25 mm and 6 mm, respectively. The thickness of the tensile specimens was the same as that of the fabricated thin plate samples. The tensile test was performed at 25 ton MTS, and the strain rate was 1 mm/min. The tensile specimens were cut with a gap between the gate and Mazes 2 and 4. For the measurement of Vickers hardness, the specimens for microstructure assessment were used. Each specimen was tested five times for hardness; the results were averaged (symbol: ■) and the maximum and minimum are also presented.

Figure 8 shows microstructures of the samples fabricated under all conditions. Table 4 shows the grain size of the primary α-Al phase. In the microstructures, the white parts indicate the primary α-Al phase and the remaining parts indicate the eutectic phase; the black parts indicate air pockets. The primary α-Al phase in the microstructure of the sample, fabricated at a melt temperature of 730 °C, was coarser than that fabricated at 710 °C, which indicates that its primary α-Al phase was coarser because the pouring temperature of 730 °C required more time to solidify than that of 710 °C. For the plunger speed, the microstructure of the sample fabricated at 3.0 m/s has more primary α-Al phases than that at 2.5 m/s; when the plunger speed was set to 3.0 m/s rather than 2.5 m/s, the melt more evenly and stably filled the cavity. The vacuumed samples had some porosity whereas the non-vacuumed ones had many gas pockets, indicating that vacuum operation is excellent for removing gases from the cavity and the melt. If the primary α-Al phase is coarse or irregular in shape, the hardness decreases; because the primary α-Al phase is soft, less primary α-Al phase also decreases the ductility. A large degree of porosity easily causes cracks and degrades mechanical properties. The microstructure of the sample fabricated under Condition 6 showed the most even distribution. The mix of the primary α-Al and eutectic phases was properly fractional, and the equivalent diameter in the primary α-Al phase was 74 μm in size and generally spherical.

To measure the porosity of samples formed under vacuum the blister test was conducted. Samples formed without vacuum were excluded due to many defects. Maze 1 was prepared for the blister test and heat-treated at 520 °C for five hours. Figure 9 shows the appearance of samples after the blister test. The red areas represent blister defects. Samples fabricated under Conditions 1 and 4 at a melt temperature of 710 °C have some blisters. In the case of Conditions 3 and 6 fabricated at a melt temperature of 730 °C, the Condition 6 sample has only one blister and the Condition 3 sample has no blisters.

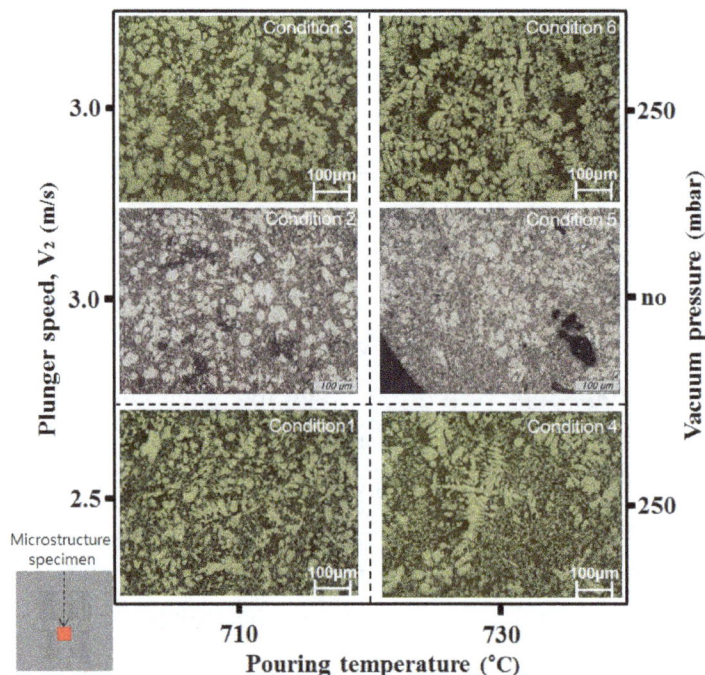

Figure 8. Microstructures of samples fabricated by vacuum die casting with different parameters: Condition 1 (T = 710 °C, V_2 = 2.5 m/s, P_{vacuum} = 250 mbar); Condition 2 (T = 710 °C, V_2 = 3.0 m/s, no vacuum); Condition 3 (T = 710 °C, V_2 = 3.0 m/s, P_{vacuum} = 250 mbar); Condition 4 (T = 730 °C, V_2 = 2.5 m/s, P_{vacuum} = 250 mbar); Condition 5 (T = 730 °C, V_2 = 3.0 m/s, no vacuum); Condition 6 (T = 730 °C, V_2 = 3.0 m/s, P_{vacuum} = 250 mbar).

Table 4. Grain size of primary α-Al phase.

Condition	1	4	2
Size	70	80	70
Condition	5	3	6
Size	68	71	74

Figure 10 presents the tensile strengths, elongations, and Vickers hardness values of the samples made under all conditions. For Conditions 2 and 5 without vacuum operation, the tensile strengths of the samples were 180–185 MPa; the elongations were 4%–5%; and Vickers hardnesses were 78–79 HV. These results are very different from those under vacuum conditions. The mechanical properties among the vacuum samples varied little. As for the effects of the melt pouring temperatures, Conditions 4 and 6 when the temperature was 730 °C showed better mechanical properties than those at 710 °C: tensile strengths were higher by 5–7 MPa, elongations were higher by 0%–1%, and Vickers hardnesses were higher by 3 HV. The primary α-Al phase of the sample fabricated at a melt temperature of 730 °C was coarser than that at 710 °C; on the other hand, the tensile strength, elongation, and hardness of the samples fabricated at 730 °C were higher than those at 710 °C. The reason for this is that samples fabricated at 710 °C have more porosities than those at 730 °C, as shown in Figure 9. For the plunger speed set to 3.0 m/s under Conditions 3 and 6, tensile strengths were higher by 8–10 MPa, elongations longer by 1.5%–2.5%, and Vickers hardnesses higher by 2 HV than those under Conditions 1 and 4 when plunger speed was set to 2.5 m/s. This is because samples fabricated at 3.0 m/s have more of the primary α-Al phase than samples at 2.5 m/s. In addition, the primary α-Al phase of sample fabricated at 2.5 m/s was more irregular in shape than for samples at 3.0 m/s. Therefore, the effect of plunger speed on the mechanical properties was greater than that of melt temperature. The mechanical properties of Condition 6, which displayed good formability and the most evenly distributed microstructure compared to the other samples, were the superior. Its tensile strength and elongation were 265 MPa and 8.5%, respectively, and its Vickers hardness was 105 HV.

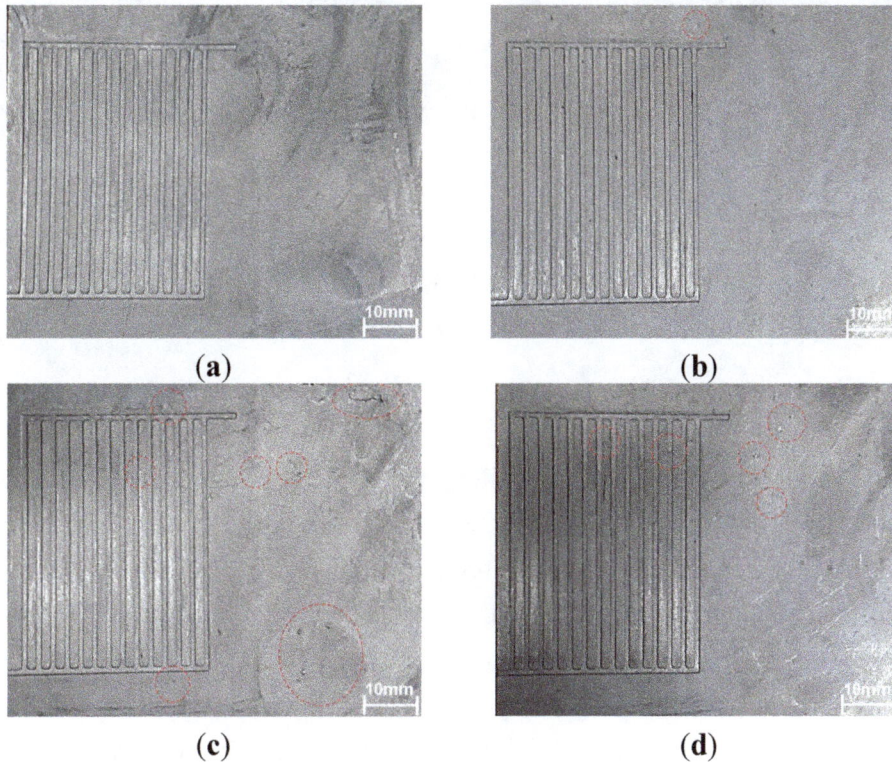

Figure 9. Appearance of samples after blister test (490 °C, five hours): (**a**) Condition 3 (T = 710 °C, V_2 = 3.0 m/s); (**b**) Condition 6 (T = 730 °C, V_2 = 3.0 m/s); (**c**) Condition 1 (T = 710 °C, V_2 = 2.5 m/s) and (**d**) Condition 4 (T = 730 °C, V_2 = 2.5 m/s). The red areas represent blister defects.

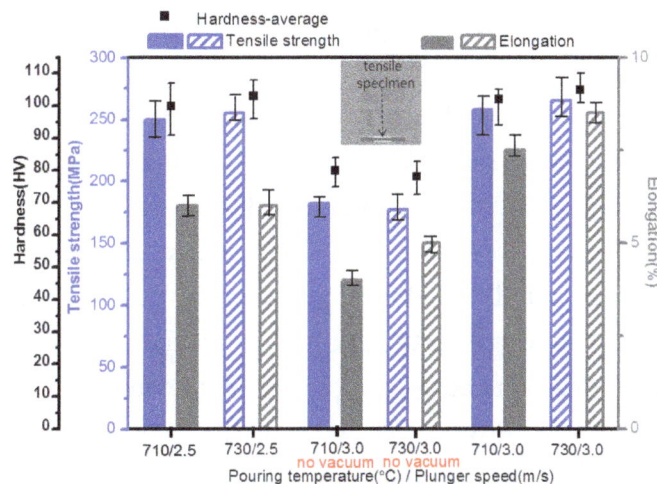

Figure 10. Mechanical properties of samples fabricated by vacuum die casting with different parameters.

The maze part of the sample fabricated under Condition 6 was cut, and its surface and cross section were measured with a digital microscope. Figure 11a presents the surface at 40× magnification and Figure 11b presents the cross section. As the cross section shows, the thicknesses ranged between 0.77 mm and 0.79 mm and the depths of the groves ranged between 0.31 mm and 0.36 mm.

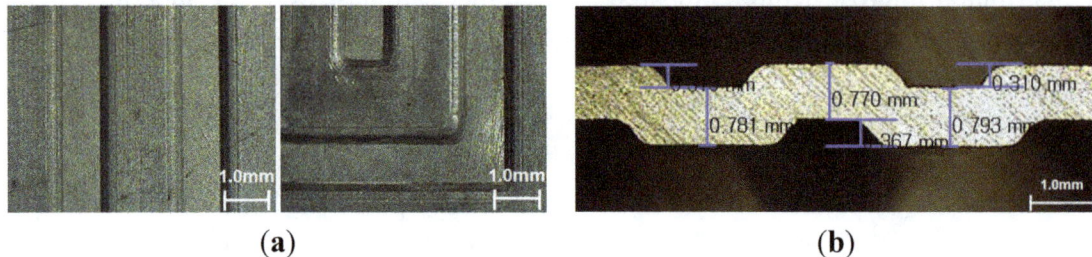

(a)   **(b)**

Figure 11. Surface and cross section in Maze 4 of a sample fabricated with a pouring temperature of 730 °C, plunger speed of 3.0 m/s, and vacuum pressure of 250 mbar (40× magnification). (**a**) Surface; (**b**) Cross section.

4. Conclusions

Using vacuum die casting with a Silafont-36 alloy, we fabricated a thin plate into samples with complex shapes containing four mazes. Findings in terms of formability and mechanical properties of samples fabricated under different melt temperature, plunger, and vacuum pressure values are summarized as follows: (1) When the melt filled the cavity, its fluidity substantially decreased and turbulence occurred while it passed over bumps in the mazes near the gate. The turbulence of the melt caused rapid solidification as it passed through the next mazes, which were partially unfilled. (2) None of the six shot conditions produced a perfect filling. However, the sample fabricated under a melt temperature of 730 °C, plunger speed of 3.0 m/s, and vacuum pressure of 250 mbar was completely filled except for the end of the product and the places where overflows were connected. (3) The microstructure of the sample fabricated at a melt temperature of 730 °C was coarser in the primary α-Al phase than in the sample fabricated at 710 °C. The microstructure of the sample formed at a plunger speed of 3.0 m/s displayed more primary α-Al phase than in the sample formed at 2.5 m/s. The non-vacuumed sample contained many air pockets. (4) The tensile strength and elongation of the sample fabricated at a melt temperature of 730 °C were higher than for those at 710 °C by 6 MPa and 0.5%, respectively. The tensile strength and elongation of the samples fabricated at a plunger speed of 3.0 m/s were higher than for those fabricated at 2.5 m/s by 9 MPa and 2%, respectively. The tensile strength and elongation of the non-vacuumed sample were lower than those of the samples formed under other conditions by approximately 74.5 MPa and 2.5%, respectively. Therefore, it appears that vacuum pressure strongly affects formability and mechanical properties, and that plunger speed rather than melt temperature is the factor that more strongly affects mechanical properties. (5) The tensile strength and elongation of the sample fabricated under a melt temperature of 730 °C, plunger speed of 3.0 m/s, and vacuum pressure of 250 mbar were 265 MPa and 8.5%, respectively.

Acknowledgments

This work was supported by the National Research Foundation of Korea (NRF) Grant funded by the Korea Government (No. 2013R1A1A2062759). This study was also supported by human resources development of the Korea Institute of Energy Technology Evaluation and Planning (KETEP) Grant funded by the Korea Government, Ministry of Knowledge Economy (No. 20104010100540). This study was also supported by the National Research Foundation of Korea (KRF) Grant funded by the Korea Government (MISP) through GCRC-SOP (No. 2012-0001204).

Author Contributions

Chu Kyu Jin designed die of vacuum die casting by performing the simulation. Chul Kyu Jin and Chang Hyun Jang conducted shot experiment and analysis the results. Chung Gil Kang maintained and examined the results of simulation and experiment. All authors have contributed to discussing and revising.

Conflicts of Interest

The authors declare no conflict of interests.

References

1. Lee, K.H.; Kwon, Y.N.; Lee, S.H. Correlation of microstructure with mechanical properties and fracture toughness of A356 aluminum alloys fabricated by low-pressure-casting, rheo-casting, and casting-forging processes. *Eng. Fract. Mech.* **2008**, *75*, 4200–4216.
2. Birol, Y. Semi-solid processing of the primary aluminium die casting alloy A365. *J. Alloys Compd.* **2009**, *473*, 133–138.
3. Atkinson, H.V. Modelling the semisolid processing of metallic alloys. *Prog. Mater. Sci.* **2005**, *50*, 341–412.
4. Dørum, C.; Laukli, H.I.; Hopperstad, O.S.; Langseth, M. Structural behaviour of Al-Si die-castings: Experiments and numerical simulations. *Eur. J. Mech. A Solids* **2009**, *28*, 1–13.
5. Fan, Z.; Fang, X.; Ji, S. Microstructure and mechanical properties of rheo-diecast (RDC) aluminium alloys. *Mater. Sci. Eng. A* **2005**, *412*, 298–306.
6. Miller, A.E.; Maijer, D.M. Investigation of erosive-corrosive wear in the low pressure die casting of aluminum A356. *Mater. Sci. Eng. A* **2006**, *435–436*, 100–111.
7. Niu, X.P.; Hu, B.H.; Pinwill, I.; Li, H. Vacuum assisted high pressure die casting of aluminium alloys. *J. Mater. Process. Technol.* **2000**, *105*, 119–127.
8. Kim, E.S.; Lee, K.H.; Moon, Y.H. A feasibility study of the partial squeeze and vacuum die casting process. *J. Mater. Process. Technol.* **2000**, *105*, 42–48.
9. Kim, Y.C.; Kang, C.S.; Cho, J.I.; Jeong, C.Y.; Choi, S.W.; Hong, S.K. Die casting mold design of the thin-walled aluminum case by computational solidification simulation. *J. Mater. Sci. Technol.* **2008**, *24*, 383–388.
10. Jin, C.K.; Kang, C.G. Fabrication process analysis and experimental verification for aluminum bipolar plates in fuel cells by vacuum die-casting. *J. Power Sour.* **2011**, *196*, 8241–8249.

11. Jin, C.K.; Kang, C.G. Fabrication by vacuum die casting and simulation of aluminum bipolar plates with micro-channels on both sides for proton exchange membrane (PEM) fuel cells. *Int. J. Hydrog. Energy* **2012**, *37*, 1661–1676.

12. Cho, C.Y.; Uan, J.Y.; Lin, H.J. Surface compositional inhomogeneity and subsurface microstructures in a thin-walled AZ91D plate formed by hot-chamber die casting. *Mater. Sci. Eng. A* **2005**, *402*, 193–202.

13. Hu, B.H.; Tong, K.K.; Niu, X.P.; Pinwill, I. Design and optimisation of runner and gating systems for the die casting of thin-walled magnesium telecommunication parts through numerical simulation. *J. Mater. Process. Technol.* **2000**, *105*, 128–133.

14. Jin, C.K.; Bolouri, A.; Kang, C.G.; Hwang, G.W. Thin-plate forming by thixo- and rheoforging. *Adv. Mech. Eng.* **2014**, *2014*, 1–12.

15. Cleary, P.W.; Ha, J.; Prakash, M.; Nguyen T. Short shots and industrial case studies: Understanding fluid flow and solidification in high pressure die casting. *Appl. Math. Model.* **2010**, *34*, 2018–2033.

16. Franke, R.; Dragulin, D.; Zovi, A.; Casarotto, F. Progress in ductile aluminium high pressure die casting alloys for the automotive industry. *Metall. Ital.* **2007**, *5*, 21–26.

17. Zovi, A.; Casarotto, F. Silafont-36, the low iron ductile die casting alloy development and applications. *Metall. Ital.* **2007**, *6*, 33–38.

3

Role of Alloying Additions in the Solidification Kinetics and Resultant Chilling Tendency and Chill of Cast Iron

Edward Fraś [1,†]**, Hugo F. Lopez** [2,*]**, Magdalena Kawalec** [1] **and Marcin Gorny** [1]

[1] Foundry Institute, AGH University of Science and Technology, Reymonta 23, Cracow 30-059, Poland;
   E-Mails: kawalec@agh.edu.pl (M.K.); mgorny@agh.edu.pl (M.G.)

[2] Department of Materials Science and Engineering, University of Wisconsin Milwaukee,
   3200 N. Cramer Street, Milwaukee, WI 53211, USA

[†] Edward Fras recently passed away on 15 January 2013.

[*] Author to whom correspondence should be addressed; E-Mail: hlopez@uwm.edu

Academic Editor: Anders E. W. Jarfors

Abstract: The present work describes the effect of the solidification processing and alloy chemistry on the chilling tendency index, CT, and the chill, w, of wedge-shaped castings made of cast iron. In this work, theoretical predictions were experimentally verified for the role of elements, such as C, Si, Mn, P and S, on the cast iron CT. In addition, inoculation and fading effects were considered in the experimental outcome. Accordingly, the graphite nucleation coefficients, N_s, b, the eutectic cell growth coefficient, μ, and the critical cooling rate, Q_{cr}, for the development of eutectic cementite (chill) were all determined as a function of the cast iron chemistry and time after inoculation. In particular, it was found that increasing the Mn and S contents, as well as the time after inoculation lowers the critical cooling rate, thus increasing the chilling tendency of the cast iron. In contrast, C, Si and P increase the critical cooling rate, and as a result, they reduce the cast iron CT and chill.

Keywords: chill; chilling tendency; gray cast iron; role of C; Si; Mn; P; Mn; inoculation effects

---

1. Introduction

One of the important indices that accounts for the quality of cast iron is its chilling tendency, that is its tendency to solidify according to the Fe-C-X metastable system. The chilling tendency, CT, depends on the physical-chemical state of the liquid iron, while the chill (the fraction of eutectic cementite in the casting) formation depends additionally on the casting cooling rate. In foundry practice, the CT for the various types of cast iron is determined from comparisons of the exhibited fraction of eutectic cementite (chill) in castings solidified under similar cooling rates. Based only on these comparisons, the difference in the chilling tendency for various cast irons can be established, but the absolute CT values for given irons cannot be disclosed [1].

It is well known that the chilling tendency of cast irons determines their subsequent performance in diverse applications. In particular, cast irons possessing a high CT tend to develop zones of white or mottled iron. Considering that these regions can be extremely hard, their machinability can be severely impaired. Alternatively, if white iron is the desired structure, a relatively small chilling tendency will favor the formation of gray iron. This, in turn, leads to low hardness and poor wear properties in as-cast components. Hence, considerable efforts [1–7] have been made to correlate the inoculation practice, iron composition, pouring temperature, *etc.*, with the cast iron CT.

In the published literature, there are only a few attempts aimed at elucidating the mechanisms responsible for the chill of cast iron [3,8–11]. Besides, none of the proposed hypotheses take into account the complexity of the solidification process. In most cases, the proposed theories assume that a single factor is the determinant in establishing the final solidification structure, while the remaining factors are ignored. In addition, various numerical models have been proposed [12,13] to predict whether a given casting or a part of it will solidify according to the stable or metastable Fe-C-X system. Yet, their application is tedious due to extensive numerical calculations. Accordingly, in this work, a simple analytical model is employed to account for the mechanism or mechanisms responsible for the CT of cast iron and, hence, the exhibited chill.

1.1. Analysis

During cast iron solidification, two processes become relevant: (a) nucleation; and (b) growth of eutectic cells, which have a direct effect on the final microstructure.

(a) Nucleation of eutectic cells

In liquid-solid transformations, nucleation is heterogeneous in nature. Accordingly, a simple model for heterogeneous nucleation of graphite in cast iron has recently been proposed [14], which describes the nucleation of eutectic cells, N, by the following expression:

$$N = N_s \exp\left(-\frac{b}{\Delta T_m}\right) \tag{1}$$

and

$$\Delta T_m = T_s - T_m \tag{2}$$

where N_s and b are the nucleation coefficients, ΔT_m is the maximum degree of undercooling at the onset of graphite eutectic solidification, T_s is the graphite eutectic equilibrium temperature and T_m is the minimal temperature at the onset of graphite eutectic solidification (see Figure 1).

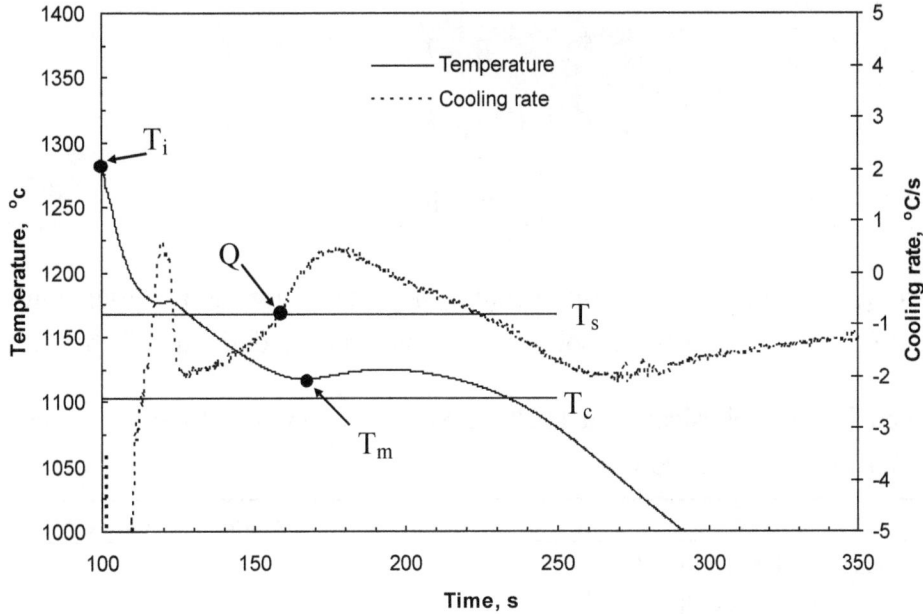

Figure 1. Cooling curves and cooling rate curves for cast iron. T_i = initial temperature of the metal in the mold cavity just after pouring; T_s = graphite eutectic equilibrium temperature; T_c = cementite equilibrium temperature; T_m = minimal temperature; and Q = cooling rate of cast iron at the onset of graphite eutectic solidification.

It is well known that each graphite nucleus gives rise to a single eutectic cell. Hence, it can be assumed that a measure of the graphite nuclei density is given by measurements of the eutectic cell count (cells/volume). Thus, the eutectic cell count can be related to the graphite nucleation potential through Equation (1), where the melt is characterized by the nucleation coefficients, N_s, and b.

(b) The growth rate of eutectic cells can be described by [15]:

$$u = \mu \Delta T^2 \tag{3}$$

where μ is the growth coefficient of the eutectic cells and ΔT is the melt undercooling.

1.2. Onset of Graphite Eutectic Solidification

The heat generated during the solidification of cast iron depends on both the cell count and the growth rate of the eutectic cells. Thus, by combining the heat extraction from the mold with the heat generated during the solidification of eutectic cells, the minimal temperature, T_m (Figure 1), or the maximum undercooling, $\Delta T_m = T_s - T_m$, at the onset of graphite eutectic solidification can be determined from [4]:

$$T_m = T_s - \Delta T_m = T_s \left[\frac{4 c_{ef} Q^3}{\pi^3 L_e N \mu^3 f} \right] \tag{4}$$

where:

$$Q = \frac{2T_s a^2}{\pi \phi c_{ef} M^2} \qquad (5)$$

$$c_{ef} = c + \frac{L_\gamma}{T_{l\gamma} - T_s} \qquad (6)$$

$$\phi = cB + c_{ef} B_1 \qquad (7)$$

$$B = \ln \frac{T_i}{T_l}; \quad B_1 = \ln \frac{T_l}{T_s} \qquad (8)$$

where M is the casting modulus, T_i is the initial liquid metal temperature just after pouring into the mold and T_l, $T_{l\gamma}$, a, c, L_e, L_γ and f are various solidification parameters, defined in Table 1.

Table 1. Selected thermophysical data [14]. C, Si, P: % content of carbon, silicon and phosphorus in cast iron, respectively.

Parameter	Value and Units
Latent heat of eutectic graphite	$L_e = 2{,}028.8$; J/cm^3
Latent heat of austenite	$L_\gamma = 1{,}904.4$; J/cm^3
Specific heat of cast iron	$c = 5.95$; J/(cm^3·°C)
Material mold ability to absorb heat	$a = 0.11$; J/(cm^2·s$^{1/2}$·°C)
Liquidus temperature for pro-eutectic austenite	$T_l = 1{,}636 - 113(C + 0.25\ Si + 0.5\ P)$; °C
Carbon content in eutectic graphite	$C_e = 4.26 - 0.30\ Si - 0.36\ P$; %
Maximum carbon content in austenite at T_s	$C_\gamma = 2.08 - 0.11\ Si - 0.35\ P$; %
Liquidus temperature of pre-eutectic austenite when its composition is C_γ	$T_{l\gamma} = 1{,}636 - 113(2.08 + 0.15\ Si + 0.14\ P)$; °C
Weight fraction of austenite cast iron at the beginning of eutectic solidification	$g_\gamma = (C_e - C)/(C_e - C_\gamma)$
Austenite density	$\rho_\gamma = 7.51$ g/cm^3
Liquid cast iron density	$\rho_l = 7.1$ g/cm^3
Volume fraction of liquid in cast iron at the beginning of eutectic solidification	$f = \rho_\gamma\, g_l/[\rho_\gamma\, g_l + \rho_l\,(1 - g_l)]$

Combining Equations (1) and (4) yields:

$$\Delta T_m = T_s - T_m = \frac{b}{8 \text{ProductLog}[y]} \qquad (9)$$

where:

ProductLog[y] = x is the Lambert function (see http://mathworld.wolfram.com./LambertW-function.html), also known as the omega function, graphically shown in Figure 2. This function can be easily calculated by means of the instruction ProductLog[y] in the Mathematica™ program.

Additionally:

$$y = \frac{b}{8\ 2^{1/4}} \left[\frac{\pi^3\ Le\ N_s\ \mu^3 f}{c_{ef}\ Q^3} \right]^{1/8} \qquad (10)$$

Combining Equations (1) and (9), the spatial cell count can be determined from:

$$N = \frac{N_s}{\exp[8\,\mathrm{ProductLog}(y)]} \tag{11}$$

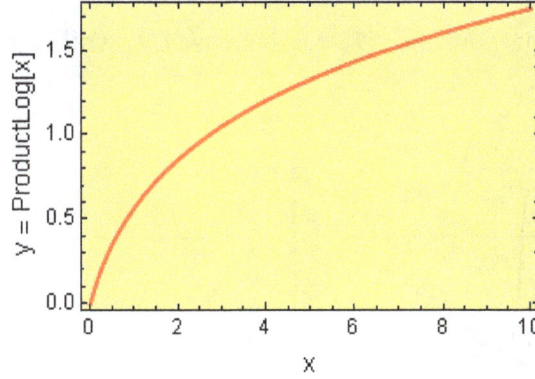

Figure 2. Graphic representation of the ProductLog[y] function for $y \geq 0$.

1.3. Chilling Tendency Index

Figure 3 shows the cooling curves with the T_m values, including the eutectic graphite and N along the wedge axis for two solidified cast irons with different physical-chemical states. The temperature range $\Delta T_{sc} = T_s - T_c$ is also given in Figure 3. Notice that in the temperature range $\Delta T_{sc} = T_s - T_c$, eutectic graphite is the only growing structure (gray cast iron). From this figure, it is apparent that increasing Q to values equal to Q_{cr} leads to an increase of eutectic cells from N to N_{cr}. Below T_c, both graphite and eutectic cementite grow simultaneously, interfering with each other and giving rise to mottled cast iron. Accordingly, T_c can be considered as the transition temperature for the solidification of eutectic cementite from eutectic graphite or the chill formation temperature. From this figure, it is apparent that an increase of the cooling rate from Q to Q_{cr} leads to a reduction in T_m to T_c and, hence, to the formation of eutectic cementite (chill development). Thus, determination of the critical cooling rate, Q_{cr}, is key in establishing the critical conditions for the development of eutectic cementite (chill) in castings.

When $Q = Q_{cr}$ or $M = M_{cr}$ and $\Delta T_m = \Delta T_{sc}$ (Figure 3), the critical cooling rate can be estimated from Equations (1) and (4) as:

$$Q_{cr} = \pi\mu \left[\frac{L_e N_s f \Delta T_{sc}^8}{4 c_{ef}} \exp\left(-\frac{b}{\Delta T_{sc}} \right) \right] \tag{12}$$

Taking into account Equations (1) and (4) (for $\Delta T_m = \Delta T_m$) and (5), the critical casting modulus, M_{cr}, under which it is possible to develop a chill yields:

$$M_{cr} = pCT \tag{13}$$

where:

$$p = \frac{a}{\pi} \left(\frac{32 T_s^3}{L_e c_{ef}^2 \phi 3} \right)^{1/6} \tag{14}$$

and:

$$CT = \left[\frac{1}{N_s f \mu^3 \Delta T_{sc}^8} \exp\left(\frac{b}{\Delta T_{sc}} \right) \right]^{1/6}$$

(15)

Notice from Equation (15) that μ, f, N_s and ΔT_{sc} all decrease CT, while b has the opposite effect.

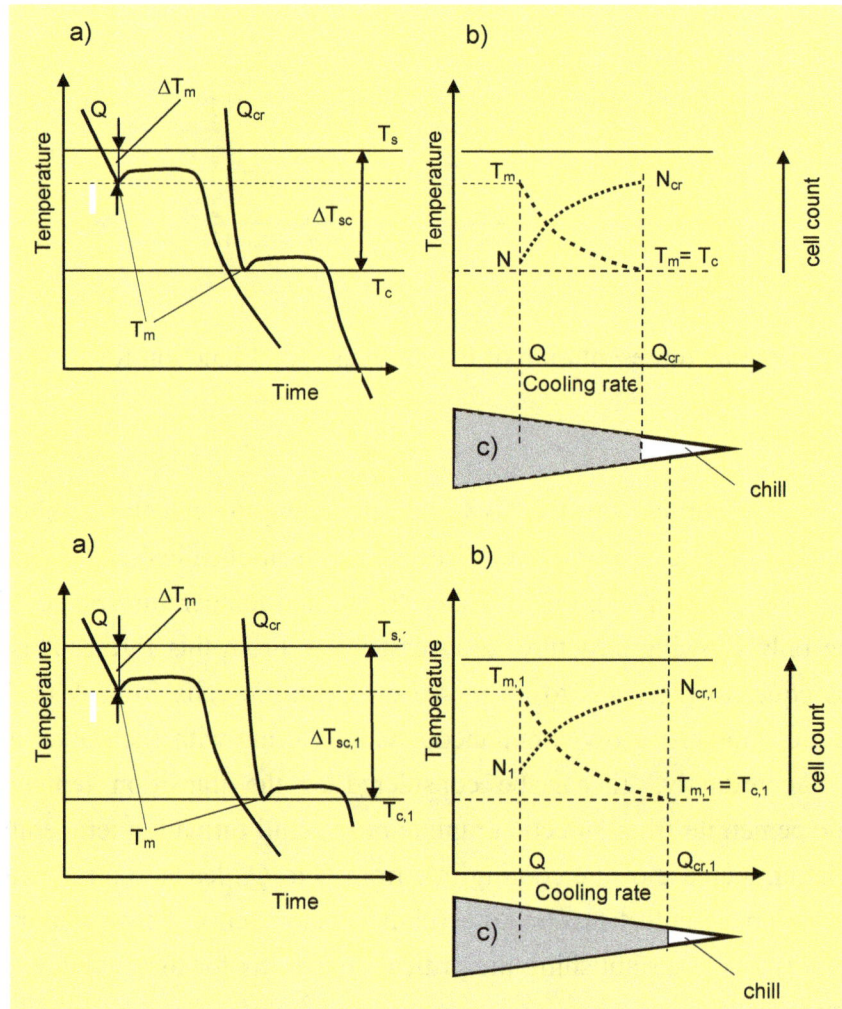

Figure 3. (**a**) Cooling curves and (**b**) effect of the cooling rate, Q, along the wedge axis on minimum solidification temperature, T_m, eutectic graphite and eutectic cell count, N, for two cast irons with different physical-chemical properties. (**c**) Scheme of a wedge section containing a chill.

1.4. Chill

In foundry practice, an assessment of the chilling tendency of cast iron is based on the chill test methods established by the ASTM A367-55T standard. In this case, wedge geometries are employed (Figure 4a). As a first approximation, it is assumed that the geometrical casting modulus, M_w, of a wedge can be estimated by:

$$M_{w} = \frac{F_{ch}}{m} = \frac{\frac{1}{2}h\frac{w}{2}}{\frac{h}{\cos(\beta/2)}} = w\frac{\cos(\beta/2)}{4} \tag{16}$$

$$M_{cr} = p; \quad CT = M_{w} \tag{17}$$

In the above expression, β is the wedge angle, F_{ch} is the half surface area chill triangle and h_{cr} and m are the critical chill height and length (Figure 4a). Taking into account Equations (14) and (17), an expression is found that relates M_{cr} to M_{w}.

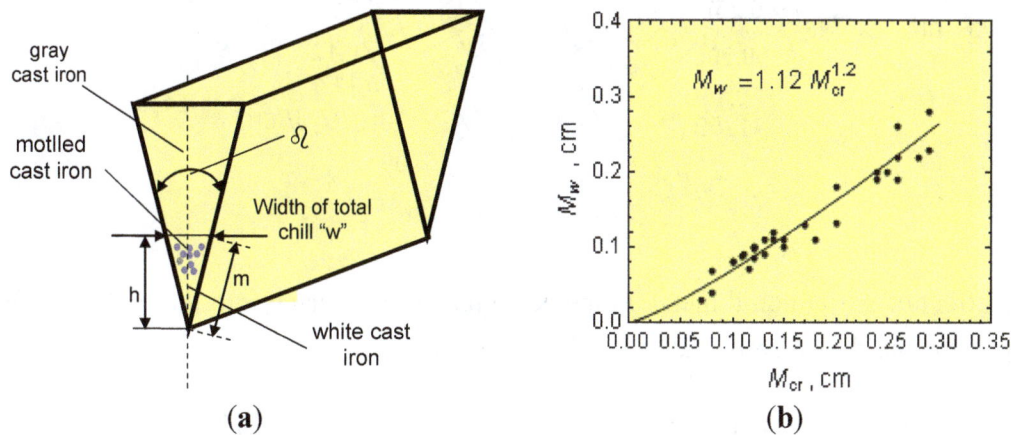

Figure 4. (a) Wedge geometry and (b) relationship between the casting modulus of the wedge with the critical casting modulus.

Nevertheless, the wedge cannot be considered as a simple body. The wedge thin sections are heated up from the thicker sections, and this is not taken into account in Equation (16). Accordingly, Equation (17) is corrected to reflect this effect. Considering the experimental w and β values and plugging them into Equation (16), the modulus M_{w} can be estimated. Figure 4b shows the relationship between the modulus M_{w} and M_{cr}. Notice from this figure that $M_{w} \neq M_{cr}$ and that the function $M_{w} = f(M_{cr})$ can be described by:

$$M_{w} = 1.12\, M_{cr}^{1.2} \tag{18}$$

Thus, by combining Equations (16) to (18), the "w" can be expressed as a function of the CT.

$$w = \frac{4.46\, p^{1.2}}{\cos(\beta/2)} CT^{1.2} \tag{19}$$

Notice that w depends additionally on the p coefficient, which includes parameters related to the cooling rate (Equation (4)), that is:

- The ability for the mold to absorb heat, a;
- The ϕ parameter, Equation (7), which depends on the B and B_1 values (Equation (8)), that is on the initial temperature, T_i, of the cast iron just after pouring the wedges.

The effect of T_i and the ability of the wedge mold to absorb heat, a, on the p parameter is shown in Figure 5. Notice that the p parameter increases as the pouring temperature T_p (and, hence, T_i) decreases, and the ability of the wedge mold to absorb heat increases. As a result, wedge removal at lower temperatures from molds with mold materials having an increasing ability to absorb heat, a (Table 2), leads to increasing chills in order to keep CT constant.

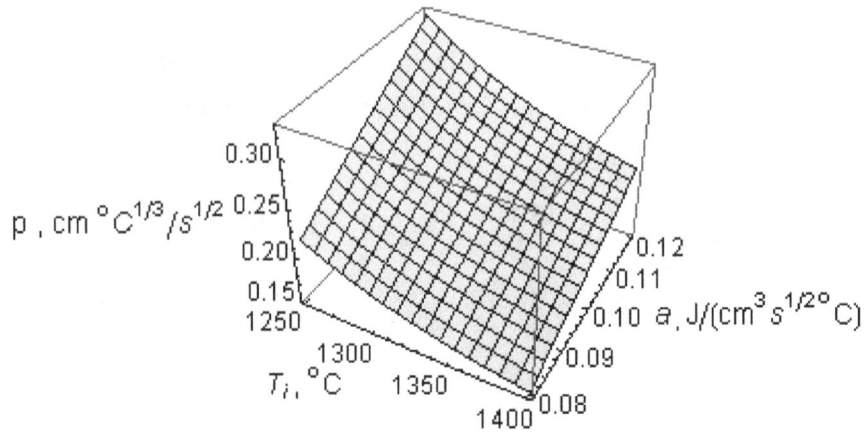

Figure 5. Effect of initial liquid metal temperature just after pouring into the mold, T_i, and the ability of the mold to absorb heat, a, on the p parameter.

Table 2. Reported a values for mold materials [11,16,17].

Mold Material	Material Mold Ability to Absorb Heat, a, $J/(cm^2 \cdot °C \cdot s^{1/2})$
Chromite	0.15
Zircon	0.13−0.15
Quartz sand	0.08−0.12
Olivine	0.10
Chemically-bonded urethane molding sand	0.10−0.12
Chemically-bonded furan molding sand	0.08−0.10
Chemically-bonded shell molding sand	0.10−0.11
LDASC	0.018−0.03

2. Experimental Section

Experimental melts were made in two stages. In the first stage, 40-mm diameter bars containing pig iron and scrap steel were prepared. The scrap composition consisted of Sorelmetal (4.27% C, 0.132% Si, 0.13% Mn, 0.026% P, 0.005% S) and steel scrap (0.2% C, 0.02% Si). In the second stage, the bars, together with various additives, such as carbon, Si, Fe-P and Fe-S, were charged into an induction furnace with a capacity of 1.5 kg. Melting was carried out at low temperatures (*i.e.*, below the equilibrium temperature for the $2C + SiO_2 = Si + 2CO$ reaction in order to avoid C losses as CO to the atmosphere [18]). In addition, low S and Mn contents were employed in order to minimize the formation of MnS compounds [19,20]. In this case, a manganese to sulfur ratio (Mn/S) of 1.7 was assumed to be "balanced" [20]. The aim of using these Mn/S ratios was to investigate the effect of the excess Mn and S on the CT and chill.

After melting and overheating to 1653 K (1380 °C), molten iron was cast into resin-bound foundry molds with bar shapes of 18 and 28 mm in diameter, as well as wedges (see Figure 6). From each melt, a sample was taken for chemical analysis.

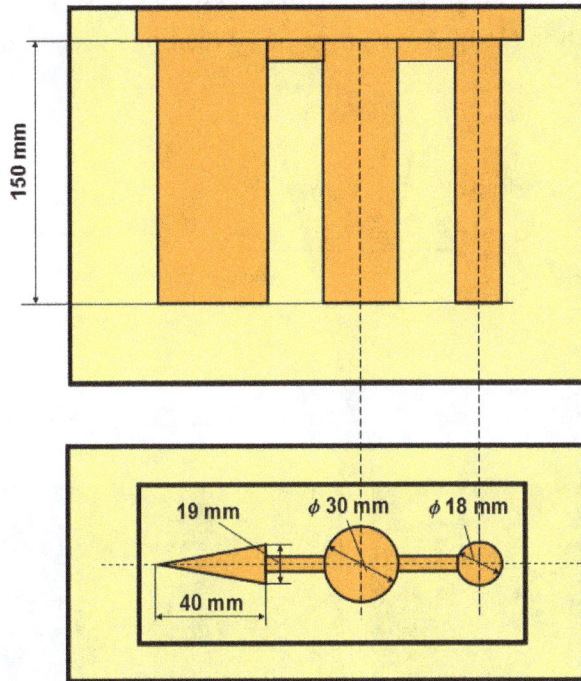

Figure 6. Mold for casting bars and wedges.

Pt-PtRh10 thermocouples were inserted in the center of the wedge and bar cavities of the sand molds. An Agilent 34970A electronic module was employed for numerical temperature recording. Figure 1 shows a typical cooling curve. The cooling curves were used for determinations of T_i just after filling the molds, T_m, $\Delta T_m = T_s - T_m$ and Q at the graphite eutectic equilibrium temperature, T_s. After solidification, specimens for metallographic examination were taken from the wedges and bars. Metallographic examinations were made on specimen cross-sections. After etching the wedges (using nital), the width (w) of the total chill was measured at the gray cast iron-chilled iron intersections (see Figure 7). The wedges and bars were polished and etched using Stead reagent to reveal the graphite eutectic cell boundaries (Figure 7). Determinations of planar cell counts, N_F, were made according to the so-called Variant II of the Jeffries method and by applying the Saltykov formula as an unbiased estimator for the rectangle of observation [21]:

$$N_F = \frac{N_i + 0.5\,N_r + 1}{F} \tag{20}$$

where N_i is the number of eutectic cells inside the measuring rectangle, N_r is the eutectic cell count that intersects the sides of the rectangle, but not its corners, and F is the surface area of the rectangle. The graphite eutectic cells have a granular morphology; hence, it can be assumed that the spatial grain configurations follow the so-called Poisson–Voronoi model [22]. A stereological formula can then be used for determinations of the volumetric cell count N, which yields the average number of eutectic cells per volume [22].

$$N = 0.568 \, (N_F)^{3/2} \tag{21}$$

The cell count, $N_{F,cr}$, measured in wedges from the rectangular surface F (Figure 7a) at the gray cast iron-chilled iron intersections is considered as the critical cell count, $N_{F,cr}$, and it is converted into a volumetric cell count, N_{cr} (Equation (19)). An average of 10 readings was used per data point.

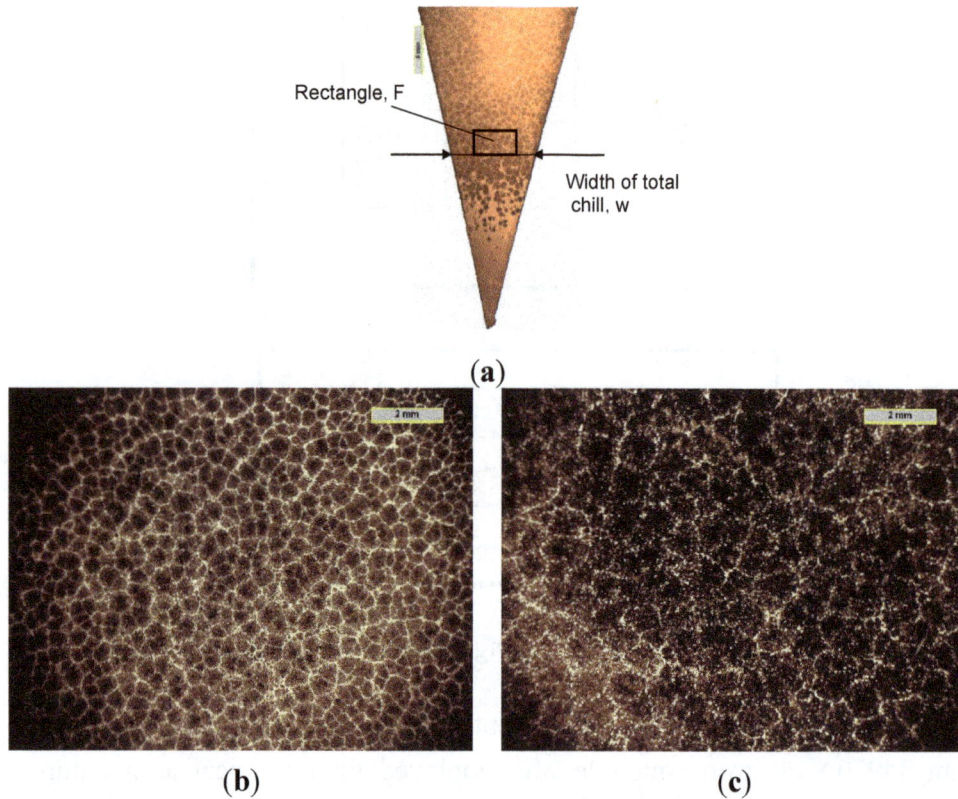

(a)

(b)   (c)

Figure 7. (**a**) Eutectic cells in a wedge-shaped casting; (**b**) bar ϕ 18 mm and (**c**) bar ϕ 28 mm.

The experimental data and using Equation (1) in the logarithmic form yields:

$$\ln N = \ln N_s - b \, \Delta T_m^{-1} \tag{22}$$

where the experimental points $(\Delta T_{m,1}, N_1)$, $(\Delta T_{m,2}, N_2)$ and $(\Delta T_{sc}, N_{cr})$ were determined for bars of 30 and 18 mm in diameter and wedges, respectively. The data are graphically plotted as a $\ln N - \Delta T_i^{-1}$ plot (where $i = 1, 2, cr$), and the data points fit a straight line (Figure 8). From this plot, experimental values of $N_s = e^{11.9} = 147,267$ cm^{-3} and $b = 97.1$ °C were obtained.

The growth coefficient for the experimental melts can be estimated as follows: for bars of diameter "d" and the length "h", the casting modulus can be given as:

$$M = \frac{d \, h}{2(d + 2 \, h)} \tag{23}$$

Thus, for 1.8- and 3.0-cm diameter bars, the casting modulus is 0.42 and 0.65 cm, respectively. In addition, considering Equations (3)–(5), μ can be determined from:

$$\mu = \frac{2 \, T_s \, a^2}{\pi^2 \varphi \, M^2} \left[\frac{4}{f_1 \, Le \, N \, c_{ef}^2 \, \Delta T_m^8} \right]^{1/3} \tag{24}$$

Furthermore, knowing M, T_i, N and ΔT_m, as well as the melt thermophysical data, the effect of C, Mn, P and S on μ can be estimated from Equation (24). In this work, the initial calculations were made separately for 18 and 30 mm bars and then averaged out.

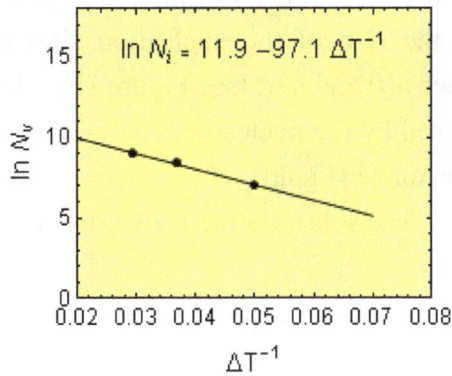

Figure 8. Relationship between undercooling, ΔT_i^{-1}, and volumetric cell count, $\ln N_i$, for melt P1.

The temperature range, ΔT_{sc}, was estimated as follows: There have been numerous experimental investigations that have involved the determinations of T_c temperatures with variable results [23,24], depending on the chemistry of the experimental iron melts. In fact, T_s is the equilibrium temperature when $Q \rightarrow 0$, making it difficult to determine. In this work, determination of the T_s temperature was possible by using selected thermodynamic data [25], as well as the work by Heine [26].

$$T_s = 1154 + 5.25\,\text{Si} - 2.36\,\text{Mn} - 14.88\,\text{P} + 35.0\,\text{S} \tag{25}$$

Furthermore, the T_c temperature can be described by [27]:

$$T_c = 1130.0 + 4.0\,\text{C} - 13.3\,\text{Si} - 50.3\,\text{P} \tag{26}$$

Hence:

$$\Delta T_{sc} = 24 - 4.0\,\text{C} + 18.6\,\text{Si} - 2.3\,\text{Mn} + 35.5\,\text{P} + 35.0\,\text{S} \tag{27}$$

It is worth mentioning that in a binary eutectic Fe-C alloy containing 4.26% C, Equations (25) and (16) yield T_s, T_c and ΔT_{sc} values of 1,427 K (1,154 °C), 1,420 K (1,147 °C) and 7 K (7 °C), respectively, in agreement with the Fe-C diagram.

3. Results and Discussion

3.1. Nucleation Coefficients N_s, b

Different particles with various sizes that act as graphite nucleation substrates are found to be present in cast iron melts. The type, number and size of these substrates are in general influenced by numerous factors, including the chemistry of the cast iron, inoculation, charge materials, melt precondition [28], temperature, bath holding times, furnace atmosphere and the type of slags. In general, the number of nucleation sites from these substrates can be described by the following size distribution function [29]:

$$n(l) = \frac{N_s}{l_a} \exp\left(-\frac{l}{l_a}\right) \qquad (28)$$

where N_s and l_a are the number of all of the potential graphite nucleation sites in the melt and the average site size, respectively; l is the size of the nucleation sites (Figure 9a). This function is schematically shown for two melt states $n(l)$ and $n_i(l)$ (see Figure 9b). The number of all of the graphite nucleation sites in the melt is determined by the nucleation coefficient, N_s (*i.e.*, the area under the $n(l)$ and $n_i(l)$ curves for the l_{min}, $l_{min,1}$ to ∞ range (Figure 9b)). However, not all of the substrate sites take an active role in the nucleation process. The substrate site size, l, from which a nucleus of graphite can grow is determined by (see Figure 9a):

$$AB = l = 2r^* \sin\theta \qquad (29)$$

where θ is the wetting angle and r^* is the critical size of a graphite nucleus:

$$r^* \geq \frac{2\,\sigma\,T_S}{\Delta H\,\Delta T} \qquad (30)$$

and σ is the nucleus-melt interfacial energy, ΔH is the latent heat of solidification and ΔT is the degree of undercooling.

Taking into account Equations (29) and (30), the graphite substrate site size at a given undercooling, ΔT, can be calculated from:

$$l \geq \frac{4\,\sigma\,T_s \sin\theta}{\Delta H\,\Delta T} \qquad (31)$$

From Equation (31), it is found that as the undercooling, ΔT, increases, the nucleation size, l, decreases. Figure 9c shows a cooling curve, $T(t)$, where the arrow indicates the maximum undercooling, ΔT_m. Notice that within the time interval t_b to t_m or $\Delta T = 0$ to $\Delta T = \Delta T_m$, all of the sites with $l_m \leq l \leq \infty$ sizes become active for nucleation purposes. At the time t_m, the degree of undercooling drops, enabling graphite particles and substrates with $l \geq l_m$ sizes to be active. In turn, this indicates that for $t > t_m$, no active sites are present in the set of graphite substrates. As a result, any nucleation events stop at t_m. Thus, the number of active nucleation sites, N, is given by the area below the $n(l)$ curve in $l_m \leq l \leq \infty$ ($0 \leq \Delta T \leq \Delta T_m$). Plugging Equation (31) into Equation (28) and integrating from l_m to ∞, the second nucleation coefficient b is obtained:

$$b = \frac{4\,T_s\,\sigma\,\sin\theta}{l_a \Delta H} \qquad (32)$$

The effect of the cast iron chemistry on N_s, σ, θ and l_a is not known, and therefore, it is not possible to determine its effect on the nucleation coefficients N_s and b.

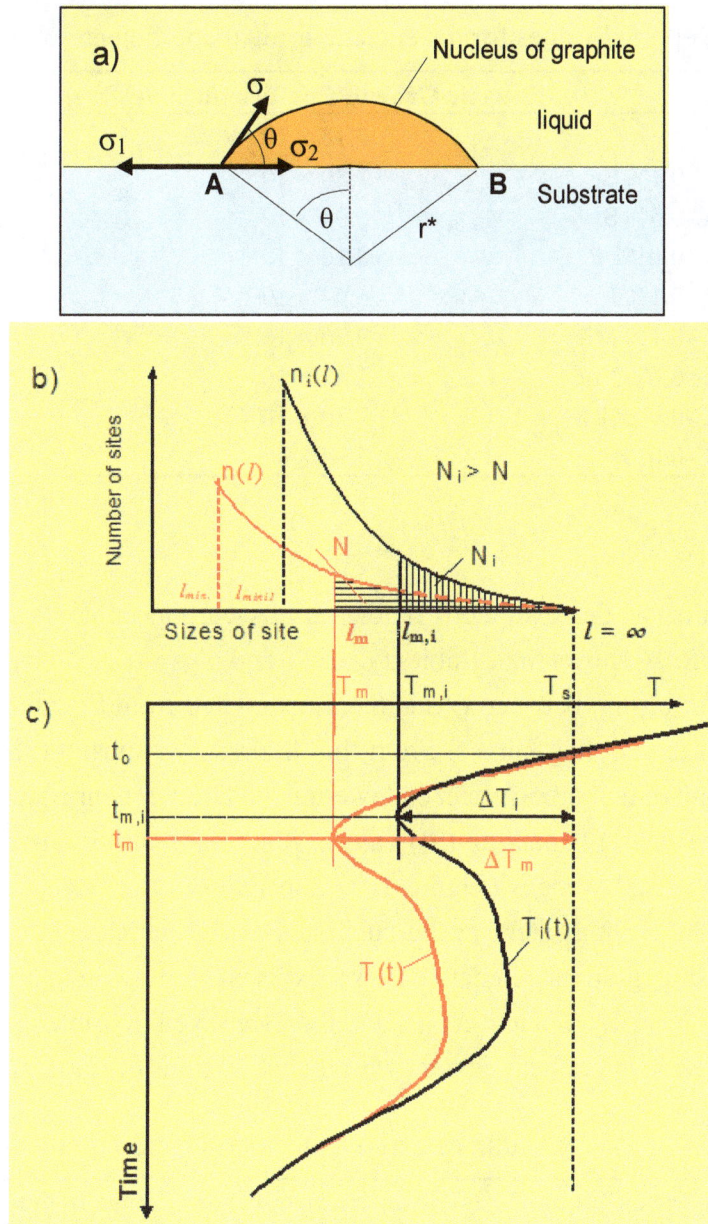

Figure 9. (**a**) Graphite nucleus on a substrate (**b**), size distribution for graphite nucleation sites and (**c**) cooling curves $T(l)$ and $T_i(l)$ for two physical-chemical states of cast iron melts.

3.2. The Growth Coefficient, μ

Little is known about μ, and reported values for μ in pure Fe-C and Fe-C-X alloys are given in Table 3 [30]. From this table, it is apparent that μ is highly sensitive to very small changes in the cast iron chemistry.

In Fe-C alloys, μ can be analytically derived [15], but the reported expressions are not simple. In general, μ is a constant depending on the composition of the cast iron [11] through: (1) the slope of the liquidus lines; (2) the volume fraction; (3) the Gibbs–Thomson coefficients for eutectic phases; (4) the contact angles between eutectic solid phases and the liquid; and (5) the liquid diffusivity. Unfortunately, there is a lack of information in the published literature on this subject. Thus, it is not possible to know the effect of the chemical composition on μ.

Table 3. Reported values for the eutectic growth coefficient, μ [30].

Alloy	Eutectic Growth Coefficient, cm/(°C²·s)
Fe-C	9.18×10^{-6}
Fe-C-0.1% Si	7.30×10^{-6}
Fe-C-0.5% Si	5.94×10^{-6}
Fe-C-0.01% P	8.16×10^{-6}
Fe-C-0.1% P	6.94×10^{-5}
Fe-C-0.1% Mn	9.76×10^{-6}
Fe-C-0.6% Mn	1.47×10^{-5}
Fe-C-0.005% S	7.30×10^{-6}
Fe-C-0.01% S	5.94×10^{-6}

3.3. Effect of Carbon

Table 4 shows the results on the exhibited T_i for various carbon contents in the experimental melts [11]. In addition, Figure 10 shows the exhibited Q, ΔT_m and N_F values.

Figure 10a shows the relationship between Q and the T_s estimated from Equation (6), as well as the experimental outcome from the cooling curves. From this figure, it is apparent that carbon increases the cooling rate. This is a result of the carbon effect on lowering the liquidus temperature, T_l, which, in turn, reduces the B_1 (Equation (8)) and ϕ (Equation (9)) values. Moreover, carbon increases the number of eutectic cells in the cast bars and wedges (Figure 10b) and reduces the undercooling (Figure 10c). Alternatively, the increase in cooling rates can be attributed to the increase in thermal conductivity associated with the increasing graphite content. These experimental correlations are described by regression equations for N_F, $N_{F,1}$ N_{cr}, ΔT_m and $\Delta T_{m,1}$, and they are given in Table 4. Notice in particular that the experimental outcome is in good agreement with the predictions of Equations (5), (9) and (11).

Table 4. Carbon content in cast iron and experimental T_i values.

No. Melt *	C, %	Initial Liquid Metal Temperature just after Pouring into the Mold, T_i, °C		
		Bar 2.8 cm	Bar 1.8 cm	Wedge
C1	3.06	1,375	1,251	1,250
C2	3.21	1,357	1,274	1,272
C3	3.42	1,335	1,260	1,272
C4	3.46	1,355	1,262	1,272
C5	3.48	1,352	1,263	1,276
C6	3.67	1,350	1,260	1,279
Cell count, cm⁻²	bar 2.8 cm	$N_F = 2{,}730.6 - 1{,}653.9\,C + 268.3\,C^2$		
Cell count, cm⁻²	bar 1.8 cm	$N_{F,1} = 7{,}001.2 - 4{,}368.1\,C + 723.4\,C^2$		
Critical cell count		$N_{cr} = -14{,}717 + 8{,}822.0\,C - 1{,}232.5\,C^2$		
Undercooling, °C	bar 2.8 cm	$\Delta T_m = 63.9 - 13.0\,C$		
Undercooling, °C	bar 1.8 m	$\Delta T_{m,1} = -347.8 + 248.4\,C - 40.3\,C^2$		
Wedge width of the chill, cm		$w = 3.05 - 0.66\,C$		
Graphite eutectic growth coefficient, μ, cm/(s·°C²)		$\mu = 10^{-5} \cdot (16.152 - 10.25\,C + 1.644\,C^2)$		
Nucleation coefficients				
°C		$b = -574.73 + 501.33\,Cz - 89.60\,C^2$		
cm⁻³		$N_s = 10^6 \cdot (-2.11 + 1.49\,Cz - 0.241\,C^2)$		

* Si = 1.40%–1.46%; Mn = 0.070%–0.076%; P = 0.025%–0.034%; S = 0.016%–0.019%.

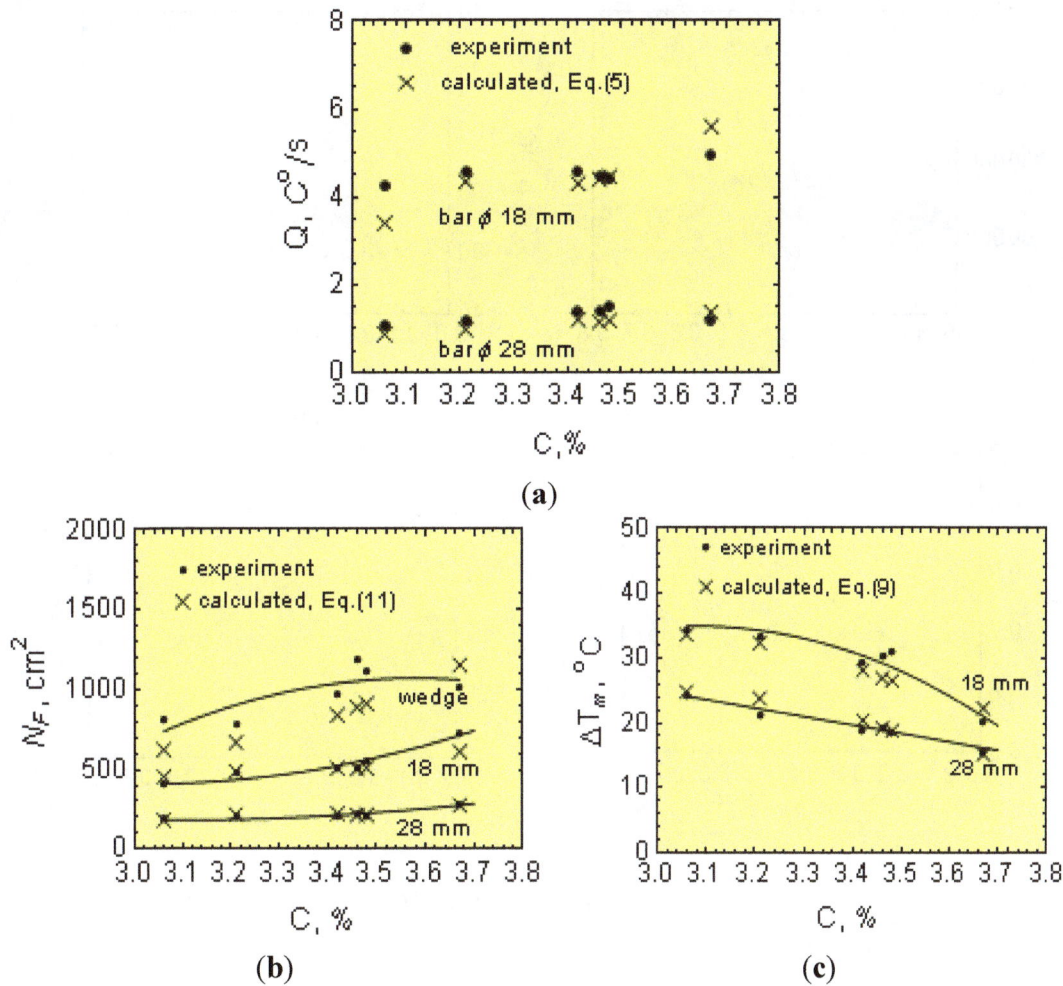

Figure 10. Effect of carbon on (**a**) the cooling rate, Q, of cast iron at the onset of solidification, (**b**) cell count and (**c**) undercooling.

Figure 11 shows the effect of carbon on the nucleation coefficients (Figure 11a,b) and on the graphite eutectic growth coefficient (Figure 11e). These correlations are described by regression expressions for μ, N_s and b in Table 4. Figure 11 also shows the volume fraction of liquid in the cast iron at the onset of eutectic solidification (Figure 11c) and the temperature range ΔT_{sc} (Figure 11d). Notice from Figure 11 that carbon increases N_s, μ and f while lowering b and ΔT_{sc}. As a result, carbon reduces the CT (Figure 11g) and increases Q_{cr} (Figure 11f), thus reducing the chill (Figure 11g). Notice that the experimentally-determined chill width and the one predicted using Equation (4) are rather similar.

From the theory on eutectic growth [15] for pure Fe-C alloys, it is found that carbon does not have any influence on μ. However, Figure 11e shows that carbon increases μ. This can be explained by considering the segregation of Si during solidification. Silicon is soluble in austenite, and it progressively decreases in the liquid phase during the solidification of the pro-eutectic austenite. This agrees with reported data (Table 3) for Si, which shows that as the Si content decreases, the graphite eutectic growth coefficient increases in magnitude.

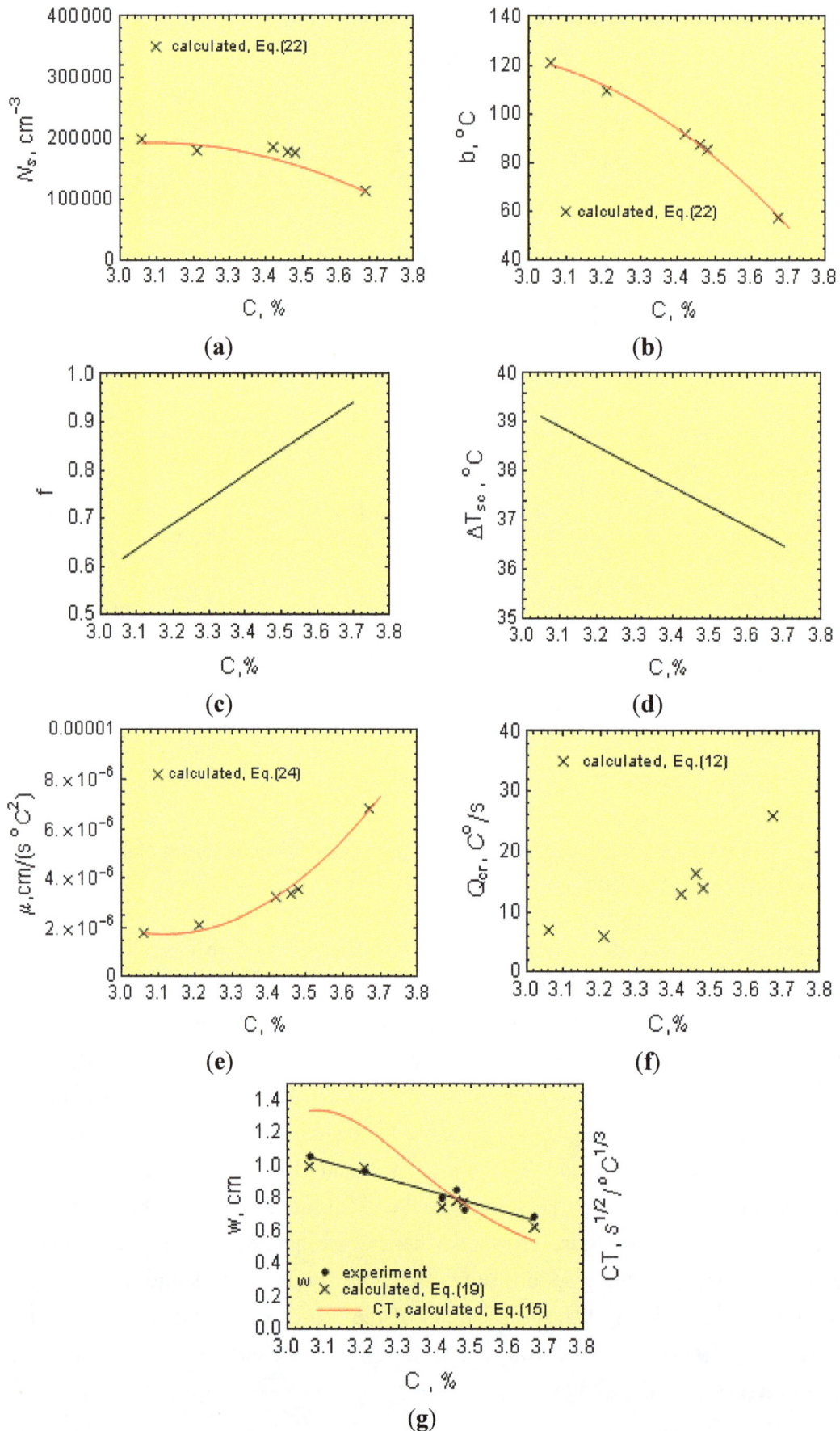

Figure 11. Effect of carbon on (**a**) N_s, (**b**) b, (**c**) f, (**d**) ΔT_{sc}, (**e**) μ, (**f**) Q_{cr} and (**g**) the chilling tendency (CT) index and w.

In order to establish the influence of carbon on the CT, the following calculations were made. The CT was estimated for a carbon content C = 3.06% using Equation (15) yielding a CT = 1.21 $s^{1/2}/°C^{1/3}$. This value was assumed as a reference value. Next, calculations were made on CT values as they are influenced by carbon through f, μ, ΔT_{sc}, N_s and b. The results of these calculations were normalized to 100%, and they are shown in Figure 12.

Notice from this figure that the intensity effect of carbon through μ on CT is very strong. When the carbon content increases, the influence of carbon on the CT through μ and N_s increases, while it decreases through b. In addition, the effect of carbon on CT through f, ΔT_{sc}, is almost balanced.

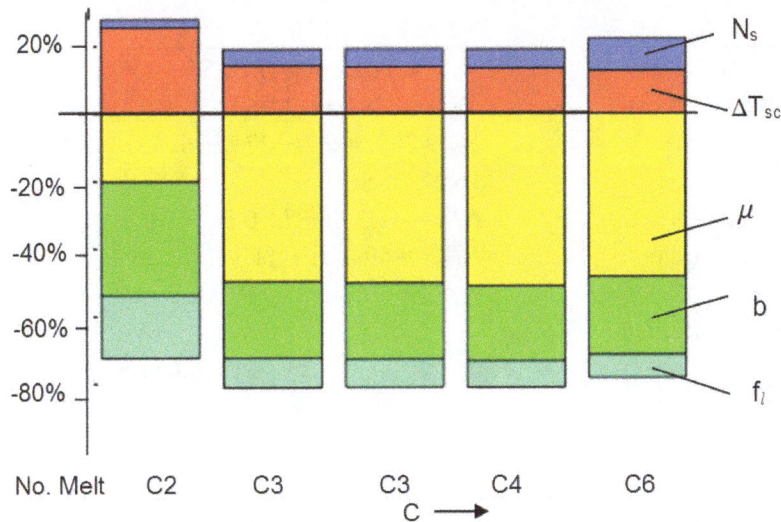

Figure 12. Carbon intensity effect on the CT index through f_l, μ, ΔT_{sc}, N_s and b (− CT decreases; + CT increases).

3.4. Effect of Silicon

The results of the experimental measurements of Si in the cast iron and the initial liquid metal temperature just after pouring into the mold, T_i, are given in Table 5, while the cooling rate, Q, of cast iron at the onset of solidification, the cell count and undercooling are shown in Figure 13. These results are reported in some detail elsewhere [31].

From Figure 13a, it is apparent that Si increases the cooling rate as a result of a reduction in the liquidus temperature, T_i. (just as with carbon). Si increases the number of eutectic cells in the experimental bars and wedges (Figure 13b), while reducing the undercooling (Figure 13c). The experimental results are described by the regression equations for N_F, $N_{F,1}$, N_{cr}, ΔT_m and $\Delta T_{m,1}$, and they are given in Table 5. Notice in particular that the experimental outcome is in good agreement with the predictions of Equations (5), (9) and (11) (see Figure 13).

Figure 14 shows the effect of silicon on the nucleation coefficients (Figure 14a,b), f (Figure 14c), ΔT_{sc} (Figure 14d) and μ (Figure 14e). Notice from these figures that in the range of up to 2.4% Si, there is an increase in the magnitude of N_s and b. At approximately 2.4% Si, N_s and b reach a maximum, and then, they decrease. In contrast, μ deceases in the Si range of up to 2.4% Si, reaches a minimum and then increases. These correlations are described by the regression equations for μ, N_s and b given in Table 5.

Table 5. Effect of silicon on the resultant T_i.

No. Melt *	Si, %	Initial Liquid Metal Temperature just after Pouring into the Mold, T_i, °C		
		Bar 2.8 cm	Bar 1.8 cm	Wedge
Si1	1.82	1,348	1,268	1,249
Si2	1.93	1,348	1,268	1,265
Si3	2.04	1,348	1,268	1,261
Si4	2.11	1,348	1,268	1,241
Si5	2.22	1,374	1,251	1,247
Si6	2.53	1,351	1,253	1,252
Si7	2.70	1,352	1,235	1,239
Si8	2.85	1,354	1,279	1,276

Cell count, cm^{-2}	bar 2.8 cm	$N_F = 4{,}040.8 - 5{,}474.5\,\text{Si} + 2{,}475.5\,\text{Si}^2 - 354.6\,\text{Si}^3$
Cell count, cm^{-2}	bar 1.8 cm	$N_{F,1} = 8{,}787.4 - 11{,}547\,\text{Si} + 5{,}165.8\,\text{Si} - 731.1\,\text{Si}^3$
Critical cell count, cm^{-2}		$N_{cr} = 200{,}666.7 - 271{,}316.3\,\text{Si} + 118{,}435.2\,\text{Si}^2 - 16{,}272.5\,\text{Si}^3$
Undercooling, °C	bar 2.8 cm	$\Delta T_m = 51.2\,\text{Si} - 11.6\,\text{Si}^2 - 36.8$
Undercooling, °C	bar 1.8 cm	$\Delta T_{m,1} = 49.4\,\text{Si} - 11.1\,\text{Si}^2 - 28.7$
wedge width of the chill, cm		$w = 2.5 - 1.45\,\text{Si} + 0.21\,\text{Si}^{-2}$
Graphite eutectic growth coefficient μ, cm/(s·°C^2)		$\mu_s = 10^{-5}(3.9 - 3.1\,\text{Si} + 0.683\,\text{Si}^2)$
Nucleation coefficients °C		$b = 822.5\,\text{Si} - 170.9$
cm^{-3}		$N_s = 7.7 \times 10^7\,\text{Si}^{72.7}\exp(-27.8\,\text{Si})$

* C = 3.20%–3.27%; Mn = 0.072%–0.077%; P = 0.018%–0.024%; S = 0.010%–0.016%.

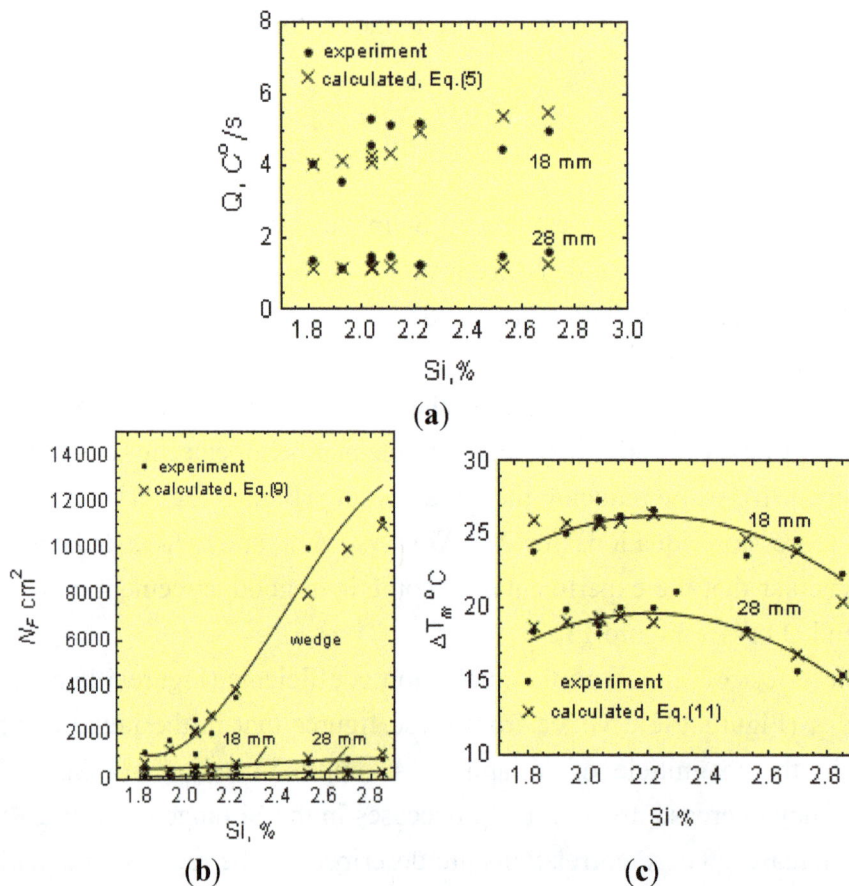

(a)

(b)

(c)

Figure 13. Effect of Si on (**a**) Q, (**b**) N_F and (**c**) ΔT_m.

Notice that the Si tendency to lower the magnitude of μ at low Si contents agrees with the data given in Table 3. Moreover, Si increases f and ΔT_{sc}. As a result, Si lowers the CT (Figure 14g) and increases Q_{cr} (Figure 14f), diminishing the chill (Figure 14g). Notice that the experimental width of the chill and the one predicted using Equation (5) are rather similar (see Figure 14g).

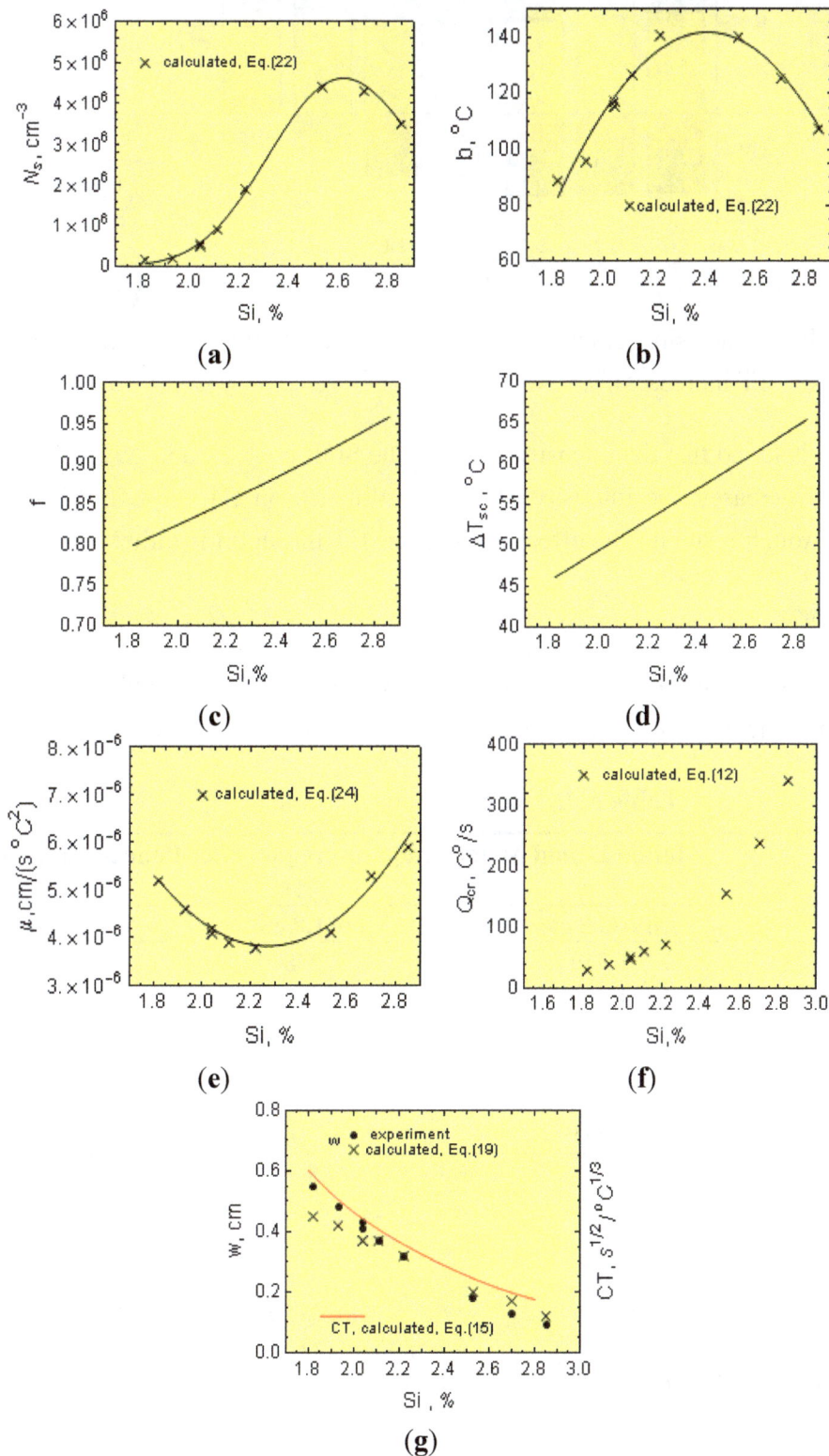

Figure 14. Effect of silicon on (**a**) N_s, (**b**) b, (**c**) f, (**d**) ΔT_{sc}, (**e**) μ, (**f**) Q_{cr} and (**g**) the CT index and w.

Similar to the effect of carbon, the influence of Si on the CT index through f, μ, ΔT_{sc}, N_s and b was determined (Figure 15).

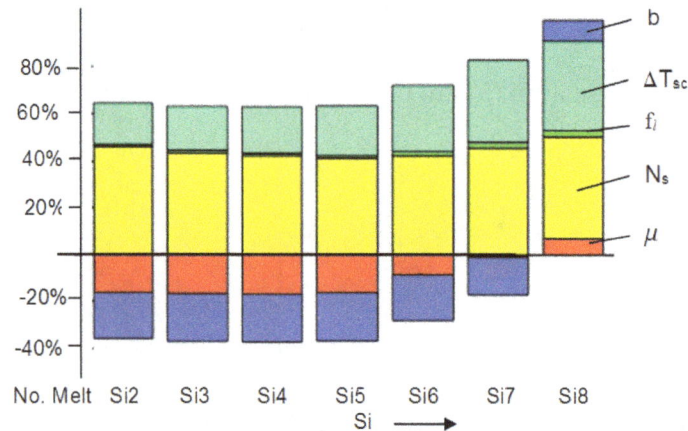

Figure 15. Silicon intensity effect on the CT index, through f, μ ΔT_{sc}, N_s and b (− CT decreases; + CT increases).

From Figure 15, it is found that the intensity effect of the Si through N_s and ΔT_{sc} on the CT is strong. When the Si content increases, the intensity of the Si influence on CT through ΔT_{sc} also increases, while it diminishes through b and μ. The effect of Si on the CT through f is rather small.

3.5. Effect of Manganese

The experimental results on the effect of Mn on the various nucleation and growth factors in the cast iron are given in Table 6. In addition, Figure 16 shows the effect of Mn on Q, N_F and ΔT_m.

Table 6. Effect of Mn on cast iron T_i.

No. Melt *	Mn, %	Initial Liquid Metal Temperature just after Pouring into the Mold, T_i, °C		
		Bar 2.8 cm	Bar 1.8 cm	Wedge
Mn1	0.078	1,350	1,248	1,242
Mn2	0.19	1,356	1,254	1,231
Mn3	0.33	1,360	1,260	1,256
Mn4	0.49	1,341	1,210	1,242
Mn5	0.73	1,348	1,264	1,244
Mn6	0.91	1,341	1,210	1,252

Cell count, cm^{-2}	bar 2.8 cm	$N_F = 115{,}587.8 - 119{,}801.3 \, Mn + 1.37 \times 10^6 \, Mn^2$
Cell count, cm^{-2}	bar 1.8 cm	$N_{F,1} = 322.3 + 457.5 \, Mn$
Critical cell count, cm^{-2}		$N_{cr} = 569.5 + 367.9 \, Mn$
Undercooling, °C	bar 2.8 cm	$\Delta T_m = 18.7 + 5.4 \, Mn$
Undercooling, °C	bar 1.8 cm	$\Delta T_{m,1} = 27.02 + 2.8 \, Mn$
Wedge width of the chill, cm		$w = 0.92 + 0.22 \, Mn$
Graphite eutectic growth coefficient, μ cm/(s·°C^2)		$\mu_s = 10^{-6} \, (4.2 - 2.1 \, Mn)$
Nucleation coefficients °C		$b = 84.6 + 64.6 \, Mn$
cm^{-3}		$N_s = 115{,}587.8 - 119{,}801.3 \, Mn + 1.37 \times 10^6 \, Mn^2$

* C = 3.29%–3.37%; Si = 1.21%–1.29%; P = 0.020%–0.03%; S = 0.01%–0.02%.

Manganese increases the number of eutectic cells in bars and wedges (Figure 16b) and the undercooling (Figure 16c). These results are described by regression equations for N_F, $N_{F,1}$, N_{cr}, ΔT_m and $\Delta T_{m,1}$, and they are given in Table 6. Notice that the experimental and calculated values from Equations (5), (9) and (11) are very close to each other.

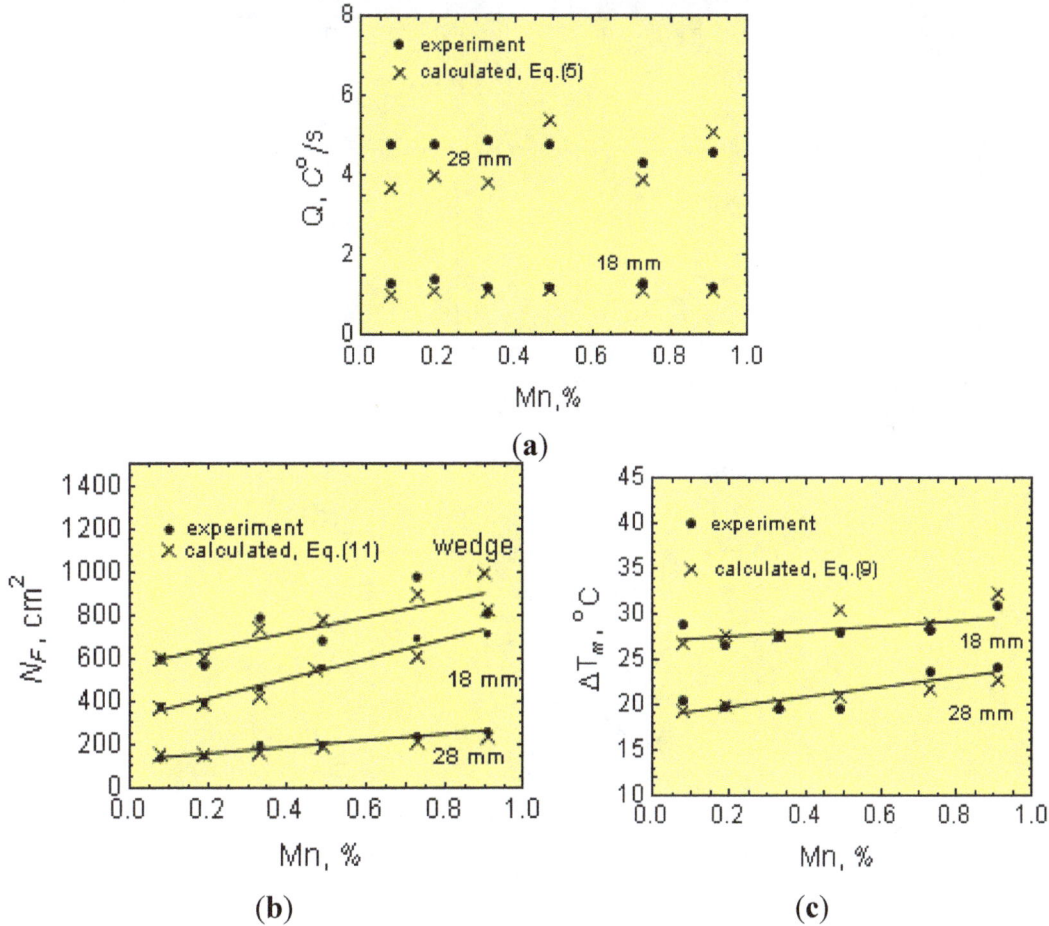

Figure 16. Effect of manganese on (a) Q, (b) N_F and (c) ΔT_m.

Manganese does not have a significant effect on f and ΔT_{sc}. Figure 17 shows the effect of Mn on the nucleation coefficients (Figure 17a,b) and on μ (Figure 17c) (see the regression equations for μ, N_s and b given in Table 6). Notice from Figure 17 that Mn increases the nucleation coefficients, N_s and b, while decreasing μ. In particular, the Mn tendency to lower μ agrees with the data given in Table 3. Mn increases the CT index (Figure 17e) and Q_{cr} (Figure 17d), reducing the chill (Figure 17e). Notice that there is good agreement between the experimental outcome and the predictions of Equation (5) for w (see Figure 17e).

The manganese intensity effect on the CT index through N_s, b and μ is shown in Figure 18. Notice that the most significant effect is on N_s. In addition, the intensity of the influence of Mn on the CT tendency index, through μ and b, is rather similar.

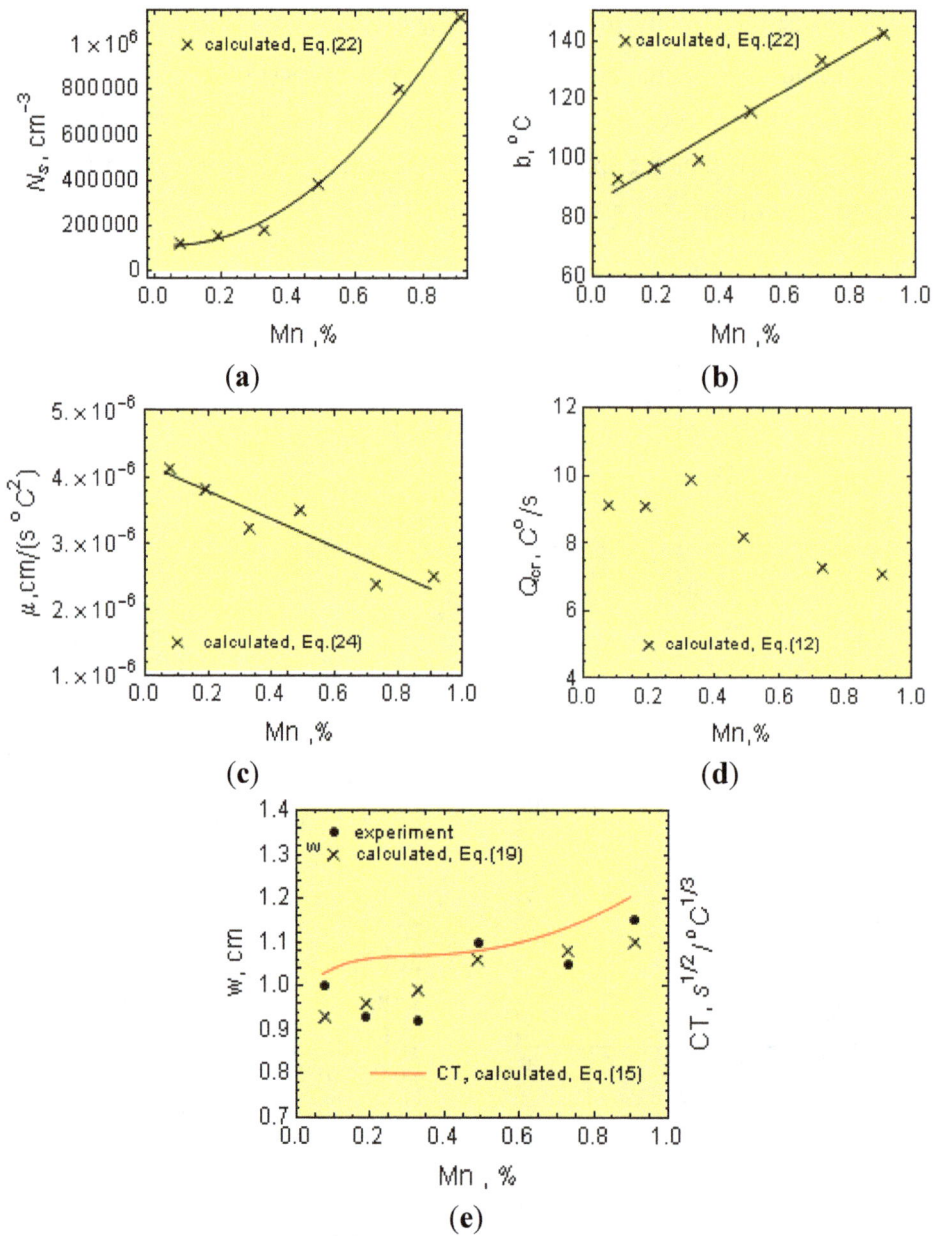

Figure 17. Effect of manganese on (**a**) N_s, (**b**) b, (**c**) μ, (**d**) CT index and (**e**) w.

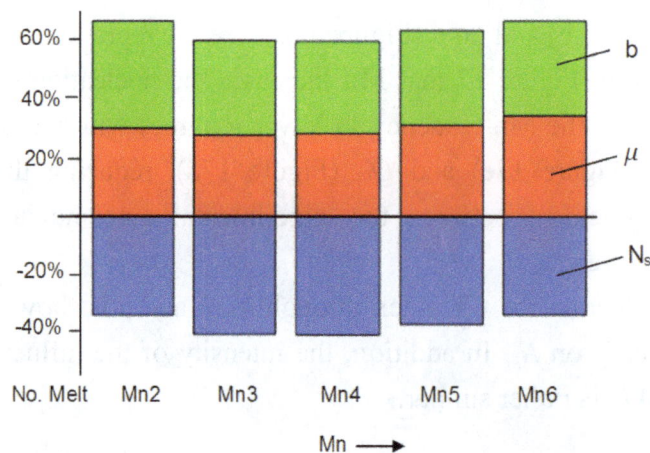

Figure 18. Manganese intensity effect on the CT index through μ, N_s and b (− CT decreases; + CT increases).

3.6. Effect of Phosphorous

The experimental results on the effect of P on the various nucleation and growth factors in the cast iron are given in Table 7. In addition, Figure 19 shows the effect of P on Q, N_F and ΔT_m.

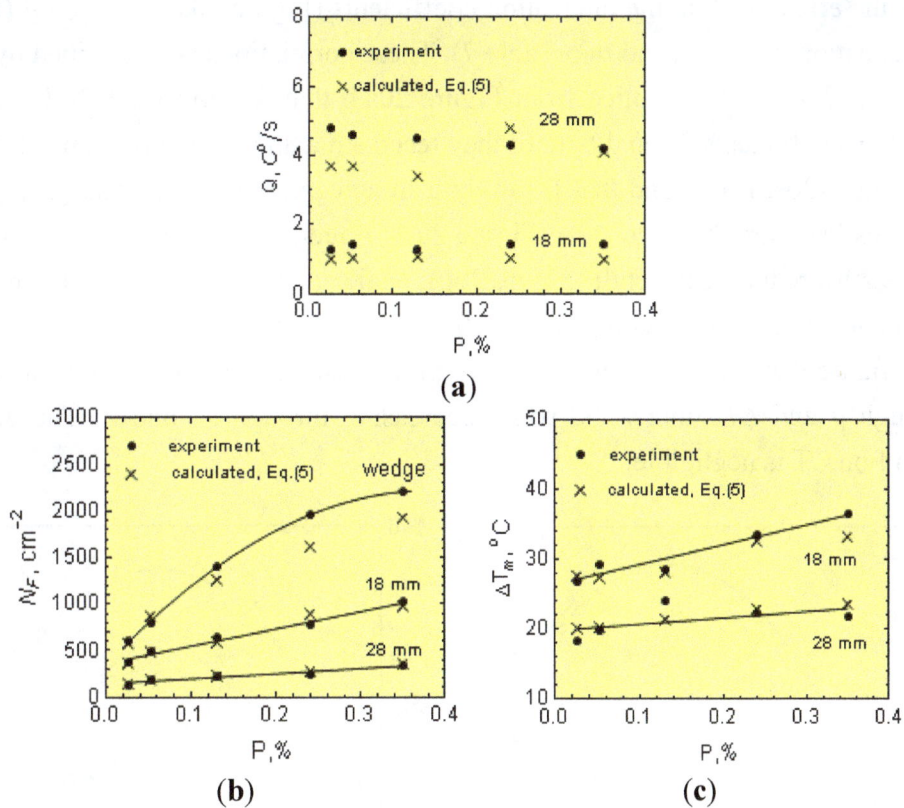

Figure 19. Effect of phosphorous on (a) Q, (b) N_F and (c) ΔT_m.

Table 7. Effect of P on cast iron T_i.

No. Melt *	P, %	Initial Liquid Metal Temperature just after Pouring into the Mold, T_i, °C		
		Bar 2.8 cm	Bar 1.8 cm	Wedge
P1	0.026	1,350	1,248	1,247
P2	0.052	1,340	1,251	1,249
P3	0.13	1,351	1,273	1,262
P4	0.24	1,351	1,213	1,244
P5	0.35	1,368	1,243	1,268

Cell count, cm^{-2}	bar 2.8 cm	$N_F = 143.1 + 541.6\,P$
Cell count, cm^{-2}	bar 1.8 cm	$N_{F,1} = 362.8 + 1{,}844.9\,P$
Critical cell count, cm^{-2}		$N_{cr} = 341.7 + 9{,}797.2\,P - 12{,}769.5\,P^2$
Undercooling, °C	bar 2.8 cm	$\Delta T_m = 19.8 + 8.98\,P$
Undercooling, °C	bar 1.8 cm	$\Delta T_{m,1} = 26.5 + 28.3\,P$
wedge width of the chill, cm		$w = 1.02 - 1.66\,P + 1.94\,P^{-2}$
Graphite eutectic growth coefficient, μ cm/(s·°C^2)		$\mu_s = 10^{-6}\,(6.12 - 6.02\,P^{0.25})$
Nucleation coefficients °C		$b = 91.8 + 358.3\,P - 770.4\,P^2$
cm^{-3}		$N_s = 10^6\,(7.7\,P - 16\,P^2 - 0.04)$

* C = 3.29%–3.37%; Si = 1.21%–1.29%; P = 0.020%–0.03%; S = 0.01%–0.02%.

Phosphorous increases the number of eutectic cells in in the experimental bars and wedges (Figure 19b) and the undercooling (Figure 19c). These results are described by regression equations for N_F, $N_{F,1}$, N_{cr}, ΔT_m and $\Delta T_{m,1}$, and they are given in Table 7. Notice that the experimental and calculated values from Equations (5), (9) and (11) are rather similar.

Figure 20 shows the effect of P on the nucleation coefficients (Figure 20a,b) and on μ (Figure 20e) (see the regression equations for μ, N_s and b in Table 7). These correlations are described by regression equations for μ, N_s and b (Table 7). Notice from Figure 20a,b that for up to 0.25% P, phosphorous increases the nucleation coefficients, N_s and b, until they reach a maximum at approximately P = 0.25%. Furthermore, from Figure 20e, it is found that P lowers μ in agreement with the data given in Table 3. In addition, P increases f (Figure 20c) and ΔT_{sc} (Figure 20d). P reduces the CT index (Figure 20g) and increases Q_{cr} (Figure 20f), reducing the chill (Figure 20g). Notice that there is good agreement between the experimental outcome and the predictions of Equation (5) for w (see Figure 20g).

Notice from this figure that increasing the P content in the cast iron, the intensity influence of P on the CT index through μ and ΔT_{sc} increases, while decreasing through N_s and b. The effect of the phosphorus through f on CT is negligible.

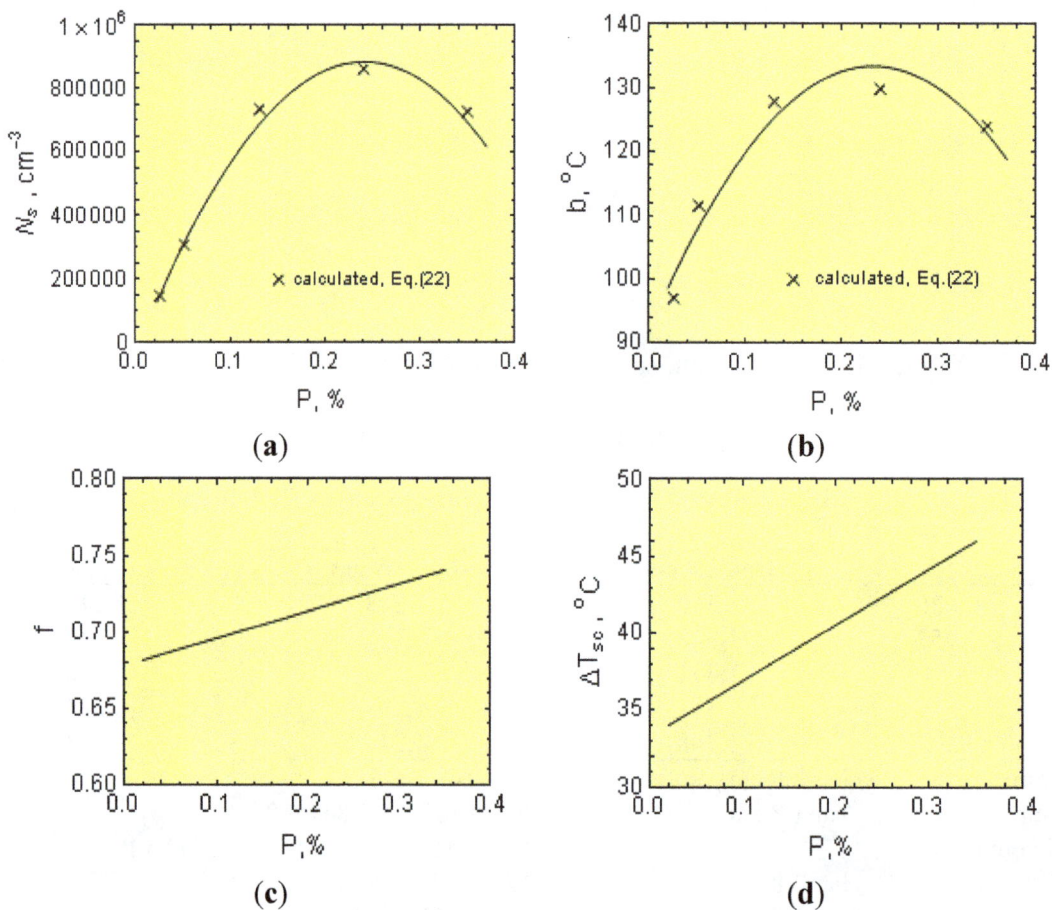

Figure 20. *Cont.*

(e)

(f)

(g)

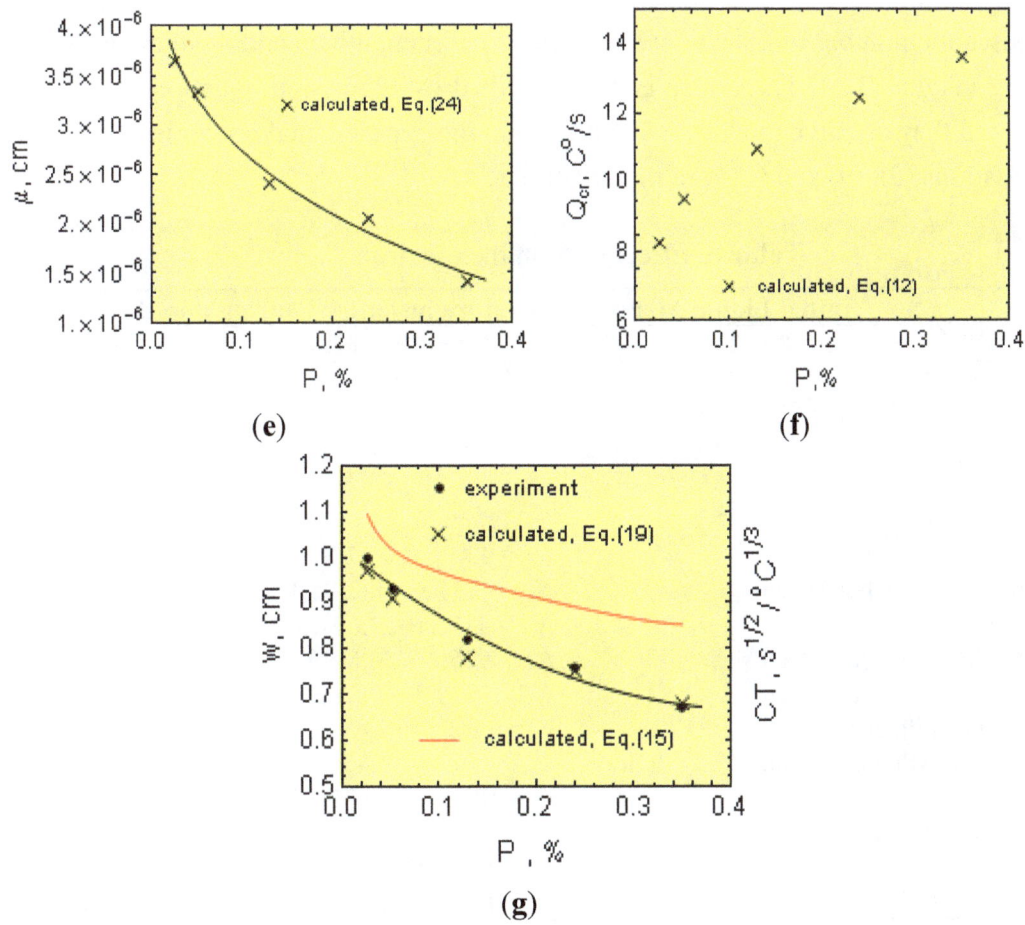

Figure 20. Effect of phosphorous on **(a)** N_s, **(b)** b, **(c)** f, **(d)** ΔT_{sc}, **(e)** μ, **(f)** Q_{cr} and **(g)** w.

The phosphorous intensity effect on the CT index through f, μ ΔT_{sc}, N_s and b is shown in Figure 21.

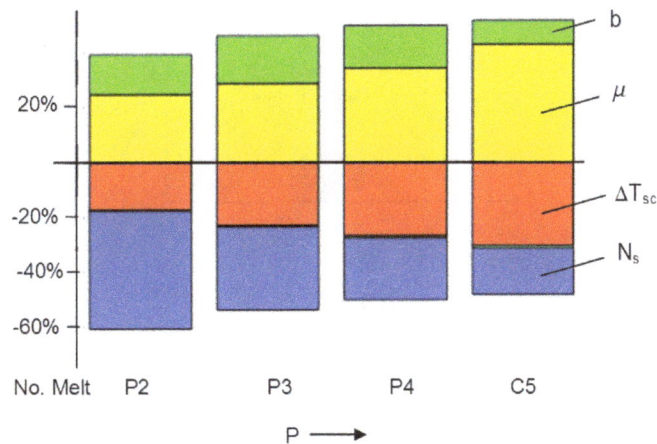

Figure 21. Phosphorous intensity effect on the CT index, through μ, N_s, b, ΔT_{sc} and f ($-$ CT decreases; $+$ CT increases).

3.7. Effect of Sulfur

The experimental results on the effect of S on the various nucleation and growth factors in the cast iron are given in Table 8. In addition, Figure 22 shows the effect of S on Q, N_F and ΔT_m. These results are reported in some detail elsewhere [32,33].

Sulfur increases the number of eutectic cells in the experimental bars and wedges (Figure 22b) and the undercooling (Figure 22c). These results are described by regression equations for N_F, $N_{F,1}$, N_{cr}, ΔT_m and $\Delta T_{m,1}$, and they are given in Table 8. Notice that the experimental outcome and the calculated values from Equations (5), (9) and (11) are rather similar.

Table 8. Effect of S on the cast iron T_i.

No. Melt *	S, %	Initial Liquid Metal Temperature just after Pouring into the Mold, T_i, °C		
		Bar 2.8 cm	Bar 1.8 cm	Wedge
S1	0.044	1,368	1,300	1,275
S2	0.079	1,354	1,245	1,247
S3	0.098	1,349	1,248	1,245
S4	0.121	1,355	1,251	1,242
S5	0.131	1,357	1,224	1,241

Cell count, cm^{-2}	bar 2.8 cm	$N_F = 9.4 + 4{,}267.2\ S$
Cell count, cm^{-2}	bar 1.8 cm	$N_{F,1} = 231.1 + 11{,}595.1\ S$
Critical cell count		$N_{cr} = 417.2 + 52{,}694.6\ S$
Undercooling, °C	bar 2.8 cm	$\Delta T_m = 15.2 + 55.8\ S$
Undercooling, °C	bar 1.8 cm	$\Delta T_{m,1} = 19.2 + 96.1\ S$
wedge width of the chill, cm		$w = 0.2 + 1.8\ S$
Graphite eutectic growth coefficient, μ $cm/(s\cdot°C^2)$		$\mu_s = 10^{-6}\,(6.62 - 72.5\ S + 242.0\ S^2)$
Nucleation coefficients		
°C		$b = 78.8 + 470.5\ S$
cm^{-3}		$N_s = 10^6\,(0.214843 - 1.888\ S + 223.206\ S^2)$

* C = 3.29%–3.35%; Si = 2.21%–2.27%; Mn = 0.077%–0.082%; P = 0.020%–0.03%.

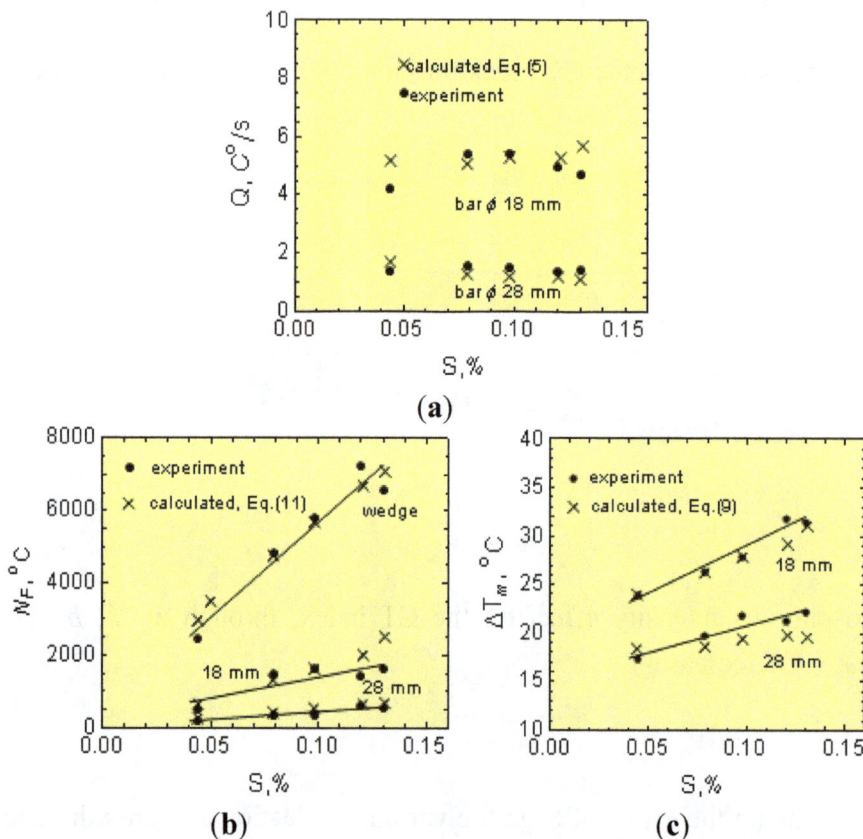

(a)

(b)

(c)

Figure 22. Effect of sulfur on (a) Q, (b) N_F and (c) ΔT_m.

Sulfur does not have a significant effect on f nor on ΔT_{sc}. Figure 23 shows the S effect on N_s, b (Figure 23a,b) and μ (Figure 23c). These correlations are described by regression equations for μ, N_s and b in Table 8. Notice from Figure 23 that S increases N_s and b, while lowering μ. The S tendency to lower μ is in agreement with the data given in Table 3. Sulfur increases the CT index (Figure 23e) and lowers Q_{cr} (Figure 23d), thus diminishing the chill (Figure 23e). Notice that there is good agreement between the experimental and the predictions of Equation (5) for w (see Figure 23e).

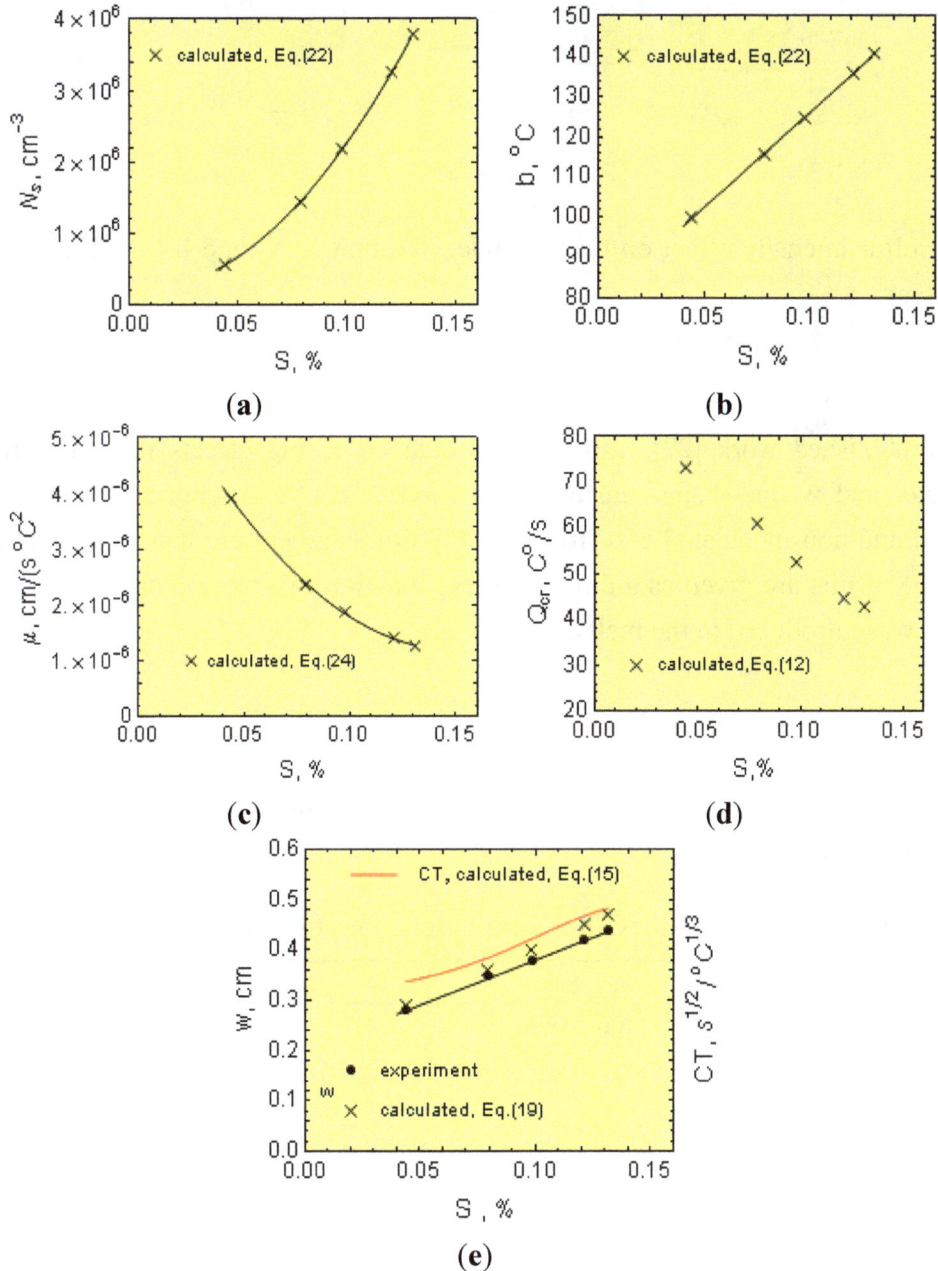

Figure 23. Effect of sulfur on (**a**) N_s, (**b**) b, (**c**) μ, (**d**) CT index and (**e**) w.

Figure 24 shows the intensity effect of S on the CT index through μ, N_s and b. Notice that the most significant effect of S is through μ. When the S content in the cast iron increases, the intensity influence of S on the CT index increases through μ, while it is reduced through N_s and b.

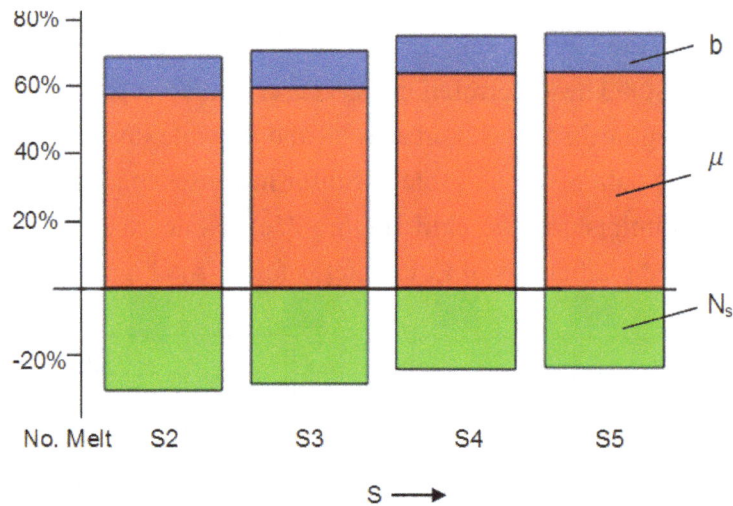

Figure 24. Sulfur intensity effect on the CT index through μ, N_s and b ($-$ CT decreases; $+$ CT increases).

3.8. Inoculation

In a previously published work [27], experimental data on fading effects in cast iron had been reported using plate- and wedge-shaped castings of various sizes. The reported experimental work included inoculated and non-inoculated cast irons and the time-temperature history of the inoculation effects (fading). The results are given as a dimensionless function of time calculated from the instant when the inoculant was introduced in the melt as:

$$t_r = \frac{t_a}{t} \qquad (33)$$

where t_a is given in Table 9 and t is a reference time during which changes in the cell count are negligible (approximately 25 min).

Table 9. Absolute and relative times after inoculation in cast iron and their effects on N_F.

No. Melt		1	2	3	4	5	6	7
		Base Iron			Inoculated Iron			
Absolute time after inoculation, t_a, min		-	1.5	5	10	15	20	25
Relative time after inoculation, t_r		-	0.06	0.2	0.4	0.6	0.8	1.0
Cell count, cm^{-2}	plate 1.0 cm	$N_{F,1} = 2{,}619.0 - 4{,}302.2\, t_r + 2{,}108.4\, t_r^2$						
Cell count, cm^{-2}	plate 2.2 cm	$N_F = 358.8 - 414.6\, t_r + 128.3\, t_r^2$						
Undercooling, °C	plate 1.0 cm	$\Delta T_{m,1} = 22.4 + 3.0\, t_r + 7.3\, t_r^2$						
Undercooling, °C	plate 2.2 cm	$\Delta T_m = 16.2 - 0.06\, t_r + 8.8\, t_r^2$						
Wedge width of the chill, cm								
base cast iron		$w = 0.8$						
inoculated cast iron		$w = 0.32 + 0.17\, t_r + 0.05\, t_r^2$						
Graphite eutectic growth coefficient, cm/(s·°C^2)								
base cast iron		$\mu = 2.64 \times 10^{-6}$						
inoculated cast iron		$\mu = 1.79 \times 10^{-6}$						

Table 9. *Cont.*

No. Melt	1	2	3	4	5	6	7
	Base Iron			Inoculated Iron			
Nucleation coefficients							
base cast iron	$b = 76.8$, °C $N_s = 1.6 \times 10^6$, cm^{-3}						
inoculated cast iron	$b = 96.9 + 122.6\, t_r - 59.2\, t_r^2$, °C $N_s = 10^6 \cdot [6.5{-}0.8\, t_r - 5.3\, t_r^2]$, cm^{-3}						
Chemistry							
base cast iron	C = 3.25%; Si = 1.17%; Mn = 0.13%; P = 0.08%; S = 0.047%						
inoculated cast iron	C = 3.18%; Si = 1.91%; Mn = 0.13%; P = 0.09%; S = 0.064%						
inoculant	Si = 73%–75%; Al = 0.75%; Ca = 0.75%–1.25%; Ba = 0.75%–1.25%						

Figure 25 shows that the cell count (Figure 25a) diminishes while the undercooling increases (Figure 25b) with t_r. In addition, Figure 25c,d shows the effect of t_r after inoculation on the nucleation coefficients. Notice from these figures that N_s increases, while b decreases with t_r. As a result, the CT index increases (Figure 25f), and the Q_{cr} is reduced (Figure 25e), lowering the chill (Figure 25f).

Figure 26 shows the t_r intensity effect on the CT index through N_s and b. Notice that t_r does not have an effect on f, ΔT_{sc} and μ (*i.e.*, the chemistry of the cast iron is a constant). Apparently, the CT index is affected by the nucleation parameters Ns and b with t_r (increasing through N_s and decreasing through b).

From the graphite nucleation potential curves, the following arguments can be made for the nucleation of graphite cells. Consider two melts (base and inoculated irons). For the base melt, the nucleation potential is given by the curve n(*l*), which is relatively small when compared with that for inoculated iron, n$_i$(*l*) (Figure 9b). Although, ΔT_m for the base melt is higher than for the inoculated melt, ΔT_{mi}, the area below the curve n(*l*) for $l_m \leq l \leq \infty$ ($0 \leq \Delta T \leq \Delta T_m$) is smaller than the area below the curve n$_i$(*l*) in the range $l_{m,i} \leq l \leq \infty$ ($0 \leq \Delta T \leq \Delta T_{m,i}$). In turn, this indicates that during solidification, the inoculated melt contains a higher nuclei density than the base melt.

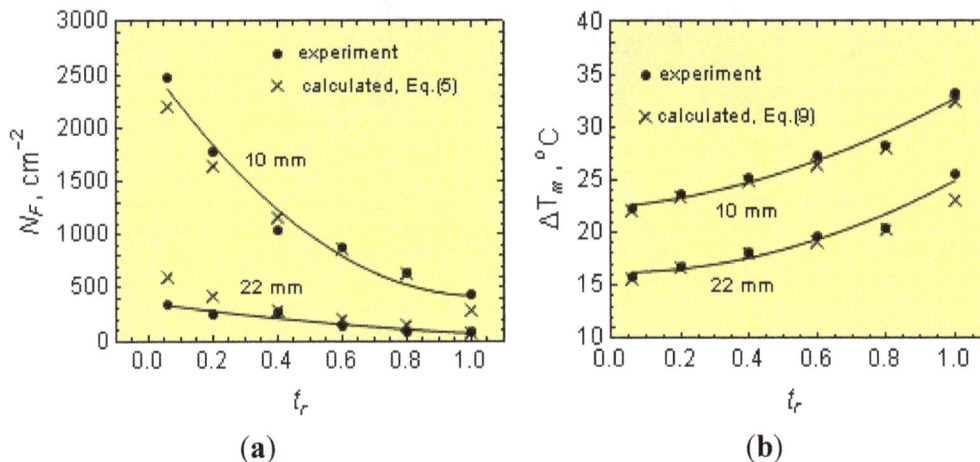

(a)   (b)

Figure 25. *Cont.*

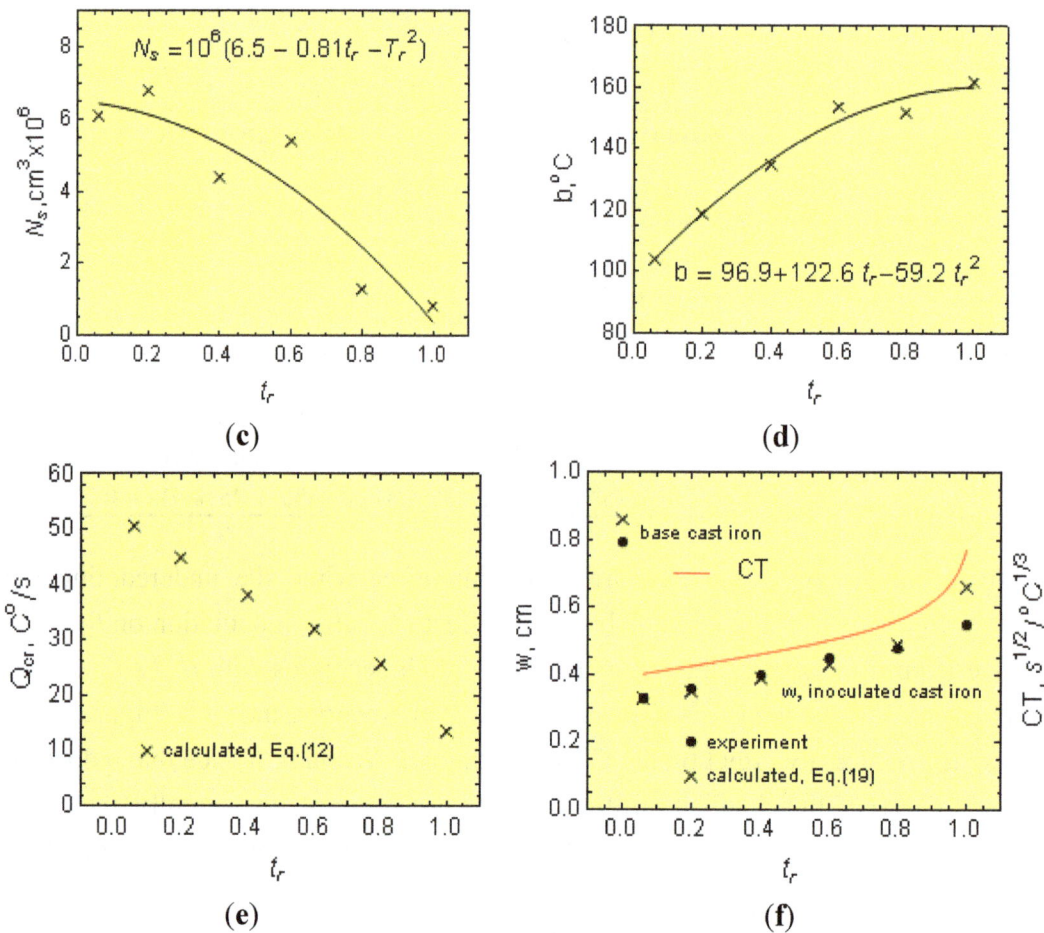

Figure 25. Effect of t_r after inoculation on (**a**) N_F, (**b**) ΔT_m, (**c**) N_s, (**d**) b, (**e**) the CT index and (**f**) w.

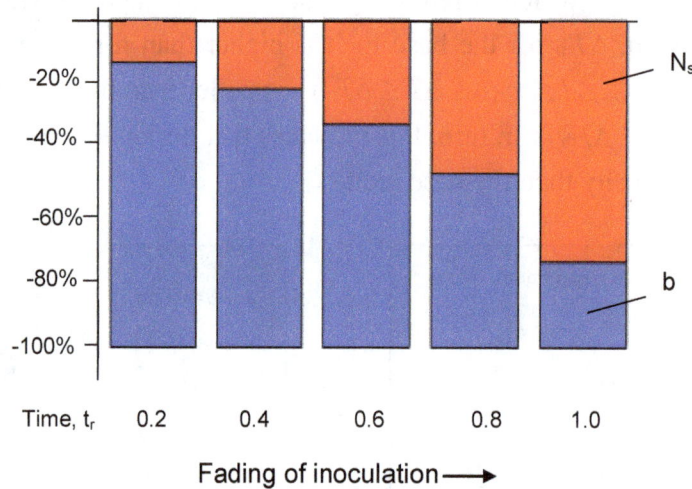

Figure 26. Intensity effect of t_r on the CT index through N_s and b (− CT decreases).

Finally, in the present work, the role of various technological parameters on the CT index and the chill of cast iron can be explained by means of a common analytical model of general validity. This theory was experimentally verified using C, Si, Mn, P and S, and it includes fading effects as examples. Consequently, general relations can be found between Q_{cr}, the CT index and the chill (see Figure 27).

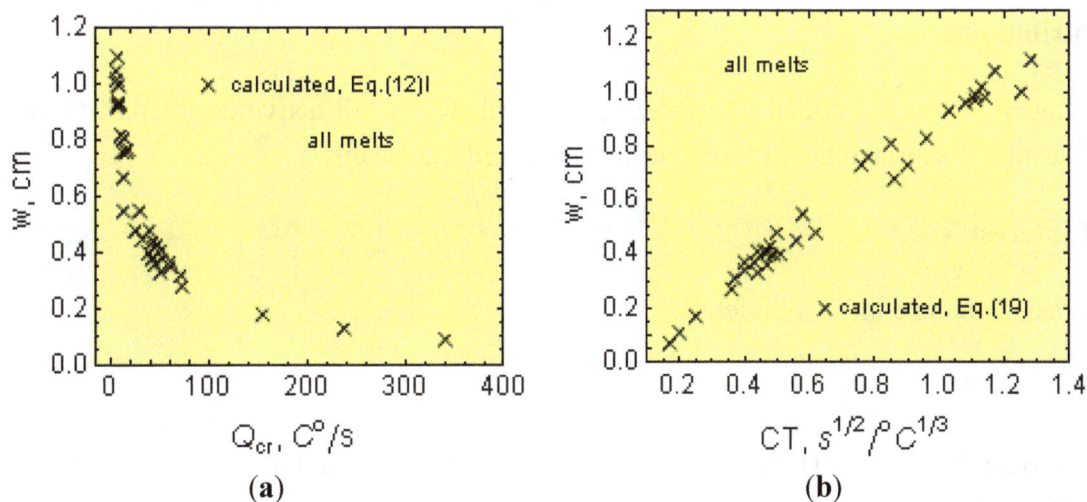

Figure 27. Relationship between (**a**) Q_{cr} and the chill, w, and (**b**) the CT and w for all melts. Experimental and calculated values were found using Equation (19).

4. Conclusions

In this work, experimental and theoretical results on cast iron solidification and chill tendencies, CT, are found to be in good agreement with each other, and they are summarized as follows:

- Carbon decreases N_s, b and ΔT_{sc} and increases μ, f and Q_{cr}; as a result, ΔT_m decreases, Q_{cr} and N_F increase, while the CT and the w decrease.
- Silicon increases f and ΔT_{sc} for the entire range of silicon content; it also increases N_s and b for up to 2.4% Si followed by a reduction in its effect. Furthermore, Si lowers μ in the range of 0–2.4% Si, and this is followed by an increase in μ. As a result, N_F increases, while ΔT_m changes from increasing to decreasing for Si contents beyond 2.4%. Thus, Q_{cr} increases, while the CT index and w both decrease.
- Manganese increases N_s and b and decreases μ; consequently, it increases ΔT_m and N_F and decreases Q_{cr}, resulting in an increase in the CT and the w.
- Phosphorous increases f and ΔT_{sc} and decreases μ for the entire P content range. It also increases N_s and b for up to 0.25% Si, and then, its effect diminishes. Consequently, there is an increase in N_F and ΔT_m, including Q_{cr}, while the CT and the w both decrease.
- Sulfur increases N_s and b and decreases μ; this leads to an increase in ΔT_m and N_F and a reduction in Q_{cr}, resulting in an increase in the CT, as well as the w.
- Inoculation increases N_s and b and decreases μ; consequently, inoculation decreases the ΔT_m and increases N_F, leading to a decrease in the CT, as well as the w.
- The time after inoculation decreases N_s and b; consequently, ΔT_m increases, while N_F decreases over time, leading to a reduction in Q_{cr} and an increase in the CT, as well as the w.

Acknowledgments

The authors are grateful to Edward Fras from the University of Science and Technology in Krakow, Poland, who recently passed away. Fras has a long history of impressive contributions to the field of solidification, both in technical publications and in published books.

Author Contributions

Hugo F. Lopez wrote and edited the paper and contributed to all activities. Marcin Gorny and Magdalena Kawalec performed the experiments and analyzed the results.

Conflicts of Interest

The authors declare no conflict of interest.

References

1. Fraś, E.; Górny, M.; Lopez, H. Graphite nodule and eutectic cell count in cast iron. *Arch. Foundry Eng.* **2007**, *7*, 47–52.
2. Fuller, A.G. Effect of superheating on chill and mottle formation. *BCIRA J. Res. Dev.* **1961**, *9*, 693–708.
3. Boyes, J.W.; Fuller, A.G. Chill and Motlle formation in cast iron. *BCIRA J. Res. Dev.* **1964**, *12*, 424–431.
4. Girshovitz, N. *Solidification and Properties of Cast Iron*; Mashinostroyene: Moscow-Leningrad, Russia, 1966. (In Russian)
5. Merchant, H.D. Solidification of cast iron. In *Recent Research on Cast Iron, Gordon and Breach*; Merchant, H., Ed.; Science Publishers: New York, NY, USA, 1968; pp. 1–100.
6. Dawson, J.V.; Maitra, S. Recent research on the inoculation of cast iron. *Br. Foundrym.* **1976**, *4*, 117–127.
7. Fraś, E.; Serrano, J.L.; Bustos, A. *Fundiciones de Hierro*; ILAFA: Valparaiso, Chile, 1990.
8. Kubick, E.J.; Javaid, A.; Bradley, F.-J. Investigation on effect C, Si, Mn, S and P on solidification characteristics and chill tendency of gray iron—Part II: Chill tendency. *AFS Trans.* **1997**, *103*, 579–586.
9. Oldfield, W. The chill-reducing mechanism of silicon in cast iron. *BCIRA J. Res. Dev.* **1962**, *10*, 17–27.
10. Hillert, M.; Subba Rao, V.V. Grey and white solidification of cast iron. In *The Solidification of Metals*; The Institute of Metals: London, UK, 1968; pp. 204–212.
11. Magnin, P.; Kurz, W. Competitive growth of stable and metastable Fe-C-X eutectic: Part I experiments. *Metallurg. Trans. A* **1988**, *19*, 1955–1963.
12. Frederiksson, H.; Svenson, I.L. Computer simulation of the structure formed during solidification of cast iron. In *The Physical Metallurgy of Cast Iron*; Fredrickson, H., Hillert, M., Eds.; North Holland: New York, NY, USA, 1985; pp. 273–284.
13. Nastac, L.; Stefanescu, D.M. Prediction of grey-to-white transition in cast iron by solidification modelling. *AFS Trans.* **1995**, *103*, 329–337.
14. Nastac, L.; Stefanescu, D.M. Modelling of stable-to-metastable structural transition in cast iron. In *Physical Metallurgy of Cast Iron V*; Lesoult, G., Lacaze, J., Eds.; Scitec Publications: Zürich, Switzerland, 1997; pp. 469–484.
15. Fras, E.; Górny, M.; López, H.F. The transition from grey to white cast iron during solidification: Part I. Theoretical background. *Metallurg. Mater. Trans. A* **2005**, *36*, 3007–3082.

16. Magnin, P.; Kurz, W. An analytical model of irregukar eutectic growth and its application to Fe-C. *Acta Metallurg.* **1987**, *35*, 1119–1128.

17. Showman, R.E.; Aufderheide, R.C. A process for thin-wall sand castings. *AFS Trans.* **2003**, *111*, 567–578.

18. Midea, T.; Shah, J.V. Mold material thermophysical data. *AFS Trans.* **2002**, *110*, 121–136.

19. Popescu, M.; Thompson, J.; Zavadil, R.; Sahoo, M. Summary of AFS Research project on restoring techniques iron-phase I. *AFS Trans.* **2002**, *110*, 1047–1065.

20. Sommerfeld, A.; Tonn, B. Nucleation in cast iron. In *Proceedings of the 5th Decennial International Conference on Solidification Processing*, Sheffield, UK, 23–25 July 2007; pp. 467–471.

21. Goodrich, G.M.; Oakwood, T.G.; Gundlach, R.B. Manganese, sulfur and manganese-sulfur ratio effects in gray cast iron. *AFS Trans.* **2003**, *111*, 783–812.

22. Rys, J. *Stereology of Materials*; Fotobit: Krakow, Poland, 1995.

23. Osher, J.; Lorz, U. *Quantitative Gefuengenanalysie*; DVG Leipzig-Stuttgard: Frankfurt, Germany, 1994.

24. Doepp, R.; Schwenkel, S. Contribution to the influence of chemical composition and cooling conditions on the eutectic solidification of Fe-C-X melts. *Mater. Sci. Eng. A* **2005**, *413*, 334–338.

25. Kanno, T.; Kikuchi, T.; Kang, I.; Nakae, H. Effect of alloying elements on the eutectic temperature in cast iron. *AFS Trans.* **2005**, *113*, 825–833.

26. Neuman, F. The influence of additional elements on the physic-chemical behaviour of carbon in saturated molten iron. In *Recent Research on Cast Iron*; Gordon and Breach: New York, NY, USA, 1998; pp. 659–705.

27. Heine, R. The carbon equivalent Fe-C-Si diagram and its application to cast iron. *AFS Cast Iron Res. J.* **1971**, *79*, 49–54.

28. Fras, E.; Górny, M.; López, H.F. The transition from grey to white cast iron during solidification: Part II. Experimental verification. *Metallurg. Mater. Trans. A* **2005**, *36*, 3083–3092.

29. Juretzko, F.R.; Dix, L.P.; Ruxanda, R.; Stefanescu, D.M. Precondition of ductile iron melts for light weight casting: Effect on mechanical properties and microstructure. *AFS Trans.* **2004**, *112*, 773–785.

30. Fras, E.; Wiencek, K.; Górny, M.; López, H.F. Graphite nodule and eutectic count in cast iron, Theoretical model based on Weibull statistic and experimental verification. *Metallurg. Mater. Trans. A* **2007**, *38*, 385–394.

31. Fraś, E.; Górny, M.; López, H.F. Mechanism of carbon on the transition from graphite to eutectic cementite in cast iron. *Metallurg. Mater. Trans. A* **2014**, *45*, 5601–5612.

32. Fraś, E.; Górny, M.; López, H.F. Mechanism for the role of silicon on the transition from graphite to eutectic cementite in cast iron. *Metallurg. Mater. Trans. A* **2012**, *43*, 4204–4218.

33. Fraś, E.; Górny, M.; López, H.F. Role of Sulfur on the Transition from Graphite to Eutectic cementite in Cast Iron. *Metallurg. Mater. Trans. A* **2013**, *44*, 2512–2522.

4

Effect of Heat Treatment on Microstructure and Hardness of Grade 91 Steel

Triratna Shrestha [1], **Sultan F. Alsagabi** [1], **Indrajit Charit** [1], **Gabriel P. Potirniche** [2] and **Michael V. Glazoff** [3,*]

[1] Department of Chemical and Materials Engineering, University of Idaho, Moscow, ID 83844-3024, USA; E-Mails: triratna.shrestha@gmail.com (T.S.); ssagabi@kacst.edu.sa (S.F.A.); icharit@uidaho.edu (I.C.)

[2] Department of Mechanical Engineering, University of Idaho, Moscow, ID 83844-0902, USA; E-Mail: gabrielp@uidaho.edu

[3] Energy Systems Integration, Idaho National Laboratory, Idaho Falls, ID 83415-3710, USA

*   Author to whom correspondence should be addressed; E-Mail: michael.glazoff@inl.gov

Academic Editor: Hugo F. Lopez

Abstract: Grade 91 steel (modified 9Cr-1Mo steel) is considered a prospective material for the Next Generation Nuclear Power Plant for application in reactor pressure vessels at temperatures of up to 650 °C. In this study, heat treatment of Grade 91 steel was performed by normalizing and tempering the steel at various temperatures for different periods of time. Optical microscopy, scanning and transmission electron microscopy in conjunction with microhardness profiles and calorimetric plots were used to understand the microstructural evolution including precipitate structures and were correlated with mechanical behavior of the steel. Thermo-Calc™ calculations were used to support the experimental work. Furthermore, carbon isopleth and temperature dependencies of the volume fraction of different precipitates were constructed.

Keywords: modified 9Cr-1Mo steel; Grade 91 steel; heat treatment; microstructure; differential scanning calorimetry; Thermo-Calc™; precipitates

1. Introduction

The Next Generation Nuclear Plant (NGNP) is expected to address the growing energy demand by producing electricity and, at the same time, mitigate greenhouse gas emissions by co-producing hydrogen from the process heat. The Very High Temperature Reactor (VHTR) is a Gen-IV reactor system at the heart of the NGNP. VHTRs are designed to operate at temperatures much higher than those of currently operating reactors. Moreover, they are designed for longer service periods (60 years or more) compared to the current operating reactors [1]. The operating temperature of the reactor pressure vessel (RPV) in VHTR can vary between 300 and 650 °C. Furthermore, the RPVs will be more than double the size of a typical RPV as found in a Light Water Reactor (LWR) [2].

The modified 9Cr-1Mo steel (Grade 91) is a material of choice in fossil-fuel-fired power plants with increased efficiency, service life, and reduction in emissions of CO_2, NO_x, and SO_2. The efficiency of fossil-fired power strongly depends on the temperature and pressure of steam. One percent increase in net efficiency reduced the emission of CO_2, NO_x, SO_2, and particulates by 2.4 metric ton, 2000 ton, 2000 ton, and 500 ton, respectively, while also reducing fuel costs by 2.4% [3]. Temperature of steam in coal-fired power plant is expected to be in the range of 550–720 °C and pressure is expected to be above 24 MPa [3–5].

Grade 91 steels are a ferritic-martensitic (F-M) class of steel with superior creep strength [6,7]. The addition of strong carbide and carbonitride formers like vanadium (V), niobium (Nb), and titanium (Ti) lead to the precipitation of various particles, such as $M_{23}C_6$, and MX type precipitates: (Ti,Nb)(N,C), (Nb,V)(C,N), and (V,Nb)(N,C). The smaller interparticle spacing and increased volume fraction of the fine MX carbides enhance the alloy strength. These precipitates obstruct the movement of dislocations, refine grains during normalizing, and delay plastic deformation [6,8]. However, precipitates like Z-phase and Laves phase may lead to a decrease in the strength of the alloy by weakening the solid solution.

Evolution of various precipitates and microstructure can be understood from the heat treatment, differential scanning calorimetric (DSC) studies, and thermodynamic calculations using Thermo-Calc. The heat treatment study of Grade 91 steel helps in understanding its microstructural evolution and mechanical properties; as this material, whether being used for a reactor pressure vessel or boiler component, goes through a series of heating, cooling, and welding processes. With the changes involved in normalizing and tempering temperatures, different types of carbides such as $M_{23}C_6$, Z-phase, Ti/Nb/V rich MX type carbide/carbonitrides precipitate and coarsen in Grade 91 steel resulting in a change in the creep strength, ductility, hardness, and microstructure. Ahn *et al.* [9] observed a decrease in the high temperature strength of an Nb containing steel due to coarsening of Nb-rich precipitates. DSC was used to measure the heat absorbed or liberated during heating or cooling associated with carbide precipitation and other phase changes. Ganesh *et al.* [10] were able to find the phase transformation temperatures including temperature ranges for dissolution of various precipitates in different low carbon steels.

The present article aims to elucidate the effects of various heat treatment conditions (normalizing and tempering) on Grade 91 steel. There are numerous studies on the normalizing/tempering of Grade 91 steel [11–15], but these studies are limited to few temperatures and times. Thus, this paper has taken about 50 normalizing and tempering conditions to study the relevant microstructural evolution.

To that end, a combination of microstructural characterization efforts and microhardness measurements was utilized. The findings were discussed with the help of DSC studies, and Thermo-Calc™ was used to understand the microstructural evolution and precipitate stability.

2. Experimental Details

2.1. Material

The manufacturer-supplied chemical composition of ASTM A387 Grade 91 CL2 steel used in this study is shown in Table 1. The hot rolled Grade 91 plates were obtained from American Alloy Steel, Houston, TX, USA, in a normalized and tempered condition (*i.e.*, austenitized at 1040 °C for 240 min followed by air cooling, and tempered at 790 °C for 43 min). The as-received plates were 10.4 cm × 10.4 cm × 1.27 cm in size. Heat treatment specimens were cut out from these plates using a diamond wafering blade. At room temperature, the as-received Grade 91 steel had an average yield strength of 533 MPa, ultimate tensile strength of 683 MPa and elongation to fracture of 19%.

Table 1. Chemical composition (in wt.%) of Grade 91 steel.

Element	Nominal	Measured
Cr	8.00–9.50	8.55
Mo	0.85–1.05	0.88
V	0.18–0.25	0.21
Nb	0.06–0.10	0.08
C	0.08–0.12	0.10
Mn	0.30–0.60	0.51
Cu	0.4 (max.)	0.18
Si	0.20–0.50	0.32
N	0.03–0.07	0.035
Ni	0.40 (max.)	0.15
P	0.02 (max.)	0.012
S	0.01 (max.)	0.005
Ti	0.01 (max.)	0.002
Al	0.02 (max.)	0.007
Zr	0.01 (max.)	0.001
Fe	Balance	Balance

2.2. Heat Treatment

Heat treatment was performed using an electric resistance furnace, capable of reaching a temperature exceeding 1200 °C. Normalizing was first done on each sample within a temperature range of 1020–1100 °C for 2, 4, and 8 h. Normalizing temperature and time were so chosen that austenitization had been complete before tempering was carried out. Normalized samples were air cooled down to room temperature before tempering at various temperatures and times. Samples that have been normalized at 1040 °C for 2, 4 and 8 h were tempered at 690, 725, 745 and 790 °C for 2, 8 and 20 h. Tempering of samples normalized at 1040 °C for 2 h was expanded to include temperatures in the range of 635–850 °C. This created a matrix of close to 50 possible combinations of normalizing and tempering scenarios, which provided an adequate sequence to reveal the microstructural changes during heat treatment.

2.3. Metallographic Characterization

Optical microscopy was performed with Olympus optical microscope, Center Valley, PA, USA, on both the as-received and heat treated specimens for characterization of the grain structure. Optical micrographs were taken for each sample and provided visual representation of the changing phases and microstructural features. Transmission electron microscopy (TEM) studies were done on selected samples. Conventional metallographic procedures of cold mounting, grinding and polishing were followed to prepare the specimen surface to 0.5 μm finish before etching was carried out using Marble's reagent: A solution made of 50 mL distilled water, 50 mL hydrochloric acid and 10 g of copper sulfate. Subsequently, an Olympus light microscope, Center Valley, PA, USA was used to examine the metallographic specimens, and an attached CCD camera was used to record the images.

For the TEM study, the sectioned samples were mechanically polished down to ~120 μm thickness, and then 3 mm diameter disks were punched out of the samples. Those disks were then jet polished in Fischione twin-jet polisher using a solution of 80 volume% methanol and 20 volume% nitric acid solution at a temperature of about −40 °C. Dry ice bath was used to achieve low temperature. Philips CM200, San Francisco, CA, USA, and JEOL JEM-2010, Pleasanton, CA, USA, TEMs operated at an accelerating voltage of 200 kV were used to study in detail the grain and precipitate morphology of the material under both as-received and select heat treated conditions. The scope of the study did not involve extensive TEM study of these samples. The EDS system available in the same TEMs was used to estimate the chemical composition of precipitates. Hardness was measured using a Leco Vickers microhardness tester, St. Joseph, MI, USA; the applied load was 500 gf and the hold time was 15 s.

2.4. Thermodynamic Modeling

The calculations of thermodynamic equilibrium in Grade 91 steel and construction of C-isopleths were done with the help of Thermo-Calc AB software (Thermo-Calc Classic Version S, Stockholm, Sweden). Thermo-Calc TCFE6 database (Stockholm, Sweden) for iron-based alloys was used to conduct all the equilibrium calculations. Due to the significant complexity of the Grade 91 steel (iron plus 13 alloying elements), impurities such as S or P were excluded from the thermodynamic calculations.

2.5. Differential Scanning Calorimetry

Differential scanning calorimetric study of Grade 91 steel was done using a Netzsch STA 409 PC calorimeter, San Francisco, CA, USA. The DSC chamber was evacuated and purged several times with the argon gas of high purity (99.999%), and the flow rate of argon was maintained at 85 cm^3/min. The sample mass ranging between 50 and 60 mg was found suitable for an acceptable signal to noise ratio. The sensitivity curves for scan rate of 20 °C/min were calibrated using the melting point of pure indium, tin, zinc, aluminum and gold under research grade argon atmosphere. Baseline calibrations were performed by using a pair of empty crucibles corresponding to the heating rate of 20 °C/min. Non-isothermal

DSC measurements were done by heating the sample from 25 to 1400 °C, and then cooling was undertaken at the same rate down to 200 °C.

3. Results

3.1. Microstructural Characteristics

The as-received Grade 91 steel had a tempered martensitic microstructure with a lot of precipitates, as shown in Figure 1a,b. Precipitation hardening is one of the main strengthening mechanisms in highly alloyed steels like Grade 91. Alloying elements, such as C, N, Ti, Nb, V, Cr and Mo, promote the formation of precipitates like Cr-rich $M_{23}C_6$, and Ti or Nb or V-rich MX particles, where M stands for metals, *i.e.*, Cr, Mo, Ti, Nb or V. The Cr-rich $M_{23}C_6$ precipitates were elongated rod-like or block-like particles, while Ti-rich, Nb-rich and V-rich MX precipitates had nearly spherical shape. The Cr-rich $M_{23}C_6$ type precipitates were mainly observed at the lath boundaries and prior austenite grain boundaries (PAGB). The dimension of the elongated rod-like and block-like Cr-rich precipitates was measured, and found to have an average length of 285 ± 80 nm and width of 121 ± 39 nm, which is similar to that reported by Shen *et al.* [16] in 11Cr F-M steel and Anderson *et al.* [17] in Grade 91 steel. The average diameter of near-spherical MX-type precipitates was 37 ± 15 nm, similar to values reported in literature [3,16]. These types of precipitates were mainly located in the matrix and martensitic lath structure. These thermally stable fine precipitates enhance the long term creep resistance by impeding movement of mobile dislocations, prior austenite grain boundaries, martensite lath and subgrain boundaries, and restrict fine grain structure from recrystallization. For a constant normalizing time of 2 h, prior austenite grain size and martensite lath size increased with increasing normalizing temperatures of 1020 °C–1100 °C, as shown in Figure 2a–d. The as-normalized (1040 °C for 2 h) microstructure had hard martensitic lath structures with high dislocation density, as observed in Figure 2c,d. Similarly, for constant normalizing temperature of 1040 °C, martensite lath size and prior austenite grain size increased with increasing normalizing time of 2, 4 and 8 h. Figure 2e and f shows the optical and TEM micrograph of sample normalized at 1040 °C for 8 h, respectively. Figure 2g shows the EDS spectrum of a $M_{23}C_6$ type precipitate shown in Figure 2f, indicating the presence of Cr.

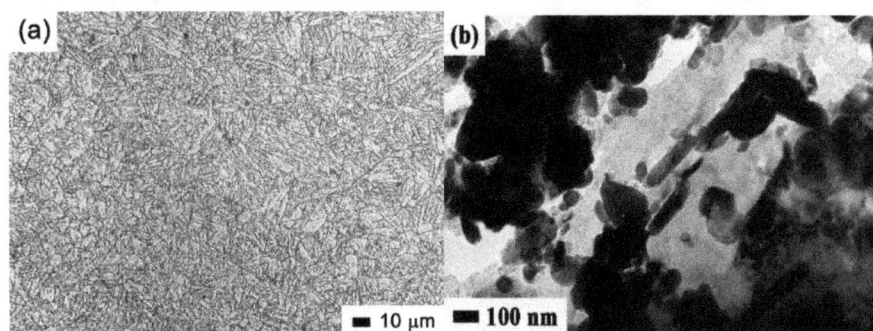

Figure 1. As-received microstructure: (**a**) optical and (**b**) TEM micrograph.

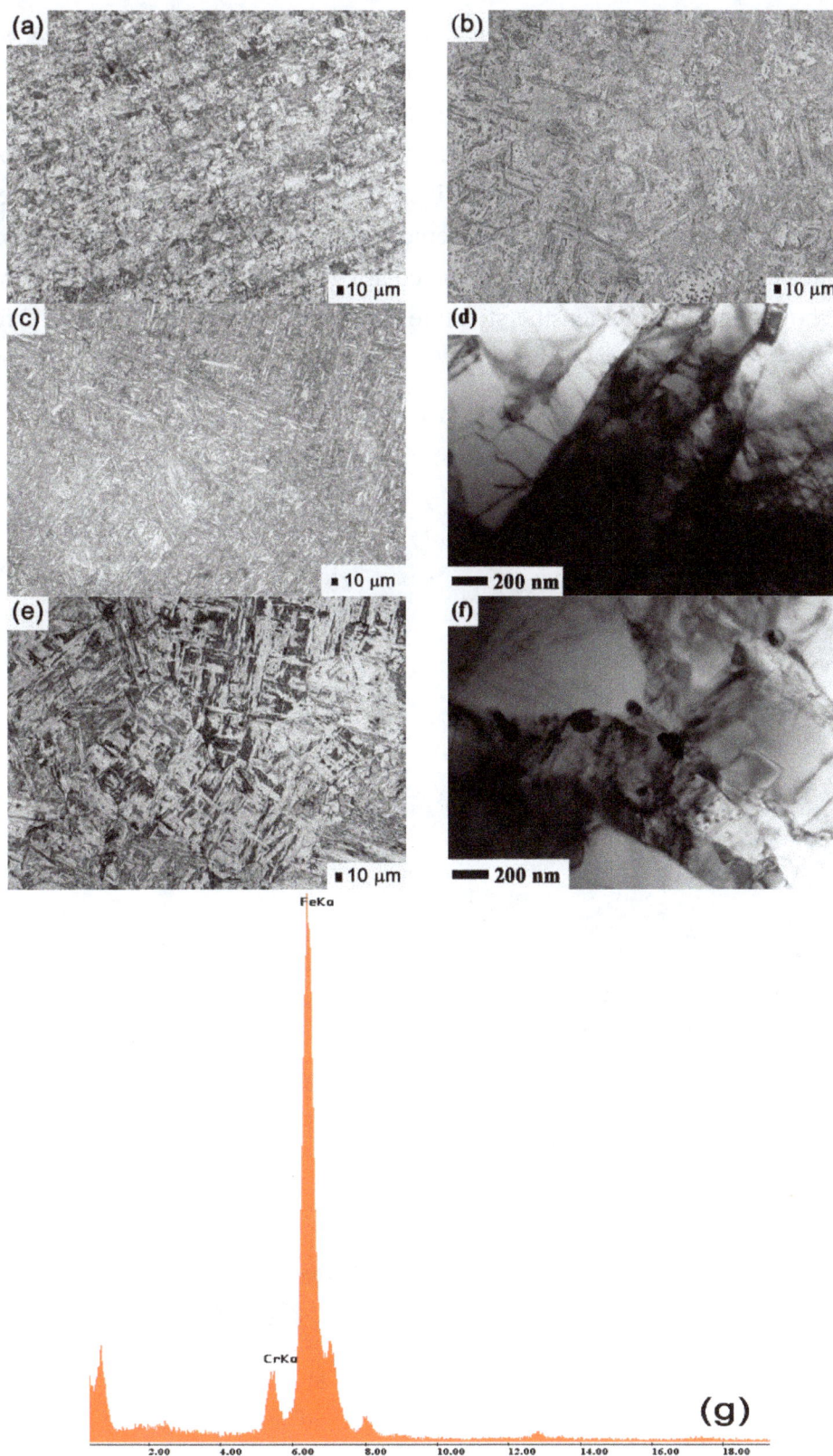

Figure 2. As-normalized optical micrograph (**a**) normalized at 1020 °C for 2 h; (**b**) normalized at 1100 °C for 2 h; (**c**) normalized at 1040 °C for 2 h; (**d**) TEM micrograph of sample normalized at 1040 °C for 2 h; (**e**) optical micrograph of sample normalized at 1040 °C for 8 h; (**f**) TEM micrograph of a sample normalized at 1040 °C for 8 h; and (**g**) the EDS spectrum of $M_{23}C_6$ type precipitate shown in Figure 2f.

The tempered microstructure of Grade 91 steel is shown in Figure 3. Figure 3a shows fine microstructure of sample normalized at 1040 °C for 2 h and tempered for 690 °C. Figure 3b shows coarse martensitic microstructure of sample normalized at 1040 °C for 2 h tempered at 745 °C for 8 h. Figure 3c and d shows the martensitic lath structure of sample normalized at 1040 °C for 2 h and tempered at 790 °C for 2 h. Grain size increased with increasing normalizing time, and tempering time. The sample normalized at 1040 °C for 2 h and tempered at 790 °C for 2 h has grain size of ~11 μm, while sample normalized at 1040 °C for 8 h and tempered at 790 °C for 20 h resulted in a grain size of ~17 μm. Figure 3e shows tempered martensitic structure of sample normalized at 1040 °C for 4 h and tempered at 725 °C for 2 h. Sample normalized at 1040 °C for 8 h and tempered at 790 °C for 20 h had coarse grain structure, as shown in Figure 3f.

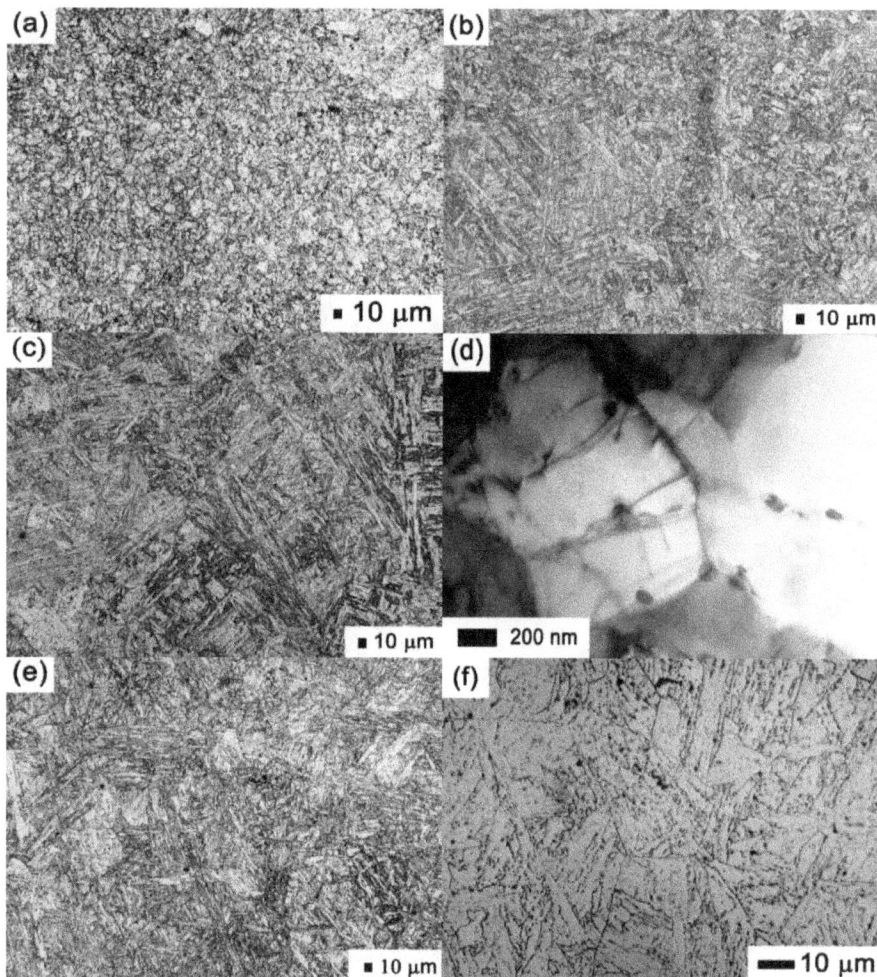

Figure 3. Tempered optical micrographs: (**a**) normalized at 1040 °C for 2 h and tempered at 690 °C for 2 h; (**b**) normalized at 1040 °C for 2 h and tempered at 745 °C for 8 h; (**c**) normalized at 1040 °C for 2 h and tempered at 790 °C for 2 h; (**d**) TEM micrograph of sample normalized at 1040 °C for 2 h and tempered at 790 °C for 2 h; optical micrographs (**e**) normalized at 1040 °C for 4 h and tempered at 725 °C for 2 h; and (**f**) normalized at 1040 °C for 8 h and tempered at 790 °C for 20 h.

3.2. Hardness of Heat Treated Steel

Normalizing treatment is intended for austenitization and homogenization of the solid solution. However, the right combination of normalizing temperature and time has to be found to optimize the material mechanical properties. Hardness measurements were performed on the alloy samples normalized at 1020, 1040, 1050, 1060, 1080 and 1100 °C while keeping the time constant at 2 h. The hardness decreased with increasing normalizing temperature, as shown in Figure 4a. The hardness value was about the same for normalizing carried out at 1020 and 1040 °C, but after that, the hardness gradually decreased until reaching 1100 °C. The effect of change in normalizing time was studied by heat treating Grade 91 steel at 1040 °C for 2 h, 4 h and 8 h. The hardness of the alloy decreased with increasing normalizing time, as shown in Figure 4b.

Alloy normalized at 1040 °C for 2 h was tempered at 690, 725, 745 and 790 °C for 2 h, 8 h and 20 h. Hardness of the alloy decreased with increasing tempering temperature and time, as shown in Figure 4c. Alloy tempered at 690 and 725 °C showed significant drop in hardness with increasing tempering time, but for 745 and 790 °C the decrease in hardness was gradual. Similar profiles were observed for samples normalized at 1040 °C for 4 h then tempered, and samples normalized at 1040 °C for 8 h then tempered at aforementioned temperatures, as shown in Figure 4d,e, respectively. Figure 4g shows the hardness of all tempered samples. Among all samples normalized at 1040 °C and tempered, samples normalized for 2 h and tempered at 690 °C for 2 h had the highest hardness, while sample normalized for 8 h and tempered at 790 °C for 20 h had the lowest hardness. The tempering temperature of the sample normalized at 1040 °C for 2 h was expanded to temperatures slightly above Ac_1 temperatures. Hardness of the alloy normalized at 1040 °C for 2 h and tempered for 2 h decreased with increasing tempering temperature up to 745 °C, stabilized till 820 °C, then increased, as shown in Figure 4f. Tempering above Ac_1 temperature was done to study the role of re-precipitation in Grade 91 steel.

3.3. Differential Scanning Calorimetry

Phase transformations of as-received (normalized and tempered) Grade 91 steel were studied using DSC. The phase transformations and reactions associated with heating and cooling of the alloy, as shown in Figure 5, are listed in Table 2. The listed temperatures for phase transformation and precipitate in Table 2 were obtained from the DSC curves, particularly with the help of its derivative in which inflection points are noticeably defined. The nature of the transformations was primarily corroborated by the Thermo-Calc calculations. The transformation of martensite to ferrite started at 550 °C, and the Curie temperature (T_c) signifying the change in ferromagnetic property was observed at 741 °C. The austenite phase started to form at 820 °C (Ac_1), peaked at 848 °C (Ac_p), and the transformation was complete at 870 °C (Ac_3), which was confirmed by differential thermal analysis (DTA) and the results of our thermodynamic calculations. The dissolution of $M_{23}C_6$ precipitate was complete at 870 °C. Dissolution of (V,Nb)(N,C) type precipitates bottomed out at 998 °C, but beyond that temperature they re-precipitated out. With the help of the first derivative of the heating curve, the onset of the second dissolution reaction of fine (V,Nb)(N,C) type precipitates was found to be at ~1140 °C. From cooling curve and its first derivative, the martensite start temperature (M_s) was calculated as 380 °C. Using Ac_1 as the onset temperature and Ac_3 as the end temperature, the DSC plot was used to calculate the enthalpy associated with the α-*ferrite→austenite* phase transformation. The phase transformation enthalpy ($\Delta^\circ H^{\alpha\rightarrow\gamma}$) of 10 J/g was calculated.

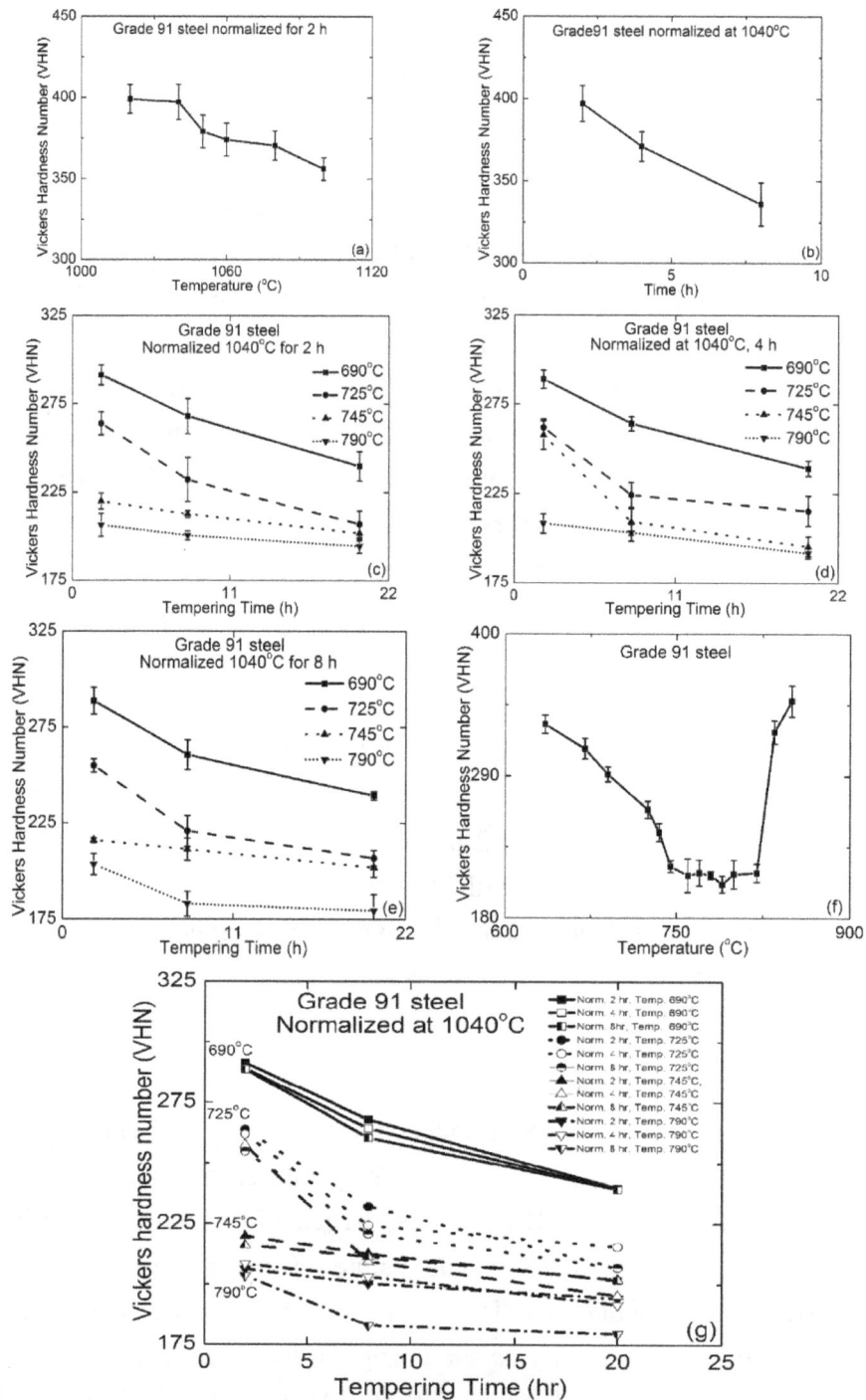

Figure 4. Variation in Vickers microhardness for various heat treatment conditions. (**a**) Variation in normalizing temperature; (**b**) samples normalized at 1040 °C for various times; (**c**) variation with tempering temperature and time for sample normalized at 1040 °C for 2 h; (**d**) Variation with tempering temperature and time for sample normalized at 1040 °C for 4 h; (**e**) variation with tempering temperature and time for sample normalized at 1040 °C for 8 h; (**f**) sample normalized at 1040 °C for 2 h and tempered at various temperatures for 2 h; and (**g**) normalized at 1040 °C and tempered samples.

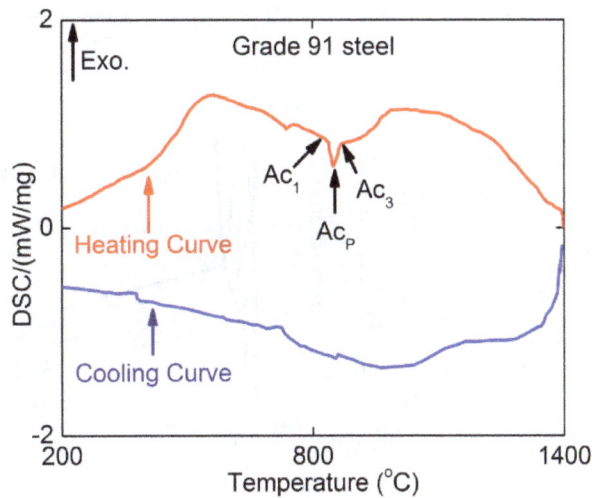

Figure 5. Differential scanning calorimetric plot of as-received Grade 91 steel.

Table 2. Various transformation temperatures of Grade 91 steel.

Description of Phase Change	Temperature (°C)
$\alpha' + M_{23}C_6 + (V,Nb)(N,C) + Z\text{-phase} \rightarrow \alpha + M_{23}C_6 + (V,Nb)(N,C) + Z\text{-phase}$	550
T_c (Curie Temperature)	741
$\alpha + M_{23}C_6 + (V,Nb)(N,C) + Z\text{-phase} \rightarrow \alpha + M_{23}C_6 + (V,Nb)(N,C) + (Nb,V)(C,N)$	770
$\alpha + M_{23}C_6 + (V,Nb)(N,C) + (Nb,V)(C,N) \rightarrow \alpha + \gamma + (V,Nb)(N,C) + (Nb,V)(C,N) + M_{23}C_6$	820 (A_{c1})
$\alpha + \gamma + (V,Nb)(N,C) + (Nb,V)(C,N) + M_{23}C_6 \rightarrow \gamma + (V,Nb)(N,C) + (Nb,V)(C,N)$	870 (A_{c3})
$\gamma + (V,Nb)(N,C) + (Nb,V)(C,N) \rightarrow \gamma + (Ti,Nb)(N,C) + (Nb,V)(C,N)$	1045
$\gamma + (Ti,Nb)(N,C) + (Nb,V)(C,N) \rightarrow \delta + \gamma + (Ti,Nb)(N,C)$	1270
$\delta + \gamma + (Ti,Nb)(N,C) \rightarrow \gamma + \delta$	1397
Martensite start (M_s)	380

3.4. Thermo-Calc Calculations

The carbon isopleth for the Fe-Cr-Mn-Mo-Nb-Ni-Ti-V-N-C full system is shown in Figure 6a. It is a property diagram, which represents the analog of the binary Fe-C phase diagram, but constructed for the full alloy system. It was calculated using the previously mentioned TCFE6 thermodynamic database. The bcc ferrite phase starts dissolving at ~810 °C and exists up to ~855 °C and then reappears above 1270 °C, as noted in Table 2. In between 850 and 1270 °C, the fcc austenite phase becomes stable. Other phases that are featured in the diagram in Figure 6a–c are the Z-phase; the $M_{23}C_6$-phase; (with metal M being mostly Cr); the Ti-rich (Ti,Nb)(N,C) precipitates; the Nb-rich and the Nb-poor (Nb,V)(C,N) precipitates; and the V-rich (V,Nb)(N,C) precipitates. All of the four phases have variable compositions but the same fcc crystalline lattice. The following notation was adopted for the latter four phases: In those cases when there was more carbon than nitrogen on a particular sublattice, the (C,N) notation was used; in the opposite case, the (N,C) notation was used. Similarly, the first position on the metal sublattice is assigned to a chemical element the concentration of which is the highest, e.g., (Nb,V); in the opposite case, the (V,Nb) notation was used.

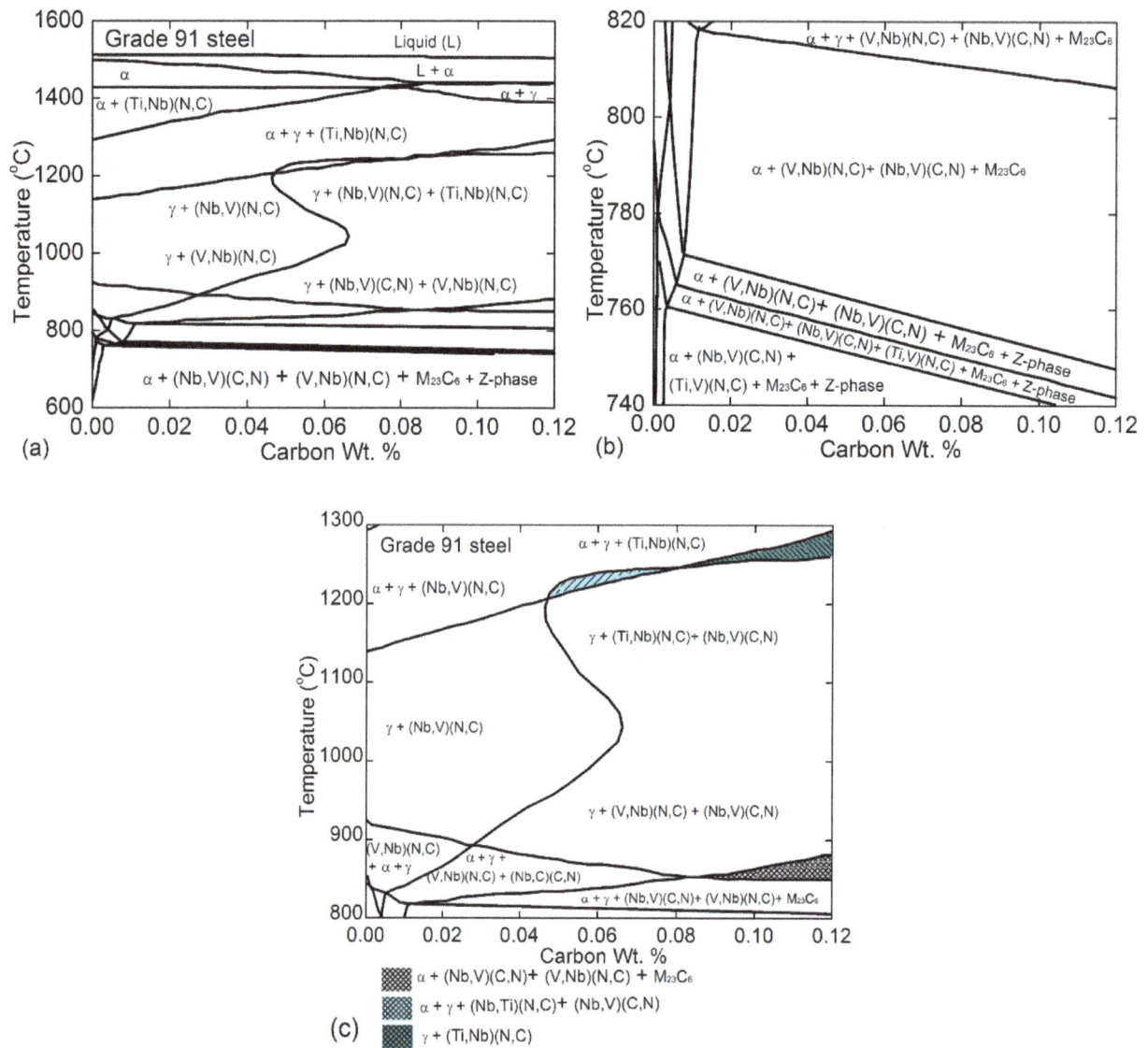

Figure 6. Carbon isopleth for Grade 91 steel. (**a**) Complete diagram: 600–1600 °C; (**b**) 740–820 °C; (**c**) 800–1300 °C.

It was established that the Ti-bearing precipitates (Ti,Nb)(C,N) were more stable than the (Nb,V)(C,N), and (V,Nb)(N,C) precipitates in the higher temperature range. Figure 7a shows a TEM micrograph of the as-normalized (normalized at 1040 °C for 2 h) sample with Ti-rich MX precipitates. The EDS spectrum of the Ti-MX particle is shown in Figure 7b. This example demonstrates the significance of simulations for the computational thermodynamics driven alloy design.

The evolution of the ferrite phase and its volume fraction with respect to temperature is shown in Figure 8a, and the evolution of $M_{23}C_6$ precipitate is shown in Figure 8b. The volume fraction of Z-phase, (Nb,V)(N,C) + (Ti,Nb)(N,C), and (V,Nb)(N,C) precipitates are illustrated by Figure 8c,d. Thermo-Calc calculations predicted that the volume fraction of $M_{23}C_6$ precipitate was higher than that of V-rich and Nb-rich MX type precipitates, owing to the higher wt.% of Cr and Mo compared to Nb and V. Higher resolution plot of the evolution of (V,Nb)(N,C) precipitate is shown in Figure 8d. Special attention was paid to the possibility of a phase separation and/or atomic clustering. Indeed, phase separation reaction and the miscibility gap were observed for the Fe-Cr system, so this possibility could not be excluded for the Fe-9Cr-1Mo steel.

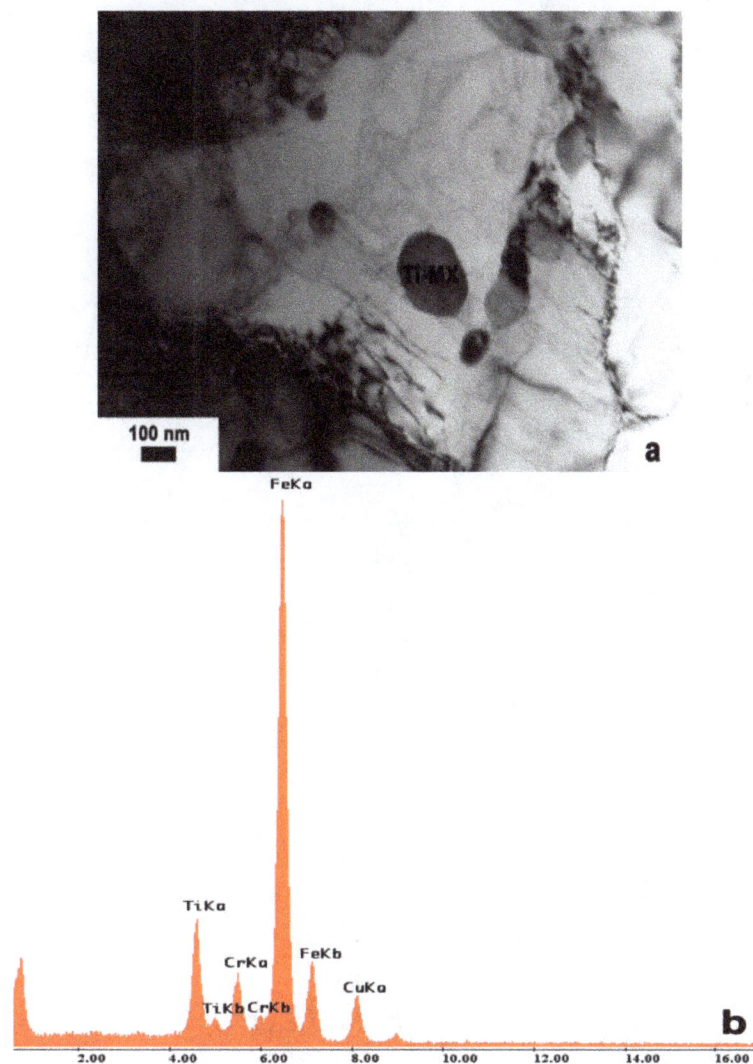

Figure 7. (**a**) A TEM micrograph of as-normalized (at 1040 °C for 2 h) sample; and (**b**) EDS of a Ti-rich MX precipitate.

4. Discussion

4.1. As-Normalized Microstructure

Hardness of the Grade 91 steel normalized for 2 h at 1020, 1040, 1050, 1060, 1080 and 1100 °C decreased with increasing temperature. Similarly, hardness of the as-normalized sample at 1040 °C decreased with increasing normalizing time. As-normalized microstructure had hard martensitic lath structures with high dislocation density, as shown in Figure 2d. Yoshino *et al.* [11] and Das *et al.* [18] reported increase in the prior austenite grain size with increase in normalizing temperatures from 1050 to 1250 °C. However, hardness remained constant in the same temperature range. However, in the current study, for normalizing carried out for 2 h in the temperature range of 1020–1100 °C, hardness decreased with increasing normalizing temperature. The decrease in the hardness value and increase in prior austenite grain size was due to a decrease in the volume fraction of precipitates with increased normalizing temperature and time. Average diameter of MX precipitates increased from 35 nm after normalizing at 1050 °C to 315 nm after normalizing at 1200 °C [11], leading to a decrease in the number density of precipitates.

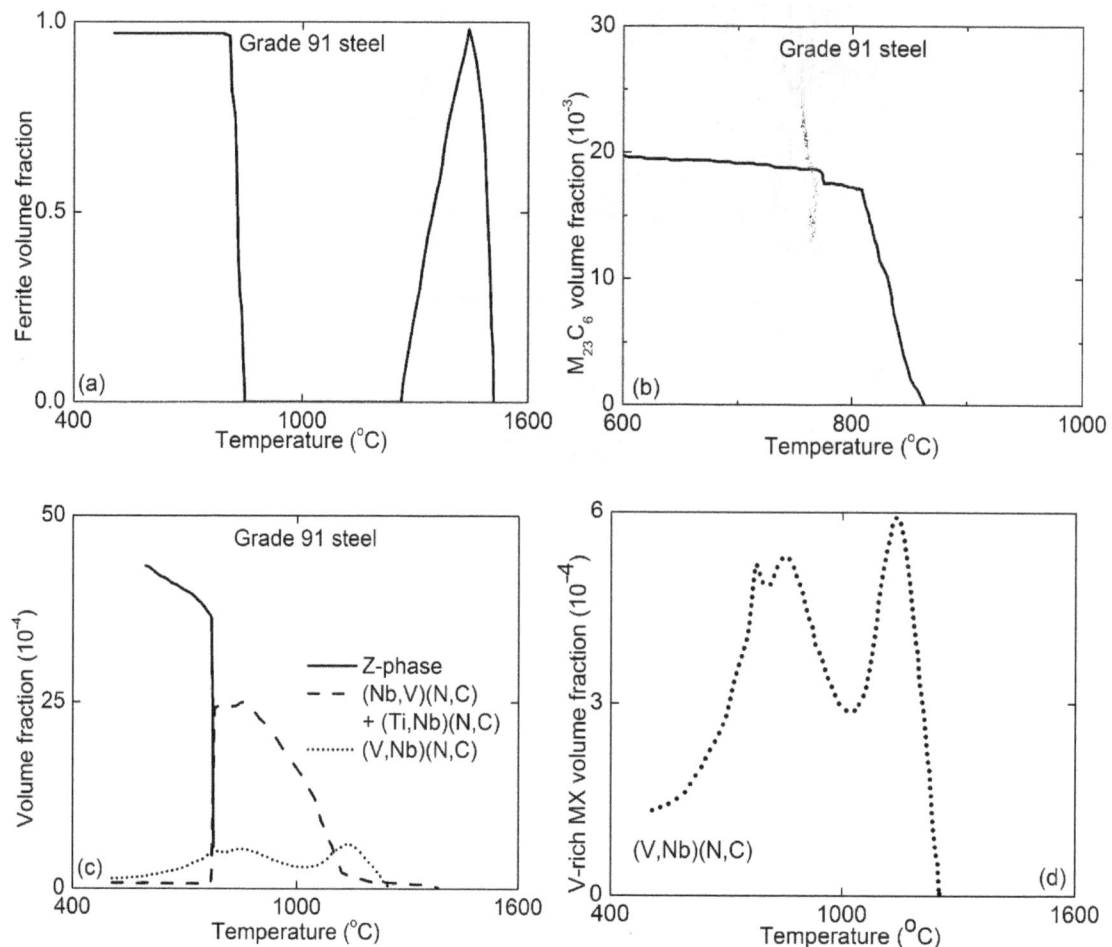

Figure 8. Evolutions of (**a**) the ferrite phase; (**b**) $M_{23}C_6$ precipitate; (**c**) Z-phase, Nb-rich MX, V-rich MX, and Ti-rich MX precipitates; and (**d**) high resolution plot of (V,Nb)(N,C) precipitate as a function of temperature.

Given the higher carbon solubility in austenite than in alpha ferrite, carbide precipitates dissolve in austenite. With increased diffusion at a higher temperature and longer time, the precipitates coarsen while the lath size and grain size also increase. Grade 91 steel normalized at 1040 °C for 8 h had a martensitic lath width of 473 ± 105 nm, and the size of the needle shaped $M_{23}C_6$ precipitates was 136 ± 75 nm in length and 28 ± 20 nm in width. While the sample normalized at 1040 °C for 2 h had a martensitic lath size of 335 ± 115 nm, while the $M_{23}C_6$ precipitates were 78 ± 15 nm in length and 13 ± 4 nm in width. At higher normalizing temperature, considerable grain growth occurred, but at lower normalizing temperature the austenite grain structure was finer. Moreover, normalizing below 900 °C resulted in partially transformed ferrite with patches of martensitic structure [12]. However, the change in the normalizing temperature in the range of 1020–1100 °C can result in different precipitate size distribution(s) that could possibly affect the mechanical behavior of the alloys.

4.2. Tempered Microstructure

The as-normalized hard martensitic lath structure with high dislocation density changed to stress-free, relaxed tempered martensitic structure upon tempering. Upon tempering, the normalized steel is reheated to a temperature just below Ac_1, which relieves stress in martensitic structure via carbon diffusion and carbide formation. With increased tempering time, the diffusion of carbon is increased and simultaneously became stress relieved. For example, a P92 steel normalized at 1050 °C showed a decrease in hardness with increasing tempering temperature from 525 up to 720 °C. Similarly, the tensile strength and yield strength decreased with increasing tempering temperature [13]. Moreover, for Grade 91 steel tempered at 765, 730, and 680 °C, specimens tempered at the lowest temperature had the highest creep strength at 600 °C, as shown by Sawada et al. [14]. Tempering above Ac_1 temperature decreased the toughness of the material due to the formation of fresh martensite as found by Silwal et al. [19]. In the present work, the hardness decreased with increasing tempering temperature till Ac_1 temperature (820 °C). The decrease in hardness, as seen in Figure 4g, is due to the coarsening of precipitates, dissolution of $M_{23}C_6$ carbides, and breakdown of martensitic lath structure.

Coarser precipitates are not effective in impeding the mobility of dislocations, and affect the strength of the steel [6]. Decrease in hardness with increased temperature can also in part be associated with the increased molybdenum in carbides. In Fe-Mo-C steels tempered at 700 °C, Shtansky and Inden [20] observed 4% increase in the molybdenum content in $M_{23}C_6$ precipitates after increasing the tempering time from 25 to 500 h. Increased molybdenum content in precipitates implies removal of Mo from the matrix, thus weakening the solid solution. Moreover, the $M_{23}C_6$ precipitates located on the prior austenite grain boundaries penetrate into the grains with increasing tempering time. The stability of hardness at temperatures around 800 °C reflects the maximum precipitation rate for 0.1 wt.% C in the alloy. However, above 820 °C, the hardness increased because of secondary hardening resulting from increased volume fraction of (V,Nb)(N,C) and (Nb,V)(N,C) precipitates, as noted in Figure 8c for the respective precipitates. The tempered microstructure still consisted of precipitates and subgrain structures but with reduced dislocation density, as seen in Figure 3d. MX type precipitates were observed inside the subgrain structure and martensite lath structure. $M_{23}C_6$ precipitates were mainly located on the prior austenite grain boundaries and martensitic lath boundaries. The microstructure of Grade 91 steel tempered at the lowest temperature had much finer martensite lath size and higher dislocation density. The as-received material had a prior austenitic grain size of ~11 μm, while that of the sample normalized at 1040 °C for 8 h and tempered at 790 °C for 20 h was 17 μm. The martensitic lath size increased and dislocation density decreased upon tempering. Dudko et al. [13] noted that when tempered at 720 °C for 3 h, the lath size increased to 300 nm from 250 nm, and the dislocation density decreased to 6.2×10^{14} m^{-2} from 7.0×10^{14} m^{-2} compared to the as-normalized microstructure. In 9%–12% Cr heat resistant steels, the dislocation density, hardness, and creep rupture life were generally reduced, but ductility and subgrain size increased with increasing tempering temperature [3].

The transformation of martensite to alpha-ferrite is dependent on the tempering temperature and holding time. The tempering temperature and holding time of tempering can be described by a single empirical parameter known as the Hollomon-Jaffe parameter (P)

$$P = T (C + \log t) \qquad (1)$$

where T (K) is temperature, C (40) constant, and t (s) time. Generally, in the Hollomon-Jaffe parameter, 20 is used as the constant (C), but in this study C (40) gave the best fit. Figure 9 shows the Hollomon-Jaffe parameter for Grade 91 steel normalized at 1040 °C/2 h, and tempered in the temperature range of 635–820 °C for 2, 8 and 20 h. Figure 9 can be used to describe the variation of hardness in the tempered Grade 91 steel as a function of tempering temperature/time (as shown in

Figures 3 and 4). In the plot in Figure 9, two distinct regions could be observed: A region with a steep decrease in hardness with increasing Hollomon-Jaffe parameter followed by a region where the hardness did not vary much. The lower Hollomon-Jaffe parameter signifies lower tempering temperature/time where microstructural evolution takes place gradually and did not achieve quite the equilibrium microstructure. However, at higher Hollomon-Jaffe parameter the microstructure achieves close-to-equilibrium structure and the microstructural parameters did not change so much to affect hardness to an appreciable extent. Figure 8 shows the Thermo-Calc calculations indicating a decrease in the volume fraction of precipitates around A_{C1} temperature. The data related to tempering condition above A_{C1} have not been included in Figure 9 as the Hollomon-Jaffe parameter does not apply to heat treatment above A_{C1} temperature.

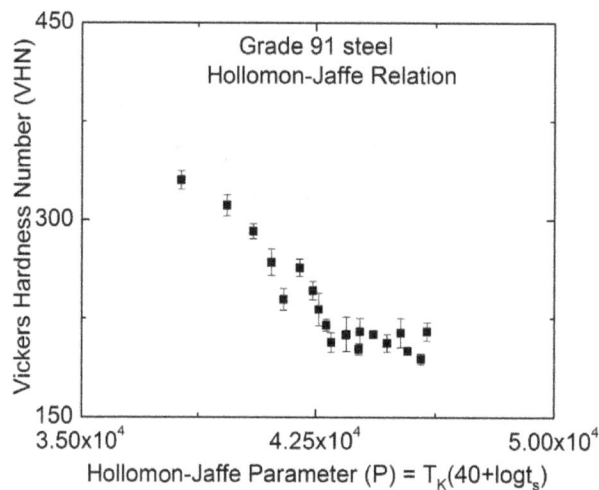

Figure 9. Hollomon-Jaffe relation of Grade 91 steel normalized at 1040 °C for 2 h and tempered at various temperatures and times.

4.3. DSC Study

The DSC plot shown in Figure 5 tracked the phase changes and reactions happening during heating and cooling of Grade 91 steel. Various temperatures associated with phase changes and reactions are listed in Table 2. DSC plot can slightly change depending on the heat treatment history of the material. Masuyama [15] reported that for various tempering temperatures, A_{C1} temperature was between 820 and 851 °C and the martensite transformation was ~400 °C. In this study, the A_{C1} temperature was 820 °C and the M_s temperature was 380 °C. DSC study of Grade 91 steel showed the change in the enthalpy associated with α-ferrite to austenite phase transformation is dependent on the starting microstructure produced by thermal aging [21]. A list of enthalpy change associated with α-ferrite to austenite phase transformation ($\Delta°H^{\alpha \rightarrow \gamma}$) is included in Table 3. The $\Delta°H^{\alpha \rightarrow \gamma}$ for or pure iron was 16 $J \cdot g^{-1}$, but decreased with increased addition of alloying elements. Table 3 includes the enthalpy values of phase transformation (alpha-ferrite to austenite) for various steels. Substitutional solutes (such as W, Mo, V, Ti, Ta, Si) present in the steels in one or other combination are all ferrite-stabilizers. Even though steels contain carbon (an austenite stabilizer), the amount of C is fairly low. Thus, the addition of ferrite stabilizers to pure iron decreases the enthalpy, requiring more energy to transform to austenite and leading to a higher A_{C1}/A_{C3} temperature. The volume fraction of (V,Nb)(C,N) precipitates peaked initially at 790 °C and 850 °C, then started to dissolve in the matrix with the dissolution bottoming out at ~1000 °C and then peaked at ~1140 °C; these temperatures were captured with the help of the first derivative of the DSC heating curve. Furthermore, these temperatures were predicted by the Thermo-Calc calculations as shown in Figure 8c,d. The activation energy of Nb diffusion in the alpha-iron lattice is larger than those of Mo and Cr diffusion, and self-diffusion of iron [22]. Thus, the dissolution of Nb rich precipitates takes place at a higher temperature compared to the $M_{23}C_6$ precipitates. The MX type precipitates rich in V dissolve in the matrix before Nb, given the smaller atomic radius of V compared to Nb.

Table 3. Comparison of enthalpy associated with α→γ phase change.

Steel Composition (wt.%)	$\Delta^o H^{a \to \gamma}$ (J·g^{-1})	Reference
Pure Iron	16	[21]
9Cr-1Mo	15	[10]
9Cr-1Mo-0.001V-0.1C	15	[21]
9Cr-1Mo-0.1C-0.42Si	13	[10]
9Cr-1W-0.23V-0.06Ta-0.1C	12	[23]
9Cr-1Mo-0.21V-0.08Nb-0.1C-0.002Ti	10	Present study

4.4. Thermo-Calc Calculations

In the present study, the normalizing temperatures (1020, 1040, 1050, 1060, 1080, and 1100 °C) and the temperature range of the subsequent tempering (690, 725, 745, and 790 °C) were of most interest. Figures 6 and 8 show the evolution of various phases as a function of temperature. These phase and precipitates play an important role in the formation of microstructure and strength of the heat treated alloy. The Thermo-Calc calculations predicted the existence of the Z-phase, but it was not observed in the TEM studies. During cooling, the austenite phase undergoes partial dissolution (the onset of transformation) at temperature ~850 °C and disappears completely at ~810 °C, in almost complete agreement with the findings of the DSC experiments (temperatures 870 and 820 °C, respectively). The Thermo-Calc does not make any distinction between α-ferrite and the δ-ferrite because they have the same bcc lattice. It would be appropriate to say that the α-ferrite dissolved at ~850 °C, but the δ-ferrite appeared at ~1270 °C peaked at ~1440 °C and then gradually dissolved as seen in Figure 8a.

The mass fraction of $M_{23}C_6$ precipitates remained constant up to the Ac$_1$ temperature but decreased drastically as austenite phase started forming. Given that austenite has higher affinity for C than α-ferrite, the $M_{23}C_6$ precipitates exist up to 865 °C, and then dissolve into the austenitic matrix. The mass fraction of MX type precipitates was constant till Ac$_3$ temperature, but then gradually decreased with increased temperature and totally dissolved at ~1200 °C [24]. In the present study, the volume fraction of Nb-rich MX increased at ~770 °C, started to decrease rapidly after ~850 °C, then gradually at ~1120 °C till 1260 °C. Since there was no clear distinction between the evolution of (Ti,Nb)(N,C) precipitate and (Nb,V)(N,C) precipitate, one could speculate that 1120 °C marks the onset of the phase separation reaction. As indicated in Figure 8c,d, the line indicating both (Nb,V)(N,C) and (Ti,Nb)(N,C) precipitates has a tail to it. The tail indicates the presence of Ti-rich MX which is stable up to ~1400 °C, as shown in Figure 6a. At around 470 °C, the dissolution of the M_6C precipitates was observed in the Thermo-Calc calculations, not the Laves phase (absent at any temperature for x(C) = 0.001 wt.%). Laves phase was not observed in the as-received 9% Cr alloys containing 0.1% and 0.05% C with 0.03% Ti [3]. The Z-phase disappeared as the Nb-rich MX precipitates started to form. It remains to be established whether at such a low temperature there is enough time for the formation of such precipitates (M_6C). However, they were identified in the computed phase diagram quite reliably for different alloy compositions.

The temperature of 1045 °C corresponded to the terminus of the carbon solubility in the V-MX + Nb-MX phase field, as shown in Figure 6a. At this temperature range, a phase transformation reaction accompanied by the formation of the three types of Nb-bearing FCC precipitates: Nb-rich, V-rich and Ti-rich was obtained. Whether this could be considered as a transformation of the V-rich precipitates of the V-MX type (containing ~47 wt.% V and ~32% of Nb) into a new FCC phase in which the amounts of Nb and *V* are practically reversed, and the evolution of Ti-rich precipitate remains to be studied. Hong *et al.* [25] reported dendritic Nb-rich carbonitrides in high strength low alloy steel. On reheating the HSLA steel from 1050 to 1400 °C, the Nb content decreased with increasing temperature till 1250 °C. However, eventually, Ti started to dissolve into the austenite phase after 1300 °C. A similar observation was made in Grade 91 steel where the (Ti,Nb)(N,C) precipitate was stable at higher temperature compared to (Nb,V)(C,N) and (V,Nb)(N,C) precipitates. The solubility of (V,Nb)(C,N) was higher than that of (Ti,Nb)(C,N) and (Nb,Ti)(C,N) in both ferrite and austenite phases [26].

5. Conclusions

The heat treatment study of Grade 91 steel was performed, and Thermo-Calc calculations were used to predict the stability of various phases and provide guidance for future research in this direction. The hardness of the alloy decreased with increasing normalizing and tempering temperatures and times. The decrease in hardness was attributed to increase in the grain size, martensite lath size, and decrease in dislocation density and precipitate coarsening. Differential scanning calorimetry indicated the various phase changes and precipitate reactions taking place during heating and cooling of Grade 91 steel. The Thermo-Calc predicted the evolution of ferrite and austenite phases, and various precipitates. With the help of differential scanning calorimetry study and Thermo-Calc, various temperatures important to heat treatment of Grade 91 steel were identified. Normalizing of Grade 91 steel should be carried out in the temperature range of 1040–1270 °C, and tempering in the temperature range of 745–820 °C. Specific tempering optimization was not feasible in this study since only hardness was evaluated without any measurement of ductility and other relevant properties.

Acknowledgments

This research was performed using funding received from the DOE Office of Nuclear Energy's Nuclear Energy University Programs (NEUP) through the US Department of Energy Grant No. 42246 release 59. We would like to thank Zack Wuthrich and Tshering Sherpa for their assistance with the optical microscopy work.

Author Contributions

Triratna Shrestha carried out experiments under Indrajit Charit and Gabriel P. Potirniche's guidance. Michael V. Glazoff performed the ThermoCalc calculations under Indrajit Charit and Triratna Shrestha's guidance. Data analyses were done in collaboration and characterization were done by Triratna Shrestha and Sultan F. Alsagabi.

Conflicts of Interest

The authors declare no conflict of interest.

References

1. Charit, I.; Murty, K.L. Structural materials issues for the next generation fission reactors. *JOM* **2010**, *62*, 67–74.
2. Murty, K.L.; Charit, I. Structural materials for Gen-IV nuclear reactors: Challenges and opportunities. *J. Nucl. Mater.* **2008**, *383*, 189–195.
3. Rojas, D. 9%–12% Cr Heat Resistant Steels: Alloy Design, Tem Characterization of Microstructure Evolution and Creep Response at 650 °C. Ph.D. Thesis, Ruhr-University Bochum, North Rhine-Westphalia, Germany, 2011.
4. Vishwanathan, R.; Bakker, W.T. Materials for boilers in ultra-supercritical power plants. In Proceedings of the 2000 International Joint Power Generation Conference, Miami Beach, FL, USA, 23–26 July 2000; pp. 1–22.
5. Buhre, B.J.P.; Gupta, R.; Richardson, S.; Sharma, A.; Spero, C.; Wall, T. PF-Fired supercritical boilers: Operational issues and coal quality impacts. Available online: http://www.steamforum. com/pictures/Supercritical%20TN%2020%20PF%20Super%20Critical.pdf (accessed on 20 January 2015).
6. Shrestha, T.; Basirat, M.; Charit, I.; Potirniche, G.P.; Rink, K.K. Creep deformation mechanisms in modified 9Cr-1Mo steel. *J. Nucl. Mater.* **2012**, *423*, 110–119.
7. Basirat, M.; Shrestha, T.; Potirniche, G.P.; Charit, I.; Rink, K. A study of the creep behavior of modified 9Cr-1Mo steel using continuum damage modeling. *Int. J. Plast.* **2012**, *37*, 95–107.
8. Shrestha, T.; Basirat, M.; Charit, I.; Potirniche, G.P.; Rink, K.K. Creep rupture behavior of Grade 91 steel. *Mater. Sci. Eng. A* **2013**, *565*, 382–391.
9. Ahn, J.C.; Sim, G.M.; Lee, K.S. Effect of aging treatment on high temperature strength of Nb added ferritic stainless steels. *Mater. Sci. Forum* **2005**, *475*, 191–194.
10. Ganesh, B.J.; Raju, S.; Rai, A.K.; Mohandas, E.; Vijayalakshmi, M.; Rao, K.B.S.; Raj, B. Differential scanning calorimetry study of diffusional and martensitic transformations in some 9 wt-% Cr low carbon ferritic steels. *Mater. Sci. Technol.* **2011**, *27*, 500–512.
11. Yoshino, M.; Mishima, Y.; Toda, Y.; Kushima, H.; Sawada, K.; Kimura, K. Influence of normalizing heat treatment on precipitation behavior in modified 9Cr-1Mo steel. *Mater. High Temp.* **2008**, *25*, 149–158.
12. Totemeier, T.C.; Tian, H.; Simpson, J.A. Effect of normalization temperature on the creep strength of modified 9Cr-1Mo steel. *Metall. Mater. Trans.* **2005**, *37*, 1519–1525.

13. Dudko, V.; Delyakov, A.; Kaibyshev, R. Effect of tempering on mechanical properties and microstructure of a 9% Cr heat resistant steel. *Mater. Sci. Forum* **2012**, *706*, 841–846.

14. Sawada, K.; Suzuki, K.; Kushima, H.; Tabuchi, M.; Kimura, K. Effect of tempering temperature on Z-phase formation and creep strength in 9Cr-1Mo-V-Nb-N steel. *Mater. Sci. Eng. A* **2008**, *480*, 558–563.

15. Masuyama, F.; Nishimura, N. Experience with creep-strength enhanced ferritic steels and new emerging computational methods In Proceedings of the 2004 ASME/JSME Pressure Vessels and Piping Conference, San Diego, CA, USA, 25–29 July 2004; Volume 476, pp. 85–92.

16. Shen, Y.Z.; Kim, S.H.; Cho, H.D.; Han, C.H.; Ryu, W.S. Identification of precipitate phases in a 11Cr ferritic/martensitic steel using electron microdiffraction. *J. Nucl. Mater.* **2010**, *400*, 64–68.

17. Anderson, P.; Bellgard, T.; Jones, F.L. Creep deformation in a modified 9Cr-1Mo steel. *Mater. Sci. Technol.* **2003**, *19*, 207–213.

18. Das, C.R.; Albert, S.K.; Bhaduri, A.K.; Srinivasan, G.; Murty, B.S. Effect of prior microstructure and mechanical properties of modified 9Cr-1Mo steel weld joints. *Mater. Sci. Eng. A* **2008**, *477*, 185–192.

19. Silwal, B.; Li, L.; Deceuster, A.; Griffiths, B. Effect of post-weld heat treatment on toughness of heat-affected zone of Grade 91 steel. *Welding J.* **2013**, *92*, 80s–87s.

20. Shtansky, D.V.; Inden, G. Phase transformation in Fe-Mo-C and Fe-W-C steels—II. Eutectoid reaction of M23C6 carbide decomposition during austenitization. *Acta Mater.* **1997**, *45*, 2879–2895.

21. Ganesh, B.J.; Raju, S.; Mohandas, E.; Vijayalakshmi, M. Effect of thermal aging on the transformation temperatures and specific heat characteristics of 9Cr-1Mo ferritic steel. *Defect Diffus. Forum* **2008**, *279*, 85–90.

22. Oono, N.; Nitta, H.; Iijima, Y. Diffusion of Nb in α-iron. *Mater. Trans.* **2003**, *44*, 2078–2083.

23. Raju, S.; Ganesh, B.J.; Rai, A.K.; Saroja, S.; Mohandas, E.; Vijayalakshmi, M.; Raj, B. Drop calorimetry studies on 9Cr-1W-0.23V-0.06Ta-0.09C reduced activation steel. *Int. J. Thermophys.* **2010**, *31*, 399–415.

24. Hald, J.; Korcakova, L.; Danielsen, H.K.; Dahl, K.V. Thermodynamics and kinetic modeling: Creep resistant materials. *Mater. Sci. Technol.* **2009**, *24*, 149–158.

25. Hong, S.G.; Jun, H.J.; Kang, K.B.; Park, C.G. Evolution of precipitates in the Nb-Ti-V microalloyed HSLA steels during reheating. *Scr. Mater.* **2003**, *48*, 1201–1206.

26. Taylor, K.A. Solubility products of titanium-, vanadium-, and niobium-carbide in ferrite. *Scr. Mater.* **1995**, *32*, 7–12.

5

Martensitic Transformation in Ni-Mn-Sn-Co Heusler Alloys

Alexandre Deltell [†], Lluisa Escoda [†], Joan Saurina [†] and Joan Josep Suñol *

Department of Physics, University of Girona, Campus Montilivi s/n, 17071 Girona, Spain;
E-Mails: alexandre.deltell@udg.edu (A.D.); lluisa.escoda@udg.edu (L.E.);
joan.saurina@udg.edu (J.S.)

[†] These authors contributed equally to this work.

* Author to whom correspondence should be addressed; E-Mail: joanjosep.sunyol@udg.edu

Academic Editor: Kurt R. Ziebeck

Abstract: Thermal and structural austenite to martensite reversible transition was studied in melt spun ribbons of $Ni_{50}Mn_{40}Sn_5Co_5$, $Ni_{50}Mn_{37.5}Sn_{7.5}Co_5$ and $Ni_{50}Mn_{35}Sn_{10}Co_5$ (at. %) alloys. Analysis of X-ray diffraction patterns confirms that all alloys have martensitic structure at room temperature: four layered orthorhombic 4O for $Ni_{50}Mn_{40}Sn_5Co_5$, four layered orthorhombic 4O and seven-layered monoclinic 14M for $Ni_{50}Mn_{37.5}Sn_{7.5}Co_5$ and seven-layered monoclinic 14M for $Ni_{50}Mn_{35}Sn_5Co_5$. Analysis of differential scanning calorimetry scans shows that higher enthalpy and entropy changes are obtained for alloy $Ni_{50}Mn_{37.5}Sn_{7.5}Co_5$, whereas transition temperatures increases as increasing valence electron density.

Keywords: magnetic shape memory; martensitic transformation; DSC; XRD; Ni-Mn-Sn-Co

1. Introduction

Ferromagnetic shape memory (FSM) alloys are of considerable interest due to their exceptional magnetoelastic properties. Their potential functional properties include: Magnetic superelasticity [1], large inverse magnetocaloric effect [2] and large magneto-resistance change [3]. Most of these effects are ascribed to the existence of a first order martensitic transformation with a strong magneto-structural coupling. Transformation temperatures of shape memory alloys depend on the composition and their values spread to a very wide range [4]. These materials are interesting for the development of new magnetically driven actuators, sensors and coolers for magnetic refrigeration [5].

FSM behavior is found in Heusler alloys, which have a generic formula X_2YZ and are defined as ternary intermetallic systems with $L2_1$ crystalline cubic structure. The most extensively studied Heusler alloys are those based on the Ni-Mn-Ga system. However, to overcome some of the problems related with practical applications (such as the high cost of Gallium and the usually low martensitic transformation temperature), Ga-free alloys have been searched and analyzed frequently during the last few decades, specifically with the introduction of In or Sn. Martensitic transformation in ferromagnetic Heusler $Ni_{50}Mn_{50-x}Sn_x$ bulk alloys with $10 \leq x \leq 16.5$ was first reported by Sutou *et al.* [6]. Later, Krenke *et al.* studied magnetic and magnetocaloric properties and phase transformations in $Ni_{50}Mn_{50-x}Sn_x$ alloys with $5 \leq x \leq 25$ [7]. Rapid solidification techniques, such as melt-spinning, are an alternative to obtain these materials (ribbon shape) [8,9].

Another important factor affecting the magnetic behavior of Ni-Mn-Sn and Ni-Mn-Sn-Co systems is the annealing process. Some authors have found different magnetic behavior in melt-spun Ni-Mn-Sn-Co ribbons annealed at temperatures from 973 K to 1173 K [10,11].

In our work, we investigate the structural and thermal behavior of three melt-spun alloys of the Ni-Mn-Sn-Co system (by modifying Mn and Sn atomic %). These ribbons were not annealed.

2. Experimental Section

Polycrystalline Ni-Mn-Sn-Co alloy ingots were prepared by arc melting high purity (99.99%) elements under argon environment in a water-cooled quartz crucible. The ingots were melted three times to ensure a good homogeneity. Thus, ingots were melt-spun on a rotating copper wheel (Buheler, Lake Bluff, IL, USA) set by controlling process parameters as: Linear wheel speed (48 ms^{-1}), atmosphere (argon, 400 mbar), injection overpressure (500 mbar) and distance between wheel and injection quartz crucible (3 mm). The as-spun ribbon samples (Alfa Aesar, Heysham, UK) obtained were: $Ni_{50}Mn_{40}Sn_5Co_5$, $Ni_{50}Mn_{37.5}Sn_{7.5}Co_5$ and $Ni_{50}Mn_{35}Sn_{10}Co_5$ (at. %). The main difference among these alloys is the partial substitution of Mn by Sn whereas the content of Co and Mn is constant.

Thermal and structural analyses were performed by applying several techniques. Scanning electron microscopy (SEM) investigations were carried out using a Zeiss DSM 960A microscope (Zeiss, Jena, Germany) operating at 30 kV and linked to an energy dispersive X-ray spectrometer (EDX; Zeiss, Jena, Germany). X-ray diffraction (XRD) analyses were performed at room temperature with a Siemens D500 X-ray powder diffractometer (Bruker, Bullerica, MA, USA) using Cu-K$_\alpha$ radiation. Thermal analyses were performed by differential scanning calorimetry (DSC) using a DSC822e calorimeter of Mettler-Toledo (Mettler Toledo, Columbus, OH, USA) working at a heating/cooling rate of 10 K/min under argon atmosphere.

3. Results and Discussion

Heusler alloys produced by melt spinning show a typical columnar structure in the fracture cross section. Figure 1 shows the micrographs of the fracture section of alloys $Ni_{50}Mn_{40}Sn_5Co_5$, $Ni_{50}Mn_{37.5}Sn_{7.5}Co_5$ and $Ni_{50}Mn_{35}Sn_{10}Co_5$, labeled as A, B and C respectively. All ribbon flakes have a similar morphology which consists of: fully crystalline and granular columnar type microstructure.

This is a sign of the quick crystallization and fast growth kinetics of the samples. This suggests that the heat removal during rapid solidification process induces the directional growth of the crystalline phase. The ribbons' width is also similar (between 12 and 15 μm).

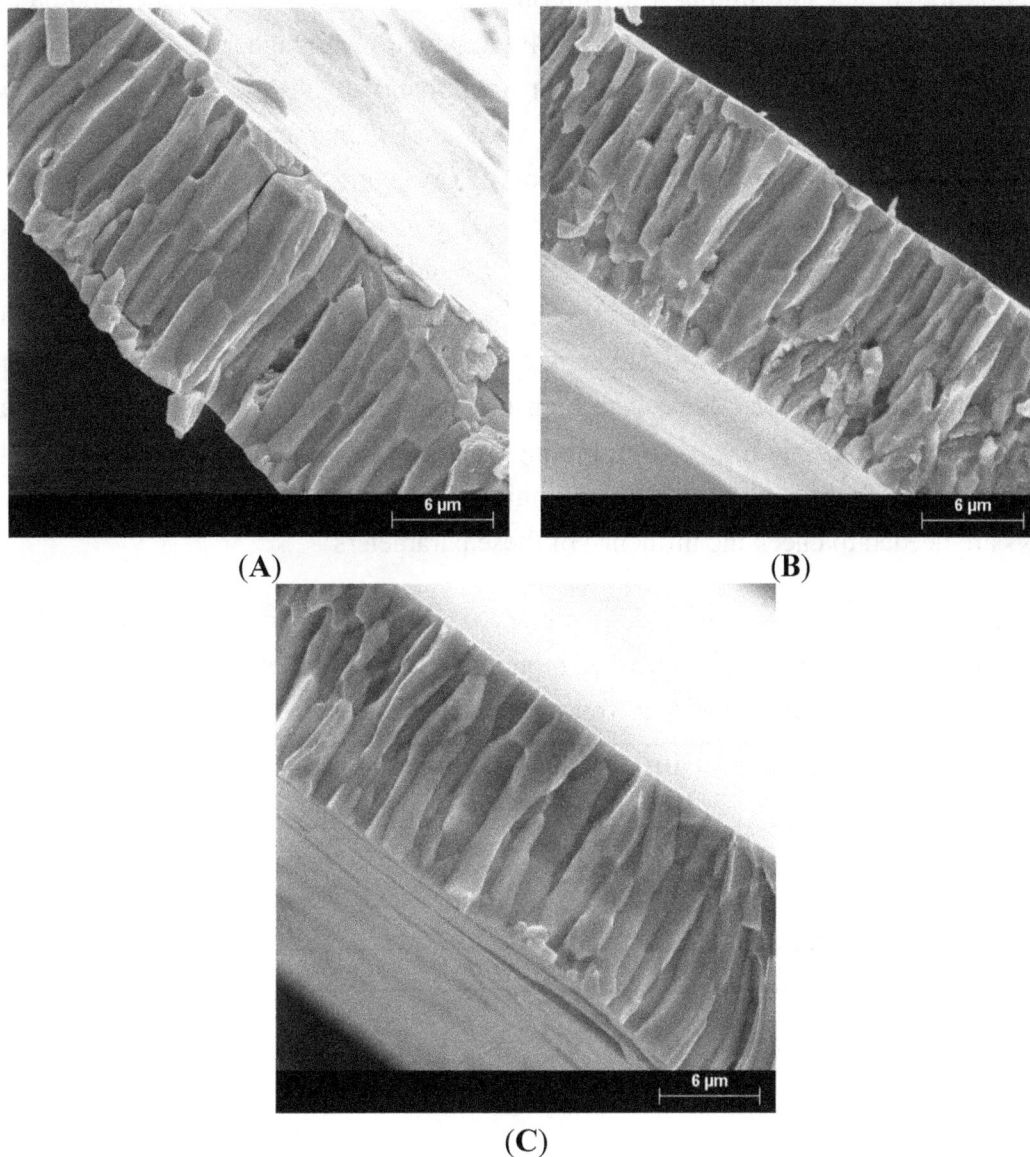

Figure 1. SEM micrographs of the cross section of alloys $Ni_{50}Mn_{40}Sn_5Co_5$, $Ni_{50}Mn_{37.5}Sn_{7.5}Co_5$ and $Ni_{50}Mn_{35}Sn_{10}Co_5$, labeled as **(A)**, **(B)** and **(C)** inside the figure.

Crystalline structures at room temperature were determined by analyzing X-Ray diffraction patterns of the three samples (see Figures 2–4 for alloys $Ni_{50}Mn_{40}Sn_5Co_5$; $Ni_{50}Mn_{37.5}Sn_{7.5}Co_5$ and $Ni_{50}Mn_{35}Sn_{10}Co_5$ respectively). X-ray diffraction analysis begins with in an indexation based on the identification proposed by other authors [7,12,13]. The lattice parameters were first calculated minimizing the global interplanar spacing (d_{hkl}) error; defined as the difference between the values calculated from the Bragg equation to every identified peak of the XRD pattern compared to the crystal system geometry equation.

It is found that, at room temperature, all alloys have martensitic structure. This martensitic structure is confirmed to be four layered orthorhombic 4O for $Ni_{50}Mn_{40}Sn_5Co_5$, four layered orthorhombic 4O and seven-layered monoclinic 14M for $Ni_{50}Mn_{37.5}Sn_{7.5}Co_5$ and seven-layered monoclinic 14M for $Ni_{50}Mn_{35}Sn_{10}Co_5$. Lattice parameters are given in Table 1 (for $Ni_{50}Mn_{37.5}Sn_{7.5}Co_5$ alloy only parameters from the main phase, 4O, are given).

In our work it is found that substituting Mn with Sn favors to the formation of the modulated 14M monoclinic structure. Thus, the martensitic structure is 4O in samples with higher Mn/Sn ratio and 14M in samples with lower Mn/Sn ratio. Opposite behavior was found in Ni-Mn-Sn bulk alloys without Co [14]. Thus, the Co addition probably influences what kind of martensitic phase is more stable. In Ni-Mn-Sn-Co ribbons, it was found that the addition of Co favors the evolution of the martensitic crystalline structure from a four-layered orthorhombic (4O) to a five-layered orthorhombic (10M) and finally to a seven-layered monoclinic (14M) [15]. Thus, the addition of Co favors the formation of the 14M structure. This effect was not found in our alloys, probably because Co content is constant. Furthermore, it has been also found that the martensitic crystal structure changes from 14M in the bulk alloy to 4O in the melt spun ribbons due to the high oriented microstructure [16]. In summary, the differences between our results and results from bibliography can be influenced by the combination of these three factors: Co constant content, Mn/Sn ratio and high oriented ribbons microstructure. More than three alloys are needed to check the influence of these parameters.

Table 1. Crystalline structure and lattice parameters of $Ni_{50}Mn_{40}Sn_5Co_5$, $Ni_{50}Mn_{37.5}Sn_{7.5}Co_5$ and $Ni_{50}Mn_{35}Sn_{10}Co_5$. The angle of the monoclinic 14M structure is 95.56°, (for $Ni_{50}Mn_{37.5}Sn_{7.5}Co_5$ alloy only parameters from the main phase are given).

Alloy	Crystalline structure	Lattice Parameter a/nm	Lattice Parameter b/nm	Lattice Parameter c/nm
$Ni_{50}Mn_{40}Sn_5Co_5$	4O	0.8828	0.5817	0.4274
$Ni_{50}Mn_{37.5}Sn_{7.5}Co_5$	4O	0.8922	0.5892	0.4281
$Ni_{50}Mn_{35}Sn_{10}Co_5$	14M	0.4298	0.5612	2.9441

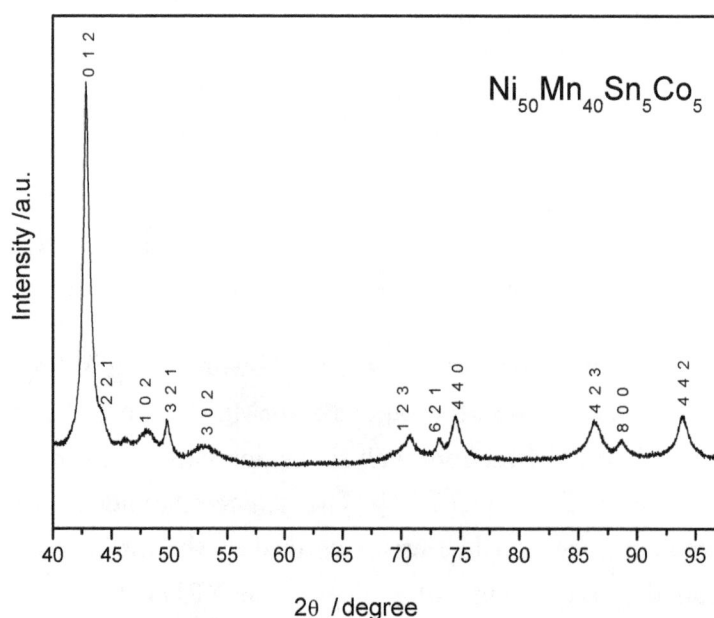

Figure 2. X-ray diffraction (XRD) pattern, at room temperature, of $Ni_{50}Mn_{40}Sn_5Co_5$ ribbon. The indexation corresponds to a four-layered 4O orthorhombic structure.

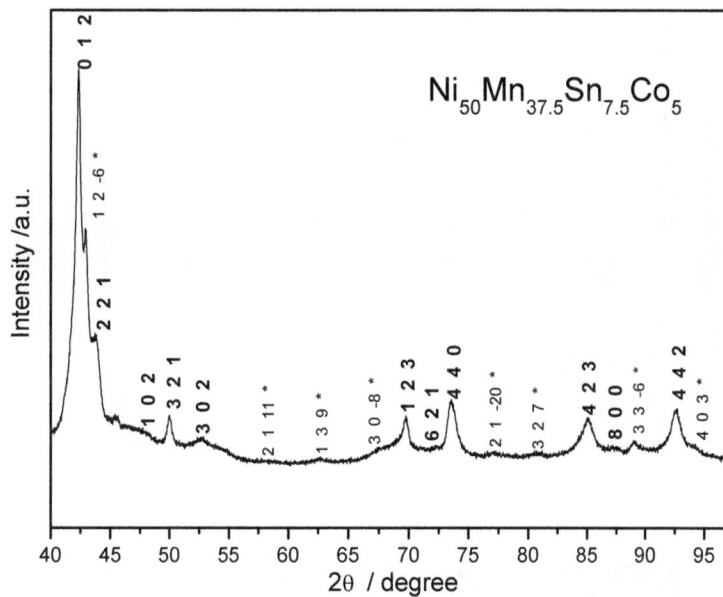

Figure 3. XRD pattern, at room temperature, of Ni₅₀Mn₃₇.₅Sn₇.₅Co₅ ribbon. The indexation of the main phase corresponds to a four-layered 40 orthorhombic structure, whereas peaks marked with * correspond to a modulated monoclinic seven-layered 14M structure.

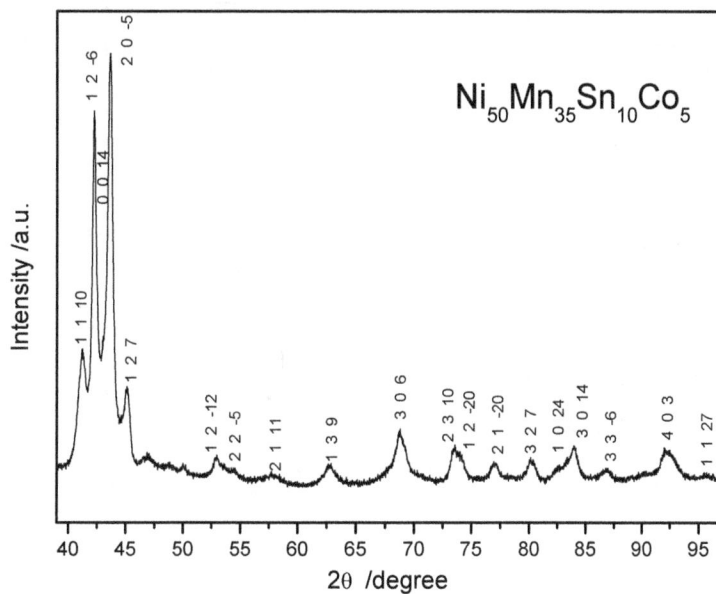

Figure 4. XRD pattern, at room temperature, of Ni₅₀Mn₃₅Sn₁₀Co₅ ribbon. The indexation of the main phase correspond to a modulated monoclinic seven-layered 14M structure.

At room temperature, XRD show that all samples have a martensitic phase. Thus, the occurrence of the martensitic transformation should be checked by DSC heating from room temperature (see Figures 5–7). The reversible austenite—martensite transformation was found in all samples. The absence of any secondary thermal process suggests that the produced ribbons are homogeneous. From DSC analysis characteristic transformation temperatures are determined. Start and finishing martensite and austenite transformation temperatures are referred as M_s, M_f and A_s, A_f respectively.

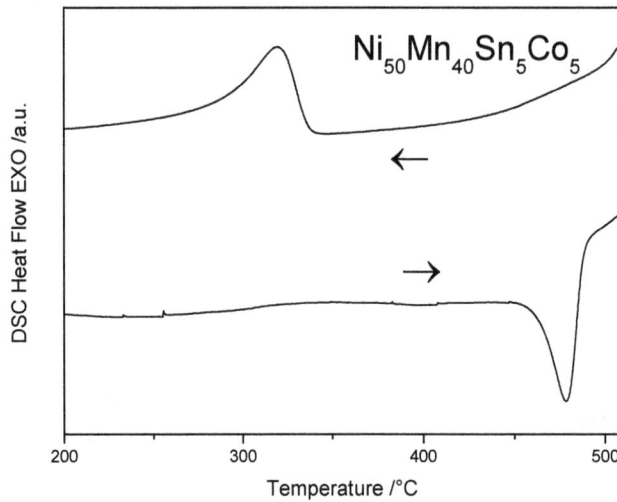

Figure 5. Differential scanning calorimetry (DSC) cyclic scan of alloy $Ni_{50}Mn_{40}Sn_5Co_5$.

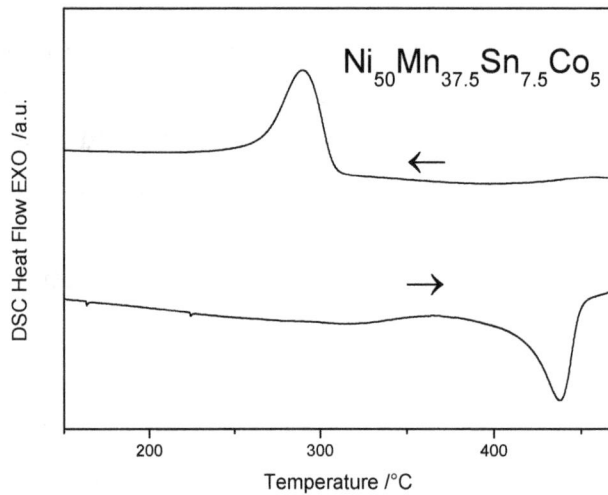

Figure 6. DSC cyclic scan of alloy $Ni_{50}Mn_{37.5}Sn_{7.5}Co_5$.

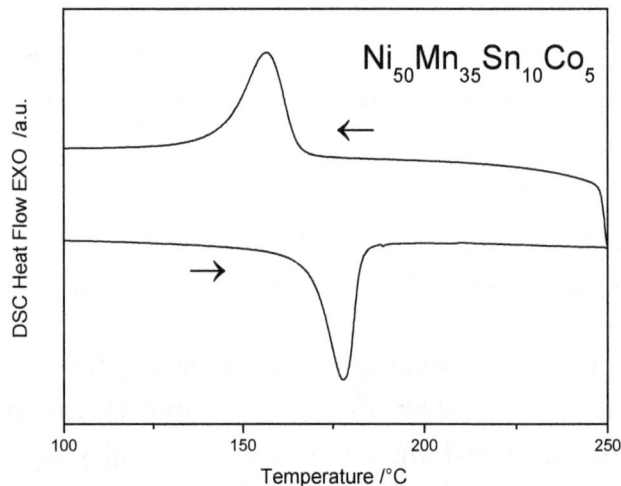

Figure 7. DSC cyclic scan of alloy $Ni_{50}Mn_{35}Sn_{10}Co_5$.

The martensitic transformation of Ni-Mn-Sn alloys is athermal in nature although a time-depending effect is observed through calorimetry interrupted measurements [17]. The thermal hysteresis, ΔT, exists due to the increase of the elastic and the surface energies during the martensitic formation. Thus, the nucleation of the martensite implies supercooling.

The equilibrium transformation temperature between martensite and austenite, T_0, is usually defined as $(M_s + A_f)/2$. All the characteristic temperatures are given in Table 2.

Table 2. Characteristic temperatures and thermal hysteresis as determined from DSC cyclic scans: $Ni_{50}Mn_{40}Sn_5Co_5$, $Ni_{50}Mn_{37.5}Sn_{7.5}Co_5$ and $Ni_{50}Mn_{35}Sn_{10}Co_5$. Start and finishing martensite and austenite formation temperatures are referred as M_s, M_f and A_s, A_f respectively. Thermal hysteresis, ΔT, and equilibrium transformation temperature between martensite and austenite, T_0.

Alloy	M_s/K	M_f/K	A_s/K	A_f/K	T_0/K	ΔT/K
$Ni_{50}Mn_{40}Sn_5Co_5$	616.8	477.2	723.8	764.8	690.8	158.1
$Ni_{50}Mn_{37.5}Sn_{7.5}Co_5$	593.3	478.0	677.7	727.8	660.6	146.7
$Ni_{50}Mn_{35}Sn_{10}Co_5$	453.2	398.2	432.1	474.9	464.1	25.5

It is found that substituting Mn by Sn favors the decrease of the phase transition temperatures. Opposite effect was found in bulk Ni-Mn-Sn alloys without Co [14]. Similarly, the addition of Co in Ni-Mn-Sn melt-spun alloys increases martensitic transformation temperatures [16]. When doping the alloys, it is important which atom is substituted. In Ni-Mn-Sn-Fe bulk alloys the partial substitution of Mn by Fe causes a diminution of the transition temperatures [12]. The same effect is observed in our alloys by substituting Mn by Co. Likewise, the partial substitution of Mn does not induce a general trend in the temperatures [13] whereas Co addition in Ni-Mn-Ga alloys increases the temperatures of the martensitic transformation [18]. Furthermore, annealing also modifies transformation temperatures and thermal hysteresis [10]. Thus, so many parameters affect structural transformation to assure which parameter determines the behavior of our samples.

Changes in enthalpy, ΔH, and entropy, ΔS, during structural transformation are calculated from the area of the DSC peaks. Figure 8 shows its evolution as a function of the average valence electron density (e/a). The shift on the characteristic temperatures and thermodynamic parameters is related to e/a [19]. The valence electrons per atom are 10 ($3d^84s^2$) for Ni, 9 ($3d^74s^2$) for Co, 7 ($3d^54s^2$) for Mn and 4 ($5s^25p^2$) for Sn, respectively.

Energy-dispersive X-ray spectroscopy microanalysis has been used to obtain the exact composition of every sample and to calculate e/a parameter. EDX elemental composition and average valence electron density are presented in Table 3.

Higher values of enthalpy and entropy are those of $Ni_{50}Mn_{37.5}Sn_{7.5}Co_5$ alloy, probably due to the coexistence of two crystalline phases.

One of the most typical ferromagnetic shape memory alloy phase diagram is the graphical representation of the martensitic start temperature as a function of the Z element content or as a function of the average valence electron density. In Figure 9 we represent M_s temperatures obtained in this work (symbols) and those obtained assuming linear relation in Ni-Mn-Sn bulk alloys [20].

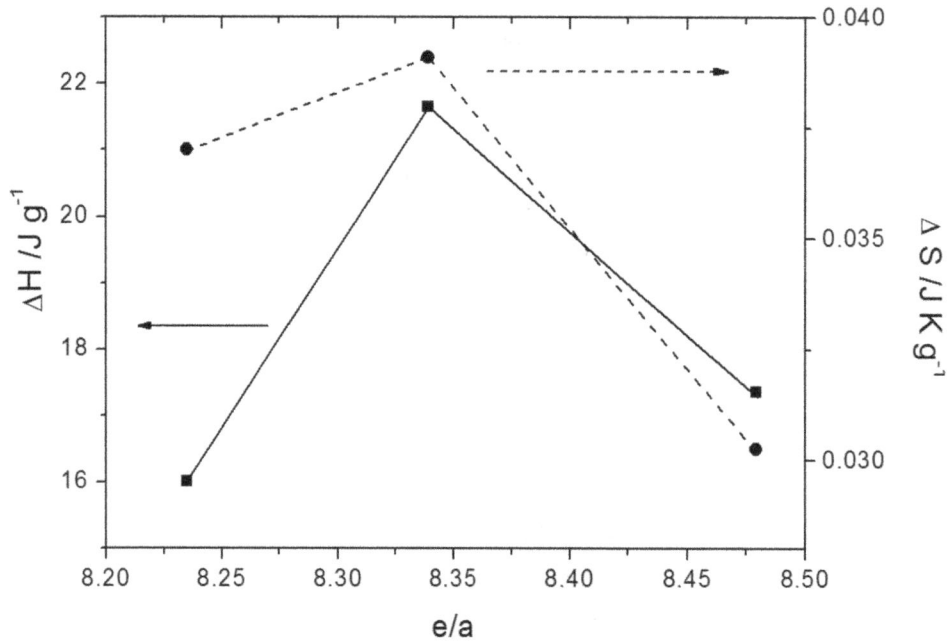

Figure 8. Enthalpy (ΔH) and entropy (ΔS) changes as a function of average valence electron density. Symbols: Enthalpy (square), entropy (circle). Error: <3%.

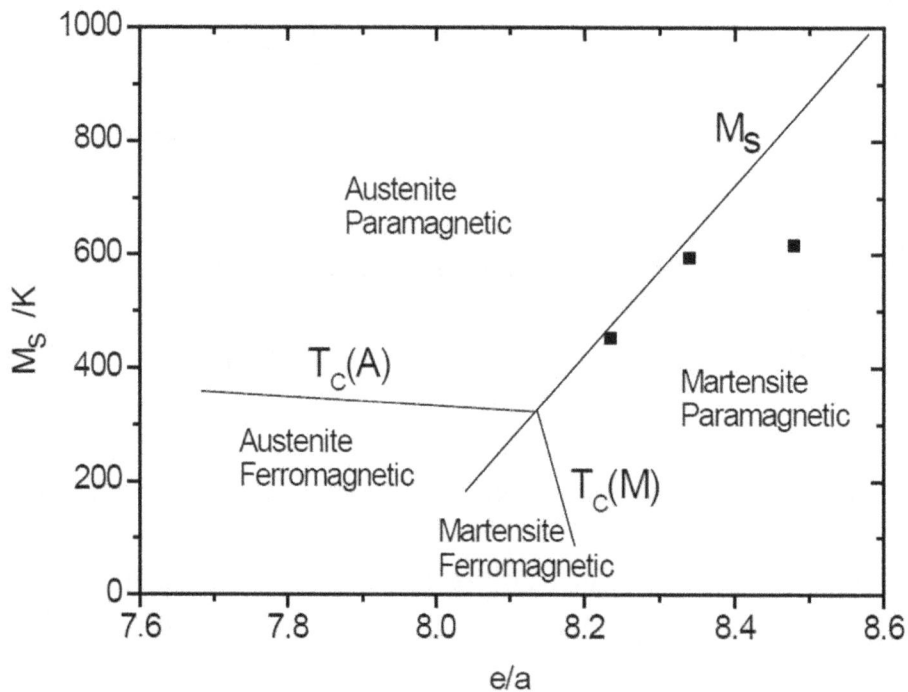

Figure 9. Martensitic start temperature *versus* average valence electron density. Lines correspond to bulk alloys [20] whereas square symbols correspond to our samples.

Table 3. Energy dispersive X-ray spectrometer (EDX) compositions of $Ni_{50}Mn_{40}Sn_5Co_5$, $Ni_{50}Mn_{37.5}Sn_{7.5}Co_5$ and $Ni_{50}Mn_{35}Sn_5Co_5$, and average valence electron density.

Alloy	Ni	Mn	Sn	Co	*e/a*
$Ni_{50}Mn_{40}Sn_5Co_5$	50.7	39.12	4.91	5.27	8.479
$Ni_{50}Mn_{37.5}Sn_{7.5}Co_5$	49.8	35.99	8.78	5.44	8.339
$Ni_{50}Mn_{35}Sn_{10}Co_5$	49.35	34.51	11.37	4.77	8.235

Our results show a diminution of the transformation temperatures. The main difference is for alloy $Ni_{50}Mn_{40}Sn_5Co_5$. In the literature it was found that the martensitic transformation, in Heusler Ni-Mn-Sn melt spun ribbons, occurs at lower temperatures than those compared to bulk alloys [15]. Moreover, a change of the martensitic crystalline structure from 14M to 4O takes place with the decrease of the martensitic transition temperature. It is proposed that the internal stress was induced due to the highly-oriented microstructure, which leads to the decrease of the transition temperature because of a refined martensite plate and the formation of dense martensitic variants with different orientations. These results were supported by high resolution transmission electron microscopy (HRTEM). Furthermore, it was also found that the partial substitution of Ni by Co shifts the martensitic transformation to lower temperatures in Ni-Mn-Sn-Co bulk alloys [21]. If our alloys have the same trend that bulk alloys (Figure 9), it is not clear the occurrence of the magnetic transformation.

4. Conclusions

Melt-spun ribbon of three alloys of the Ni-Co-Mn-Sn system has been produced: $Ni_{50}Mn_{40}Sn_5Co_5$, $Ni_{50}Mn_{37.5}Sn_{7.5}Co_5$ and $Ni_{50}Mn_{35}Sn_{10}Co_5$. The austenite to martensite reversible transformation was found in all samples. Transformation temperatures increase as Mn/Sn ratio increases.

Martensitic structure is four-layered orthorhombic 4O in samples with higher Mn/Sn ratio and monoclinic modulated seven-layered 14M in samples with lower Mn/Sn ratio. These results differ from other obtained in the bibliography. Probably, these differences are caused by the combination of three factors: constant Co content, Mn/Sn ratio and high oriented ribbons microstructure. Furthermore, substituting Mn by Sn favors to the decrease of the austenite-martensite reversible transition temperatures.

Acknowledgments

This research was supported by the projects MAT2013-47231-C2-2-P and 2014SGR1180.

Author Contributions

Alexandre Deltell is a doctorate student. The thesis works are supervised by Joan Josep Suñol and Lluisa Escoda. Joan Saurina supports DSC analysis.

Conflicts of Interest

The authors declare no conflict of interest.

References

1. Krenke, T.; Duman, E.; Acet, M.; Wasserman, E.F.; Moya, X.; Mañosa, L.; Planes, A.; Suard, E.; Ouaddiaf, B. Magnetic superelasticity and reverse magnetocaloric effect in Ni-Mn-In. *Phys. Rev. B* **2007**, *75*, 104414.
2. Caballero-Flores, R.; González-Legarreta, L.; Rosa, W.O.; Sánche, T.; Prida, V.M.; Escoda, L.; Suñol, J.J.; Batdalov, A.B.; Aliev, A.M.; Koledov, V.V.; *et al.* Magnetocaloric effect, magnetostructural and magnetic phase transformations in $Ni_{50.3}Mn_{36.5}Sn_{13.2}$ Heusler alloy ribbons. *J. Alloys Compd.* **2015**, *629*, 332–342.

3. Barandiarán, J.M.; Chernenko, V.A.; Lazpita, P.; Gutiérrez, J.; Feuchtwanger, J. Effect of martensitic transformation and magnetic field on transport properties of Ni-Mn-Ga and Ni-Fe-Ga Heusler alloys. *Phys. Rev. B* **2009**, *80*, 104404.

4. Coll, R.; Escoda, L.; Saurina, J.; Sánchez-Llamazares, J.L.; Hernando, B.; Suñol, J.J. Martensitic transformation in Mn-Ni-Sn Heusler alloys. *J. Therm. Anal. Calorim.* **2010**, *99*, 905–909.

5. Marioni, M.A.; O'Handley, R.C.; Allen, S.M.; Hall, S.R.; Paul, D.I.; Richard, M.L.; Feuchtwanger, J.; Peterson, B.W.; Chambers, J.M.; Techapiesancharoenkij, R. The ferromagnetic shape-memory effect in Ni-Mn-Ga. *J. Magn. Magn. Mater.* **2005**, *290*, 35–41.

6. Sutou, Y.; Imano, Y.; Koeda, N.; Omori, T.; Kainuma, R.; Ishida, K.; Oikawa, K. Magnetic and martensitic transformations of BiMnX (X = In, Sn, Sb) ferromagnetic shape memory alloys. *Appl. Phys. Lett.* **2004**, *85*, 4358.

7. Krenke, T.; Duman, E.; Acet, M.; Wassermann, E.F.; Moya, X.; Mañosa, L.; Planes, A. Inverse magnetocaloric effect in ferromagnetic Ni-Mn-Sn alloys. *Nat. Mater.* **2005**, *4*, 450–454.

8. Hernando, B.; Sánchez-Llamazares, J.L.; Santos, J.D.; Escoda, L.; Suñol, J.J.; Varga, R.; Baldomir, D.; Serantes, D. Thermal and magnetic field-induced mrtensite-austenite transition in $Ni_{50.3}Mn_{35.3}Sn_{14.4}$ ribbons. *Appl. Phys. Lett.* **2008**, *92*, 042504.

9. Santos, J.D.; Sánhez, T.; Álvarez, P.; Sánchez, M.L.; Sánchez-Llamazares, J.L.; Hernando, J.; Escoda, L.; Suñol, J.J.; Varga, R. Microstructure and magnetic properties of $Ni_{50}Mn_{37}Sn_{13}$ Heusler alloy ribbons. *J. Appl. Phys.* **2008**, *103*, 07B326.

10. Chen, F.; Liu, W.L.; Shi, Y.G.; Müllner, P. Influence of annealing on martensitic transformation and magnetic entropy change in Ni-Co-Mn-Sn magnetic shape memory alloy ribbon. *J. Magn. Magn. Mater.* **2015**, *377*, 137–141.

11. Ma, S.C.; Cao, Q.Q.; Xuan, H.C.; Zhang, C.L.; Shen, L.J.; Wang, D.H.; Du, Y.W. Magnetic and magnetocaloric properties in melt-spun and annealed Ni-Mn-Co-Sn ribbons. *J. Alloys Compd.* **2011**, *509*, 1111–1114.

12. Fukushima, K.; Sano, K.; Kanomata, T.; Nishihara, H.; Furutani, Y.; Shishido, T.; Ito, W.; Umetsu, R.Y.; Kainuma, R.; Oikawa, K.; *et al.* Phase diagram of Fe-substituted Ni-Mn-Sn shape memory alloys. *Scr. Mater.* **2009**, *61*, 813–816.

13. Kanomata, T.; Umetsu, R.Y.; Ohtsuki, K.; Shoji, T.; Endo, K.; Fukushima, K.; Nishihara, H.; Ito, W.; Adachi, Y.; Miura, T.; *et al.* Magnetic phase diagram of $Ni_{2}Mn_{1.44-x}Cu_{x}Sn_{0.56}$ shape memory alloys. *J. Alloys Compd.* **2014**, *590*, 221–226.

14. Zheng, H.; Wang, W.; Xue, S.; Zhai, Q.; Frenzel, J.; Luo, Z. Composition-dependent crystal structure and martensitic transformation in Heusler Ni-Mn-Sn alloys. *Acta Mater.* **2013**, *61*, 4648–4656.

15. Zheng, H.; Wu, W.; Yu, J.; Zhai, Q.; Luo, Z. Martensitic transformations in melt-spun Heusler Ni-Mn-Sn-Co ribbons. *J. Mater. Res.* **2014**, *29*, 880–886.

16. Wang, W.; Yu, J.; Luo, Z.; Zheng, H. Origin of retarded martensite transformation in Heusler Ni-Mn-Sn melt-spun ribbons. *Intermetallics* **2013**, *42*, 126–129.

17. Zheng, H.; Wang, W.; Wu, D.; Xue, S.; Zhai, Q.; Frenzel, J.; Luo, Z. Athermal nature of the martensitic transformation in Heusler alloy Ni-Mn-Sn. *Intermetallics* **2013**, *36*, 90–95.

18. Kanomata, T.; Nunoki, S.; Endo, K.; Kataoka, M.; Nishihara, H.; Khovaylo, V.V.; Umetsu, R.Y.; Shishido, T.; Nagasako, M.; Kainuma, R.; *et al.* Phase diagram of the ferromagnetic shape memory alloys $Ni_{2}MnGa_{1-x}Co_{x}$. *Phys. Rev. B* **2012**, *85*, 134421.

19. Chernenko, V.A. Composition instability of beta-phase in Ni-Mn-Ga alloys. *Scr. Mater.* **1999**, *40*, 523–527.
20. Krenke, T.; Acet, M.; Wassermann, E.F.; Moya, X.; Mañosa, L.; Planes, A. Martensitic transitions and the nature of ferromagnetism in the austenitic and martensitic states of Ni-Mn-Sn alloys. *Phys. Rev. B* **2005**, *72*, 014412.
21. Jing, C.; Li, Z.; Zhang, H.L.; Chen, J.P.; Qiao, Y.F.; Cao, S.X.; Zhang, J.C. Martensitic transition and inverse magnetocaloric effect in Co doping Ni-Mn-Sn Heusler alloy. *Eur. Phys. J. B* **2009**, *67*, 193–196.

6

Estimation of Fatigue Crack Growth Rate for 7% Nickel Steel under Room and Cryogenic Temperatures Using Damage-Coupled Finite Element Analysis

Seul-Kee Kim [1], **Chi-Seung Lee** [1], **Jeong-Hyeon Kim** [1], **Myung-Hyun Kim** [1], **Byeong-Jae Noh** [2], **Toshyuki Matsumoto** [3] and **Jae-Myung Lee** [1,*]

[1]   Department of Naval Architecture and Ocean Engineering, Pusan National University, Busan 609-735, Korea; E-Mails: kfreek@pusan.ac.kr (S.-K.K.); victorich@pusan.ac.kr (C.-S.L.); honeybee@pusan.ac.kr (J.-H.K.); kimm@pusan.ac.kr (M.-H.K.)

[2]   Hyundai Heavy Industries, Ulsan 682-792, Korea; E-Mail: bjnoh@hhi.co.kr

[3]   Research Institute of ClassNK, Tokyo 102-8567, Japan; E-Mail: rx-sec@classnk.or.jp

*   Author to whom correspondence should be addressed; E-Mail: jaemlee@pusan.ac.kr

Academic Editor: Hugo F. Lopez

Abstract: In this study, fatigue crack growth rates (FCGR) of 7% nickel steel at room and cryogenic temperatures were evaluated using damage-coupled finite element analysis (FEA). In order to perform the computational fatigue analysis effectively, methods for coupling damage to FEA are introduced and adopted. A hybrid method including the damage-coupled constitutive model and jump-in-cycles procedure was implemented into the ABAQUS user-defined material subroutine. Finally, the represented method was validated by comparing its results with the FCGR test results for 7% nickel steel under room and cryogenic temperatures. In particular, da/dN *versus* ΔK and the crack length *versus* the number of cycles were compared.

Keywords: fatigue crack growth rate; damage mechanics; 7% nickel steel; cryogenic temperature; ABAQUS user-defined material subroutine (UMAT)

1. Introduction

Industrial structures, including ships and offshore structures, are subject to periodic and arbitrary loadings. These phenomena lead to crack propagation and fracture of the structures. In order to prevent such catastrophic disaster, fatigue analysis and fatigue life assessment should be carried out prior to the design steps.

Miner's accumulative damage calculation and Goodman fatigue calculation are the representative fatigue life evaluation methods under regular and irregular cyclic loading. If there is an initial crack included in the material or structural member, the fatigue life and fatigue crack growth rate can change under the operating environments of the structure. For this reason, the fatigue crack growth rate (FCGR) test is commonly adopted to investigate the crack growth characteristics of materials [1].

The estimation of FCGR characteristics has mainly been performed by an experimental approach. Through the test, the correlation between the crack growth rate da/dN and the applied stress intensity factor range ΔK can be estimated [2].

However, it requires enormous time and cost to establish the required experimental facilities and conditions. In addition, it is extremely difficult to reproduce harsh environments such as cryogenic and high temperatures. For this reason, there are many limitations to carrying out the FCGR test effectively. As a result, computational simulation has been considered as an alternative method to solve these problems.

In computational simulation, two kinds of approaches have been principally adopted, namely, a cohesive zone model (CZM) and an extended finite element method (XFEM).

In the CZM theory, a crack is propagated through the separation surface (or cohesive zone) when the cohesive separation force is greater than the cohesive traction force. The crack propagation problems of brittle/cementitious materials, in which a nonlinear zone ahead of the crack tip is generated because of plasticity and microcracking near the cracked region, can be analyzed using the CZM [3]. There are many researchers who solve the fatigue crack growth problem using the CZM. For example, de-Andres et al. have solved a three-dimensional fatigue crack phenomenon and estimated the fatigue life of an aluminum shaft under axial loading using a three-dimensional cohesive element and an irreversible cohesive law [4]. Yang et al. have proposed a simulation method for fatigue crack initiation and propagation in quasi-brittle materials using the CZM [5], and Roe and Siegmund have studied fatigue crack growth along an interface using the actual process of material separation, and not by using Paris' law [6]. In their research, it was found that the traction-separation behavior does not follow a predefined traction-separation path as posited by the classical CZM. Hence, an improved CZM has been proposed on the basis of three types of viewpoints, namely, (i) the consideration and application of a traction-separation law considering unloading in both normal and shear separation in the cohesive zone; (ii) the development of an evolutionary damage law in order to analyze failure under subcritical loading in dependence of the accumulated separation and the current resultant traction in the cohesive zone; and (iii) the establishment of a contact model. In addition, Bouvard et al. have carried out a numerical analysis of the fatigue and creep-fatigue crack growth of single crystal superalloys using an irreversible CZM [7]. In order to calculate the fatigue damage accumulation under cyclic loads, a damage-coupled CZM has been proposed, and damage accumulation and stagnation phenomena under loading and unloading conditions have been quantitatively predicted. Ural et al. have developed a damage-based cohesive model in which the coupled model is fabricated by the combination of the linear damage-dependent traction-separation relation and the damage evolution equation [8].

On the other hand, the handicaps of the classical finite element method (FEM) for crack propagation simulation, such as an impractical crack growth shape and node singularity, have been improved by the XFEM. For using XFEM, a standard displacement-based approximation has been enriched near a crack by incorporating both the discontinuous fields and the near-tip asymptotic fields through a partition of unity method. On the basis of this technique, the entire crack is represented independently of the mesh; hence, a remeshing technique is not necessary to describe the crack growth [9]. The XFEM is most recent proposed method, and outstanding studies have been carried out since 1999. Stolarska *et al.* have developed a level set method (LSM)-coupled XFEM algorithm [10]. The LSM and XFEM were adopted to represent the crack tip location prior to the crack growth and the rate of crack growth, respectively. In addition, Sukumar *et al.* have proposed a fast marching method (FMM)-coupled XFEM technique [11]. In order to describe the crack front in an FE model, the FMM was combined with Paris' law. Moreover, the stress intensity factor for planar three-dimensional cracks was calculated, and the fatigue crack growth was simulated using the XFEM. Shi *et al.* have programmed an ABAQUS user-defined material subroutine (UMAT) and element subroutine (UEL) for the practical simulation of fatigue crack growth of a compact tension specimen using the FMM-XFEM technique [12].

The aforementioned studies using both types of methods show good agreement with the fatigue crack growth experiments using a compact tension specimen, a notched plate, *etc.* In other words, the correlation between da/dN and ΔK in the stable crack growth region, *i.e.*, region II, has been well predicted. However, these methods require experiments prior to simulation to identify the material parameters of the governing equation, which is a function of the critical stress intensity factor, the cohesive fracture energy variable, and the critical crack opening value. Although these numerical methods have been verified sufficiently for simulating fatigue crack growth, a large number of experiments are required to determine the moment and realm of crack occurrence while considering information about the initial crack such as shape and position.

In contrast with these techniques, damage mechanics has been introduced as a new method for expressing the fatigue crack growth phenomena. Damage mechanics is based on thermodynamic micromechanical approaches, and has provided governing equations that can be incorporated into the material constitutive equation to express the material degradation that causes crack initiation and propagation [13–17].

One of the distinctive factors of damaged materials under arbitrary loads is material degradation due to void and slip. In particular, the elastic stiffness variable/matrix is degraded because of these micro-defects. There is an inverse relationship between elastic stiffness and material defects. In the case of the damage mechanism, degradation of elastic stiffness and void/slip growth can be represented by an effective elastic stiffness model and a damage evolution model, respectively. On the basis of the correlation between these two models, the material/structural degradation under various loads, such as fatigue, creep, brittle, and ductile failures, can be evaluated and predicted quantitatively [18–20]. Furthermore, in order to analyze the damage/failure behavior of materials precisely, it is crucial to select a suitable damage evolution model from among the possible choices.

There are a few damage evolution models that describe the various types of material failures, such as elastic-plastic ductile damage [19,21–23], fatigue damage [24–27], creep and creep-fatigue damage [28–30], and elastic-brittle damage [31–34]. In the case of low-cycle fatigue, some research has been performed. Hamon *et al.* proposed a damage model for fatigue crack propagation. The model is based on two damage mechanisms that consider static and cyclic damage on fatigue crack propagation. In this study, da/dN and ΔK curves at different load ratios were obtained numerically (typical results for FCGR tests) and the proposed model showed good agreement with experimental results. However, there are few studies on how to predict fatigue life under high-cycle fatigue for various types of metal and nonmetal materials. Additional mechanisms such as two-scale damage are required to express crack propagation under high-cycle fatigue that occurs in an elastic regime.

Three difficulties must be overcome to computationally assess the fatigue crack grow rate using a finite element analysis (FEA) based on damage mechanics: (i) the selection of an appropriate material constitutive model to express the material's plastic behavior and evolution of the damage; (ii) a method for simulating the crack propagation based on the calculation results using the material constitutive model; and (iii) the time costs for the computation, which can be more difficult in the case of high cycle fatigue problems.

In the authors' previous study, simulations of the crack propagation were performed under monotonic loading conditions at room and cryogenic temperatures and moreover, the fracture capacities of SUS304L were also estimated computationally at both temperatures [35]. A previously proposed method based on a material constitutive model is verified by the comparison of the results, such as the strain energy release rate, G_{IC}, failure mode, and force-displacement relationship, from the experimental and computational fracture tests. This computational method could be a solution for first two difficulties mentioned above. The constitutive model used sufficiently expressed the materials' plastic behavior including strong hardening at cryogenic temperatures and the evolution of damage of SUS304L. Crack propagations, including fracture mode, are also simulated based on a calculated damage variable using the element weakening method [35,36]. Therefore, in the present study, for the numerical estimation of fatigue fracture, the aforementioned computational method using a material constitutive model including damage evolution model is employed again. In addition, the jump-in-cycles procedure and a numerical approach rearranging critical damage are employed for fatigue damage analysis, in order to reduce the computational times and to consider the fatigue fracture phenomenon differently from ductile fracture under static loading, respectively (these numerical approaches can be solution for the last difficulty as presented before). The aforementioned procedure and numerical approach for fatigue damage analysis are also implemented in ABAQUS UMAT. Finally, in the present study, a computational FCGR tests using the proposed ABAQUS UMAT for 7% nickel steel are carried out at room and cryogenic temperatures. The obtained numerical results were verified by comparing them with the experimental results.

2. Experimental FCGR Tests

Before the numerical evaluation of the fatigue crack growth rate for 7% nickel steel, experimental tests are performed. The experimental test results are compared with the numerical results in order to validate the computational method proposed in the present study. Compact tension (CT) type specimens for fatigue tests were tested at both the room temperature and the cryogenic temperature (−163 °C). The cryogenic temperature was selected in consideration of the temperature at which liquefied natural gas (LNG) is stored.

The FCGR test is conducted according to ASTM E647 [37]. The test method involves cyclic loading of notched specimens that have been acceptably precracked by fatigue. The crack size is measured, either visually or by an equivalent method, as a function of the elapsed fatigue cycles, and these data are subjected to numerical analysis to establish the rate of crack growth. Crack growth rates are expressed as a function of the stress-intensity factor range, ΔK, which is calculated from expressions based on linear elastic stress analysis.

2.1. Test Specimens

Nickel is recognized to enhance fracture toughness of materials at low temperatures. Therefore, nickel is widely employed as an ingredient in materials for constructing industrial structures, such as LNG carrier storage systems. In particular, 7% nickel steel has being recently applied in LNG carrier storage systems in order to replace 9% nickel steel, which requires a considerable amount of expensive welding materials. However, only a few studies have been carried out to assess the fracture capacity of 7% nickel steel at both the room temperature and the cryogenic temperature.

The chemical composition of 7% nickel steel, which is selected for the study of the fatigue crack growth rate test, is summarized in Table 1.

Table 1. Chemical composition of 7% nickel steel (wt. %).

Element	C	Si	Mn	P	S	Cu	Ni
weight %	0.05	0.05	0.80	0.001	0.001	0.03	7.13
Element	**Cr**	**Mo**	**Nb**	**V**	**Ti**	**Sol. Al**	
weight %	0.41	0.04	<0.003	<0.003	<0.003	0.030	

The geometry of the standard CT specimen is shown in Figure 1 based on ASTM E647 [37]. Thickness B and width W may be varied independently within the following limits. For CT specimens, it is recommended that the thickness be within the range $W/20 \leq B \leq W/4$. Specimens having thicknesses up to and including $W/2$ may also be employed.

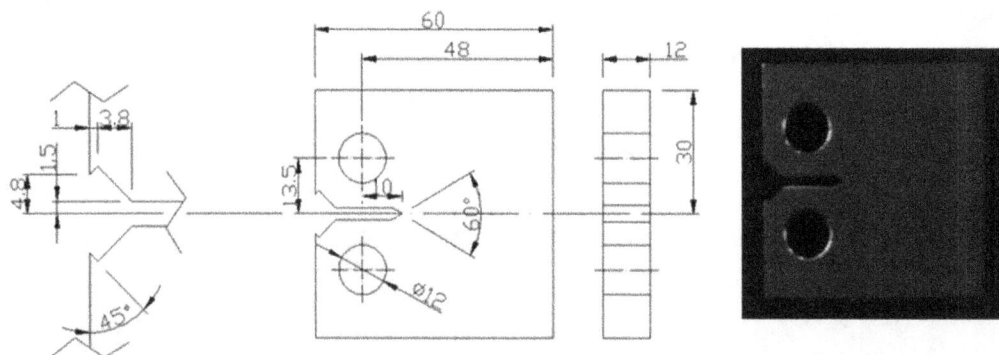

Figure 1. Geometry of the test specimen [37].

The fatigue precrack is made in the specimen using the K (stress intensity factor)-decreasing procedure based on ASTM E647 [37]. K-decreasing procedure requires that the final K_{max} during precracking shall not exceed the initial K_{max} for which test data are to be obtained. Therefore, forces corresponding to higher K_{max} values used to initiate cracking at the machined notch and the force range shall be stepped-down. Figure 2 shows the actual stepped-down force histories for precracking according to K-decreasing procedure. The frequency of fatigue loading is 10 Hz. The initial K_{max} and final K_{max} during the K-decreasing procedure at both the room temperature and the cryogenic temperature are 34.44 and 31.64 MPa $\sqrt{m}$ and 32.57 and 31.31 MPa $\sqrt{m}$, respectively.

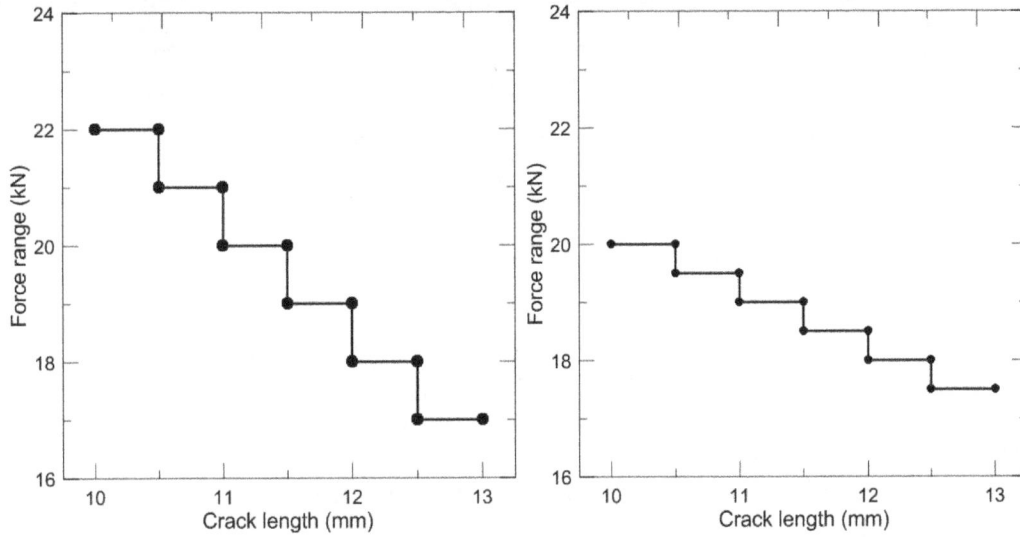

Figure 2. Actual K-decreasing by stepped force shedding at room temperature (**left**) and cryogenic temperature (**right**).

2.2. Test Facilities

Servo-hydraulic testing machines with a maximum load capacity of ±50 tons (IST-8800) were used for the fatigue crack growth rate test of 7% nickel steel. To maintain the cryogenic temperature, a special-purpose cryogenic chamber was fabricated and installed in the testing machine.

All the tests were terminated at the point where a through-width fracture occurred. A load control mode was used a 10 Hz as same as precracking procedure. The maximum loads of 18 kN are applied to specimens tested at both the room temperature and the cryogenic temperature. The stress ratios are zero for both tests.

2.3. Methods for Measurement of Crack Length

The crack length during the test is measured using a compliance method based on ASTM E647 [37]. The results of the fatigue crack growth tests are used to analyze the relationships between the crack length (a) and the number of cycles (N), as well as between da/dN and the stress intensity factor range (ΔK). The empirical formulae for calculating the crack length of the compact tension specimen are as follows ASTM E647 [37]:

$$\alpha = a/W = C_a + C_b u_x + C_c u_x^2 + C_d u_x^3 + C_e u_x^4 + C_f u_x^5 \tag{1}$$

$$u_x = \{[\frac{E\upsilon B}{P}]^{0.5} + 1\}^{-1} \quad \text{for } 0.2 \leq a/W \leq 0.975 \tag{2}$$

where a is the crack length, W is the width of the specimen, E is the elastic modulus, v is the displacement between the measurement points, B is the specimen thickness, and P is the force. $C_a \sim C_f$ are the constants determined from the measuring location, as shown in Figure 3 and Table 2.

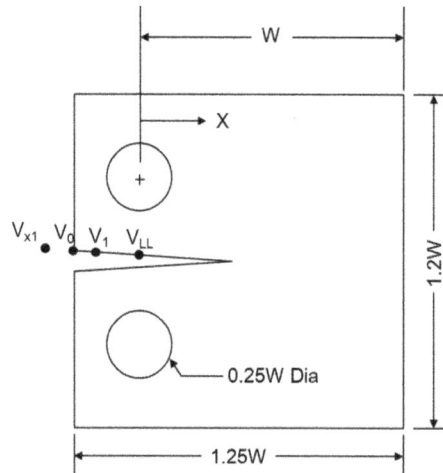

Figure 3. Normalized crack length as a function of plane stress elastic compliance for compact tension specimen ASTM E647 [37].

Table 2. Constants determined from the measuring location for CT specimen ASTM E647 [37].

Meas. Location	X/W	C_a	C_b	C_c	C_d	C_e	C_f
V_{x1}	−0.345	1.0012	−4.9165	23.057	−323.91	1798.3	−3513.2
V_0	−0.250	1.0010	−4.6695	18.460	−236.82	1214.9	−2143.6
V_1	−0.1576	1.0008	−4.4473	15.400	−180.55	870.92	−1411.3
V_{LL}	0	1.0002	−4.0632	11.242	−106.04	464.33	−650.68

Another typical result of the fatigue crack growth rate test is the relationship between da/dN and the stress intensity factor range. The linear region of this relationship in the log-log scale can be expressed using the empirical formula known as Paris' law [2]:

$$\frac{da}{dN} = C\Delta K^m \tag{3}$$

where C and m are the material parameters determined by the experimental fatigue crack growth test.

The stress intensity factor range (ΔK) in the plane stress condition can be expressed as follows ASTM E647 [37]:

$$\Delta K = \frac{\Delta P}{B\sqrt{W}} \frac{(2+\alpha)}{(1-\alpha)^{1.5}} (0.886 + 4.64\alpha - 13.32\alpha^2 + 14.72\alpha^3 - 5.6\alpha^4) \tag{4}$$

where ΔP is the force range, which is the algebraic difference between the maximum and the minimum forces in a cycle.

2.4. Test Results and Discussion

The results of the experimental fatigue crack growth rate tests are shown in Figures 4 and 5. The test results at both temperatures are presented together. Figure 4 shows the relationship between the number of cycles and the crack length. As shown in the figure, 7% nickel steel exhibits enhanced fracture resistance under fatigue loading at cryogenic temperature.

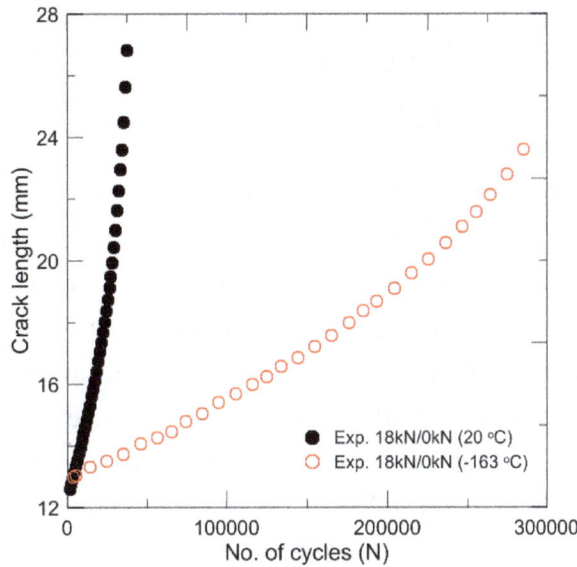

Figure 4. Relationship between the crack length and the number of cycles at the room and cryogenic temperatures.

Figure 5 shows a log-log plot of the relationship between da/dN and ΔK. The material parameters in terms of Paris' law, C and m, are determined from this result. The experimental fatigue crack growth rate test results, including the material parameters C and m, are summarized in Table 3.

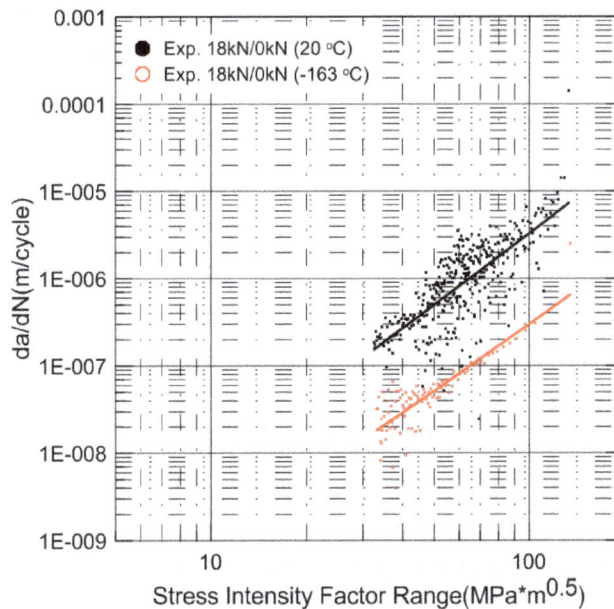

Figure 5. Relationship between da/dN and ΔK at room and cryogenic temperatures.

Table 3. Summary of experimental fatigue crack growth rate test results.

Material	Temperature	Max. Load (kN)	Min. Load (kN)	Cycle (N_f)	C	m
7% Ni	RT	18	0	40,019	2.17×10^{-11}	2.57
	CT (-163 °C)	18	0	349,541	21.61×10^{-12}	2.71

In the present study, da/dN and ΔK curves are obtained from two neighboring points of experimental results as follows:

$$da / dN = (a_{n+1} - a_n) / (N_{n+1} - N_n) \tag{5}$$

$$\Delta K = (K_{n+1} - K_n) \tag{6}$$

In Figure 5, scatters are presented around a line representing Paris' Law. The large scatter is explained by Figure 6. Here, assuming that the two neighboring points shown in the green-outlined section in the left frame have almost the same crack length (*i.e.*, ΔK, a function of ΔP and the crack length, is almost the same) at different numbers of cycles, the related point in the graph of da/dN with respect to ΔK is located in the green-outlined section in the right frame. As in the previous case, if the two neighboring points have different crack lengths at almost the same number of cycles, scatters appear in Figure 6 (blue-outlined section). The fatigue crack is not propagated at every cycle during the experiments. Therefore, some scatters in Figure 5 are negligible in view of the total fatigue-crack propagation during the FCGR tests, and the relationship between the crack length and number of cycles represented in Figure 4 is not affected by the scatters in Figure 5.

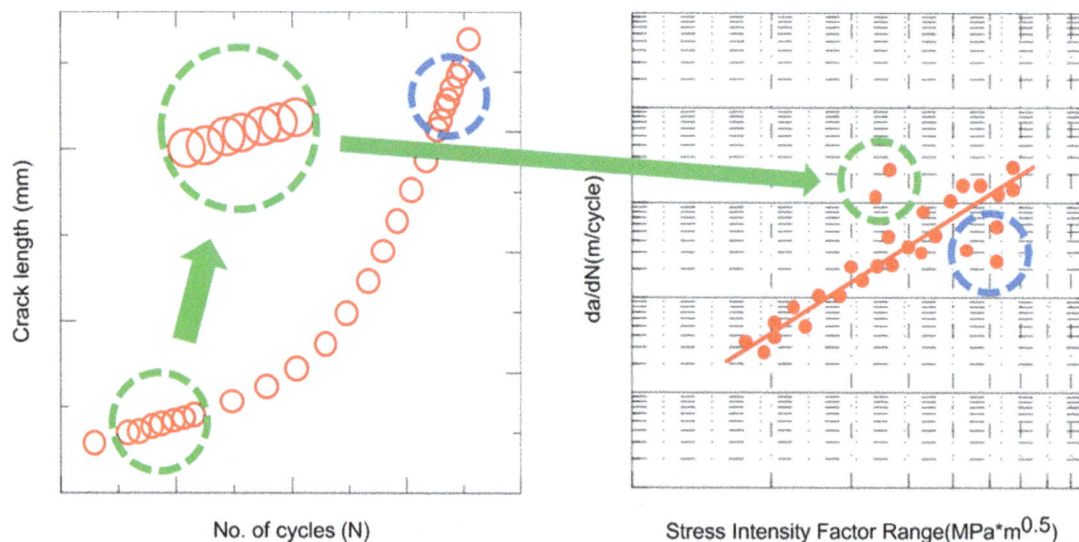

Figure 6. Schematics of scatters induced on $da/dN - \Delta K$ curve.

As shown in the test results, the fracture resistances of 7% nickel steel under fatigue loading are considerably different between room temperature and cryogenic temperature (see Figure 4). Retained austenite is known to improve the toughness and strength of steels such as nickel-alloyed steels including austenitic stainless steels [35,36,38]. The nickel decreases the temperature of the martensite transformation and thermally stabilizes the austenite. Retained austenite enhances the fracture toughness by strain-induced martensite transformation within the temperature range of room temperature to 70 K. Increasing the content of retained austenite strengthens the effect of the strain-induced martensite transformation and thereby elevates the fracture toughness. Notably, these results indicate that both the composition and microstructure can influence low-temperature strengthening mechanisms [39,40]. These phenomena reveal the importance of understanding material nonlinearities, such as temperature dependency, which result in an increased fracture-resistance capacity.

The details of the fatigue crack growth test render it difficult to perform at a low temperature. The reasons for this include the special equipment required, the special care necessary for data acquisition, the cost, and the time needed. However, prior to the industrial utilization of a new type of material, the evaluation of the fracture resistance capacity under fatigue loading is essential for safe design. Therefore, a low-cost, efficient test method is required, particularly for low-temperature conditions.

3. Coupling Damage Analysis for Computational FCGR Tests

In the present study, a method for computationally estimating the fatigue crack growth rate was proposed based on a structural damage analysis using FEA. Generally, it is known that there are two types of representative approaches for applying damage variables in conjunction with internal state variables such as the stress and strain to predict the failure of a structure: fully coupled and uncoupled approaches [36].

In an uncoupled approach, the failure prediction of a structure is carried out using a threshold value for the damage, and the damage variable is calculated using a damage evolution equation, with the stress fields obtained from a structural analysis. Although an uncoupled approach is relatively comfortable and can save calculation time, this approach cannot ensure precise results because there is no coupling between the strain and damage field during the structural analysis [17,41,42].

In a fully coupled approach, the damage variable is calculated using a damage evolution equation during a structural analysis, and the stress, strain, and damage fields are mutually affected. Therefore, the full coupling process produces more accurate results, but it requires a relatively large amount of time for the calculation, which is conducted throughout the entire FEA process [17,41,42].

The proposed computational method for estimating the fatigue crack growth rate (FCGR) was carried out by adopting both approaches in computational FCGR tests. A fully coupled analysis was primarily performed during the computational tests, and the damage variable was calculated using the material constitutive equation, which included a damage evolution equation. An uncoupled damage analysis was restrictively performed between the fully-coupled analyses of stable regions of the damage evolution.

3.1. Fully-Coupled Analysis: Damage-Coupled Constitutive Model

In this study, the Bodner-Partom material model was employed as a material constitutive model for the computational fatigue crack growth rate evaluation of 7% nickel steel [18]. This model is an elastic-viscoplastic model describing rate-dependent plasticity and creep. This means that material behavior is always being modeled as having a certain degree of inelasticity. Therefore, the Bodner-Partom material model is recognized as a unified viscoplastic model, because it can express the yield phenomenon of materials over a wide range of strain rates and temperatures quite effectively, despite the absence of yield functions. This distinguishing feature is intended so that the overall framework expressing the macroscopic response properties, such as strain rate sensitivity and temperature dependency, are consistent with the essential physics of elastic and inelastic deformation. Therefore, this model was developed in the form of macroscopic equations, and specific microscopic mechanisms were not necessary. The employed viscoplastic model is described as in the following [18].

$$\dot{\varepsilon}_{ij} = \dot{\varepsilon}_{ij}^{e} + \dot{\varepsilon}_{ij}^{P} \tag{7}$$

$$\dot{\varepsilon}_{ij}^{P} = D_0 \exp[-\frac{1}{2}(\frac{Z}{\sigma_{eff}})^{2n}]\frac{\sqrt{3}S_{ij}}{\sigma_{eff}} \tag{8}$$

where $\dot{\varepsilon}_{ij}$, $\dot{\varepsilon}_{ij}^{e}$ and $\dot{\varepsilon}_{ij}^{P}$ are the total strain, the elastic strain and the plastic strain rate tensor, respectively. D_0 is the assumed maximum plastic strain rate, Z is the total hardening variable, and n is the material parameter that controls the rate sensitivity. σ_{eff} is the effective stress. The total hardening variable is defined as follows:

$$Z = Z_1 - (Z_1 - Z_0)\exp(-m_h W_P) \tag{9}$$

$$W_P = \int dW_P = \int \sigma_{ij} d\varepsilon_{ij}^{P} \tag{10}$$

where Z_0 and Z_1 are the initial and saturated values of the isotropic hardening variable, respectively; m_h is the rate of isotropic hardening; and W_P is the plastic work. σ_{ij} is the stress tensor.

Generally, a two-scale damage model is adopted for expressing material degradation under high-cycle fatigue loading [21,43–45]. The two-scale damage model is based on the notion that initiation and evolution of the damage in brittle fractures and high-cycle fatigue fractures are governed by plastic deformation in a microscopic region of a material [21]. A two-scale damage model includes a scale transition law that links the mesoscopic and microscopic scales. Therefore, a large number of experiments are required to define a scale transition law and determine the internal state variable in a microscopic region, such as accumulated plastic strain.

In this study, a Bodner-Chan damage model was used to express the material degradation and crack growth of 7% nickel steel during the simulation of crack growth under high-cycle fatigue loading [18]. Instead of the two-scale damage model, rearrangement of the critical damage was adopted for the material degradation under high-cycle fatigue loading. As previously mentioned, the damage-coupled constitutive model employed in this study is always modeled as having a certain degree of plasticity. This feature is effective for expressing material failure under high-cycle fatigue. A scale for rearrangement of the critical damage was obtained from some simulations described in Section 4.2.

The viscoplastic model described above was incorporated in the damage model expressed below [18]:

$$\dot{\varepsilon}_{ij}^{P} = D_0 \exp[-\frac{1}{2}(\frac{Z(1-\omega)}{\sigma_{eff}})^{2n}]\frac{\sqrt{3}S_{ij}}{\sigma_{eff}} \tag{11}$$

$$\dot{\omega} = \frac{b}{h}[\ln(\frac{1}{\omega})]^{\frac{b+1}{b}} \omega \dot{Q} \tag{12}$$

$$\dot{Q} = (C_1\sigma_{max}^{+} + C_2\sigma_{eff} + C_3 I_1^{+})^{r} \tag{13}$$

where ω is the damage parameter which is the material internal state variable and S_{ij} is the deviatoric stress tensor. b and h are the material parameters that control the characteristics of damage, and $\dot{Q}$ is the multiaxial stress function proposed by Hayhurst and Leckie [46]. σ_{max}^+, σ_{eff}, and I_1^+ are the maximum tensile principal stress, effective stress, and first stress invariant, respectively. C_1, C_2, C_3 and r are constants of the material, such that $C_1 + C_2 + C_3 = 1$.

3.2. Uncoupled Analysis: Jump-in-Cycles Procedure

In the present study, the jump-in-cycles procedure is adopted to calculate fatigue damage in order to reduce the computation time. This procedure can be classified with the uncoupled method in structural damage analysis [19]. The jump-in-cycles procedure is a method to "jump" full blocks of $\overline{\Delta N}$ cycles in a damage calculation based on the postulation that damage accumulation is progressed linearly in the stable region of damage accumulation. $\overline{\Delta N}$ is expressed as follows:

$$\overline{\Delta N} = \frac{\overline{\Delta \omega}}{\left. \frac{\delta \omega}{\delta N} \right|_{N_S}} \tag{14}$$

where $\overline{\Delta N}$ is the blocks of number of cycles, and $\overline{\Delta \omega}$ is the accumulated damage for the jumped blocks. $\overline{\Delta \omega}$ is also a given value that determines the accuracy of the procedure. $\overline{\Delta \omega} \approx \omega_{cr} / 50$ (ω_{cr} is the critical damage) is a good compromise between accuracy and time cost [19]. N_S is a stabilized cycle, and $\left. \frac{\delta \omega}{\delta N} \right|_{N_S}$ is the accumulated damage increment over this single cycle. The accumulated damage is updated as

$$\omega(N_S + \overline{\Delta N}) = \omega(N_S) + \left. \frac{\delta \omega}{\delta N} \right|_{N_S} \overline{\Delta N} \tag{15}$$

3.3. Algorithm for Computational FCGR Tests

As mentioned previously, computational method for FCGR tests based on damage-coupled FEA is comprised of fully coupled and uncoupled approaches coped with material constitutive model and jump-in-cycles procedure, respectively. Each of procedures for analysis was implemented into ABAQUS user defined subroutine UMAT and algorithm for mutual calculation of damage accumulation from fully and uncoupled analyses was carried out. UMAT is well known that the commercial FE code ABAQUS allows the definition of the material state at every integration point within the FE element. In this regard, the ABAQUS user-defined subroutine UMAT is commonly used to implement a specific constitutive model in the finite element analysis. This can give a significant improvement in the FE analysis results, especially if there is no adequate material model within the ABAQUS libraries.

Figure 7 shows a schematic of the accumulated damage during the fatigue analysis using the proposed method comprised of the fully and uncoupled damage analyses (hereafter referred to as the "hybrid method"). First, the fatigue damage was calculated using the fully coupled approach (using the damage-coupled material constitutive model). This process was relevant to the section from point A to point B in Figure 7. If the damage accumulation calculated by the damage-coupled material constitutive model was stable at point B, the jump-in-cycles procedure was conducted, and the accumulated damage was updated as point C, as follows:

$$\omega_B = \omega_A + \Delta\omega_{JICint} \tag{16}$$

where $\Delta\omega_{JICint}$ is the accumulated damage during jump-in-cycles procedure and $\Delta\omega_{JICint}$ is calculated by

$$\Delta\omega_{JICint} = \left.\frac{\delta\omega}{\delta N}\right|_{N_S} \Delta N_{JICint} \tag{17}$$

where ΔN_{JICint} is the jumped blocks of number of cycles for the accumulation of damage updated by the jump-in-cycles procedure. In other words, ΔN_{JICint} is the saved computation time (number of cycles) for the fatigue damage analysis. This uncoupled approach to the damage analysis is relevant to the section from point B to point C shown in Figure 7. Subsequently, the time increment is shifted to point C' from point C, and the uncoupled process is terminated. As shown in Figure 7, $\Delta t_{CC'}$ is the shifted time increment that resulted from the jump-in-cycles procedure, and this parameter is the other value along with ΔN_{JICint}. This was because the fatigue damage analysis was performed in the real-time domain rather than using the number of cycles. In this study, the adopted damage-coupled constitutive model was formulated based on the real-time domain. ΔN_{JICint} and $\Delta t_{CC'}$ represented the saved number of cycles and saved time increments, respectively. When a ΔN_{JICint} value obtained from the jump-in-cycles procedure was 1000 cycles, there was a 100 s time increment shift in the structural analysis (assuming a time increment of 0.1 s was required for the structural analysis for a 1-cycle load). Finally, in the section from point C' to point D, the fatigue damage was again calculated based on the damage-coupled material constitutive model. The systematic process for fatigue damage analysis proposed here can be briefly summarized as follows: (1) fully coupled analysis; (2) uncoupled analysis; (3) increment shift; and (4) fully coupled analysis again.

In the present study, the parameter ΔN_{JICint} was predetermined to calculate the accumulated damage by the jump-in-cycles procedure because the ratio of the accumulated damage increment and the increment in the number of cycles in the stable region $\left.\frac{\Delta\omega}{\Delta N}\right|_{N_S}$ was difficult to obtain numerically.

Therefore, ΔN_{JICint} and $\left.\frac{\Delta\omega}{\Delta N}\right|_{N_S}$ were predetermined, and the value of $\Delta\omega_{JICint}$ was checked using $\omega_{cr}/50$.

In the present study, ΔN_{JICint} values of 1000 and 10,000 cycles were used at room and cryogenic temperatures, respectively. The increment used for the number of cycles in the stable region $\Delta N|_{N_S}$ was 100 cycles for both temperatures. It was postulated that the stable regions appeared when the accumulated damage reached 1.0×10^{-10} and 1.0×10^{-20} in the room and cryogenic temperature cases, respectively.

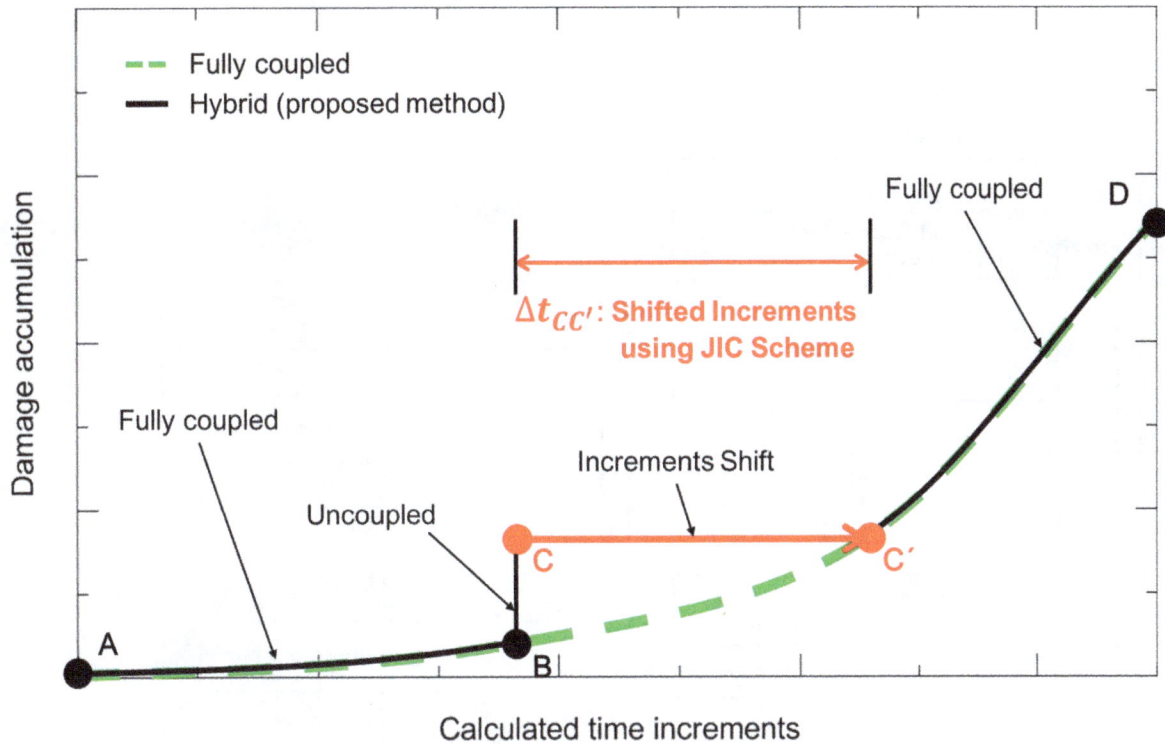

Figure 7. Schematic of accumulated damage under fatigue analysis using proposed method comprised with fully and uncoupled damage analysis approaches.

The aforementioned process for calculating the accumulated damage using jump-in-cycles procedure was implemented into ABAQUS user defined subroutine UMAT and damage-coupled material constitutive model classified as a fully coupled approach was also implemented into ABAQUS UMAT. The computational algorithm of UMAT for FCGR test simulation is presented in Figure 8. The most of times for structural damage analysis is carried out using the fully coupled method based on a damage-coupled constitutive model. The jump-in-cycles procedure is used as an uncoupled method for reducing the computation time. The jump-in-cycles procedure for damage analysis was already adopted in the author's previous study and was verified for calculating the high cycle fatigue damage, although fatigue analyses were carried out using PATRAN/NASTRAN [17].

The implicit formulation of the damage-coupled constitutive model is also carried out and implemented into ABAQUS UMAT. All of these incremental values and the damage parameter ω must be updated in the calculation procedure by the definition of the solution-dependent state variables (SDVs). SDVs are initialized in the subroutine SDVINI of ABAQUS. The value of ω progressively increases, and the subroutine FAILURE is called to reduce the corresponding element stiffness when ω reaches the predetermined critical value ω_{cr}.

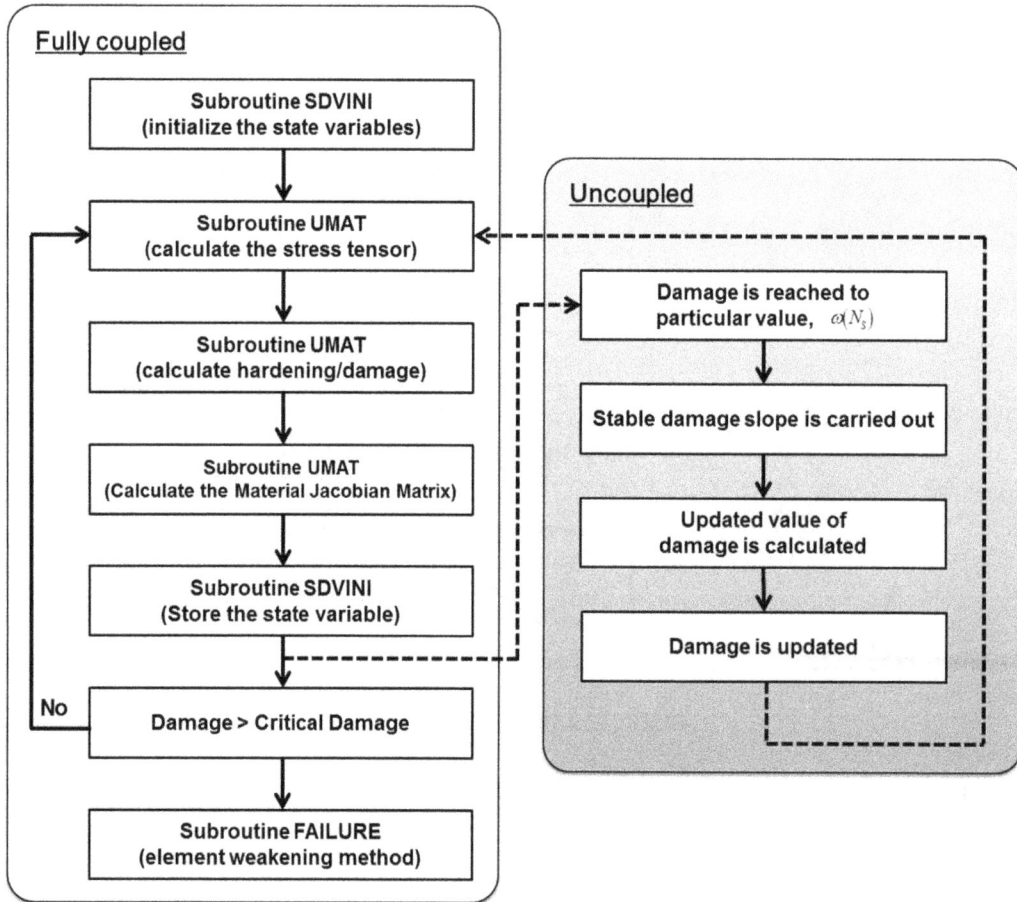

Figure 8. Computational algorithm of UMAT for FCGR test simulation.

The optimal time increment Δt and the corresponding strain increment, which are computed at the end of the previous time increment, are determined using the Jacobian matrix. The numerical integration scheme used in the present study is described below. The incremental strain is given by

$$\Delta\varepsilon_{ij} = \varepsilon_{ij}^{n+1} - \varepsilon_{ij}^{n} \tag{18}$$

The corresponding elastic trial stress tensor is computed from

$$\sigma_{ij}^{n+1(\text{trial})} = \sigma_{ij}^{n} + D_{ijkl}\Delta\varepsilon_{kl} \tag{19}$$

where D_{ijkl} is the elastic stiffness tensor. The updates of the stresses can then be obtained from

$$\sigma_{ij}^{n+1} = \sigma_{ij}^{n+1(\text{trial})} - \Delta\lambda D_{ijkl}S_{kl}^{n+1} \tag{20}$$

where λ is the plastic multiplier. The scalar $\Delta\lambda$, defined by $\Delta\lambda = \lambda\Delta t$, can be calculated on the basis of the damage using

$$\Delta\lambda = \sqrt{\frac{D_0^2}{J_2}\exp\left[-\left(\frac{Z^2(1-\omega)^2}{3J_2}\right)^n\right]}\Delta t \tag{21}$$

At the end of the increment, the time history values of each solution-dependent state variable are stored using the STATEV array. The following equations are representative state variables of the constitutive equation:

$$Z_{n+1} = Z_n + \Delta\lambda m[Z_1 - Z^I]2J_2 \tag{22}$$

$$\omega_{n+1} = \omega_n + \frac{b}{h}\left[\ln\left(\frac{1}{\omega}\right)\right]^{\frac{b+1}{b}}\omega\dot{Q}\Delta t \tag{23}$$

4. Computational FCGR Tests

4.1. Simulation of Uniaxial Tensile Tests and Determination of Material Parameters

Simulations of uniaxial tensile tests for 7% nickel steel are performed at room and cryogenic (-163 °C) temperatures in order to verify the developed UMAT and determine the material parameters for computational FCGR tests. The jump-in-cycles procedure is not considered during the simulation of tensile tests for 7% nickel steel. The material parameters of the aforementioned damage-coupled constitutive model for 7% nickel steel are determined by a comparison of the results, presented as the stress-strain relationship, between the simulation and the experiments. The determined material parameters for 7% nickel steel at room and cryogenic temperatures are employed for the computational FCGR tests. Among the material parameters, the critical damage (ω_{cr}) is rearranged for fatigue damage analysis. The FE model for simulation of uniaxial tensile tests of 7% nickel steel is adopted from author's previous study and sensitivity of mesh size is also analyzed [38]. A rod-type FE model was controlled in the y-direction and strain rate applied to the specimen is 1.0×10^{-4} /s for both temperatures. As can be observed from the Figure 9, the developed UMAT could describe the stress-strain relationship of 7% nickel steel for both the room and the cryogenic temperatures. The material parameters of the computational FCGR tests are summarized in Table 4.

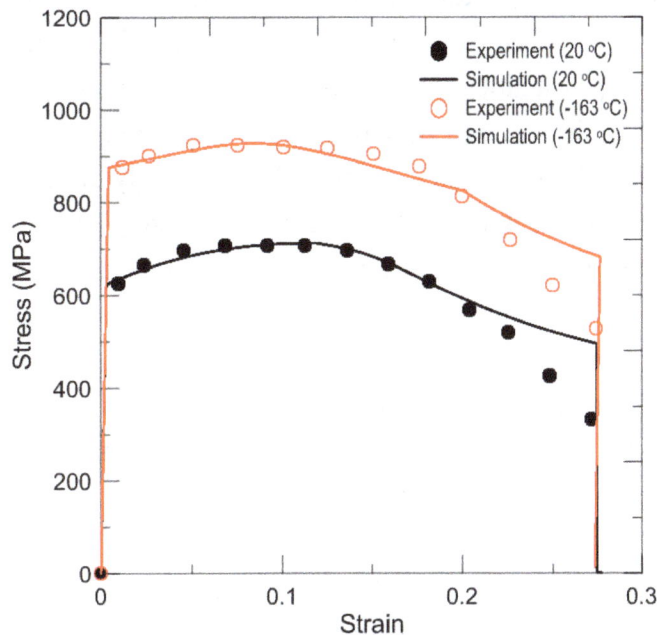

Figure 9. Comparison of uniaxial tensile tests of 7% nickel steel.

Table 4. Material parameters of damage-coupled constitutive model for computational fatigue crack growth rate tests.

	E (GPa)	v	Z_0 (MPa)	Z_1 (MPa)	D_0 (s^{-1})	m
Room Temperature (20 °C)	203	0.33	1450	1380	1.0	0.025
	n	b	h		r	ω_{cr}
	2.05	4.28	7.25×10^{18}		5.50	0.20
	E (GPa)	v	Z_0 (MPa)	Z_1 (MPa)	D_0 (s^{-1})	m
Cryogenic Temperature (−163 °C)	203	0.33	1650	1600	1.0	0.003
	n	b	h		r	ω_{cr}
	4.30	1.00	1.63×10^{19}		5.50	0.20

4.2. Rearrangement of Critical Damage (ω_{cr}) and Fatigue Precrack Simulation

As mentioned previously, a numerical approach instead of theoretical approach using a two-scale damage model for fatigue damage analysis is carried out in the present study. Critical damage (ω_{cr}) is rearranged, and fatigue precrack simulations adapting rearranged critical damage are performed at room temperature. The jump-in-cycles procedure is not considered during the fatigue precrack simulation.

The material parameters for the damage-coupled constitutive model, which are implemented into UMAT, are employed from the simulation of the material tension test at room temperature, and the rearranged values of the critical damage are also employed for the computational fatigue precrack test.

Figure 10 shows an FE model of the FCGR test specimen, which is used to perform simulations. A compact tension (CT) type specimen of the FE model is composed of 6285 elements (8 node-reduced integral elements, C3D8R in ABAQUS, using the same element type as the FE model for tension test simulation) and 8052 nodes. The crack tip radius of the FE model is 1 mm. The minimum element length is also 1 mm near the crack tip line. The problem of mesh size sensitivity for a stable solution, with regard to the simulation of crack propagation, was investigated in the authors' previous study [36]. The suitable minimum element length in the case of a 1-mm crack tip radius was carried out as 0.1 mm, although a minimum element length of 1.0 mm is employed in this study to save calculation time costs. Moreover, computational fatigue precracking, including the K-decreasing procedure, is performed in the same way as experimental precracking at room temperature in the present study. The load history according to the K-decreasing procedure was obtained from the data resulting from the experimental fatigue precrack test and was employed as a loading condition of the precracking simulation (see Figure 2). As shown in Figure 10, the red-outlined sections for the upper and lower holes were similarly controlled using MPC in the experimental tests. The lower hole was restrained in the x-, y-, and z-directions. The upper hole was restrained in the x- and z-directions and was applied cyclic loading.

In this study, the rearranged value of critical damage is assigned a value ω^*_{cr}, and the rearranged values are considered as ω_{cr}, $\omega_{cr}/100$ and $\omega_{cr}/10,000$. Figure 11 shows a comparison between the experiment and the simulation of the relationships between the crack length and the number of cycles from the fatigue precrack test, adopting each of rearranged critical damage. The effective stress contour at 27,374 cycles (the number of cycles at which the fatigue precrack test was terminated) adopting $\omega^*_{cr} = \omega_{cr}/10,000$ is also shown in Figure 11. Although there are some discrepancies between the results, it is found that rearrangement of $\omega^*_{cr} = \omega_{cr}/10,000$ is sufficient to perform the fatigue crack damage analysis. Rearranged critical damage was adopted to the material parameters for each FCGR test simulation at room and cryogenic temperatures. In addition, the accumulated damage values of each damaged element from the fatigue precrack simulation are adopted for the main FCGR test simulation as the initial accumulated damage of the FE specimen at both the room and the cryogenic temperatures.

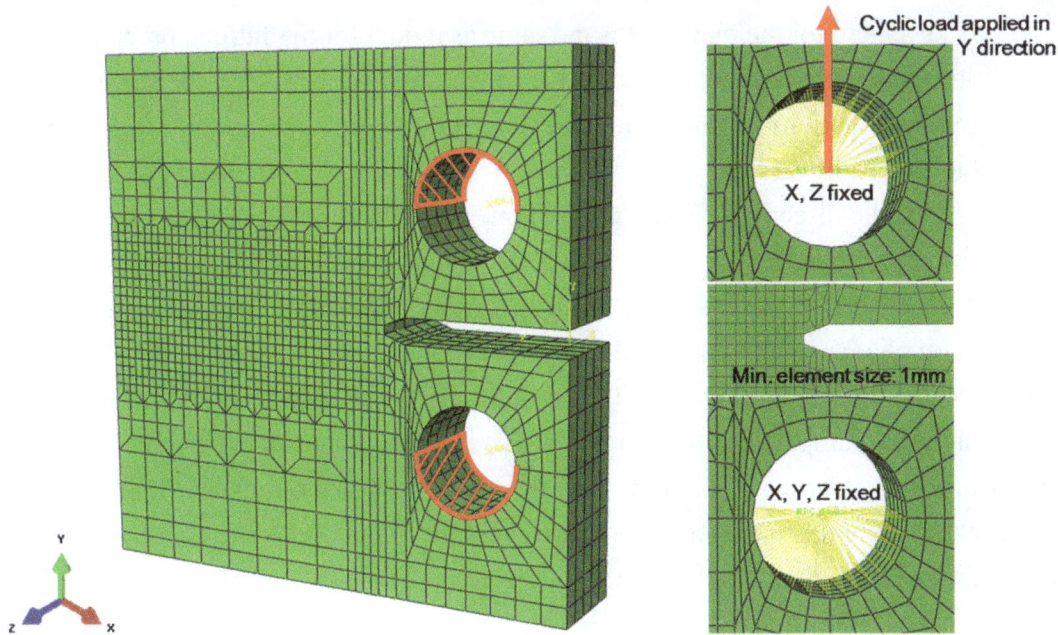

Figure 10. Finite element (FE) model of compact tension (CT) specimen for ABAQUS.

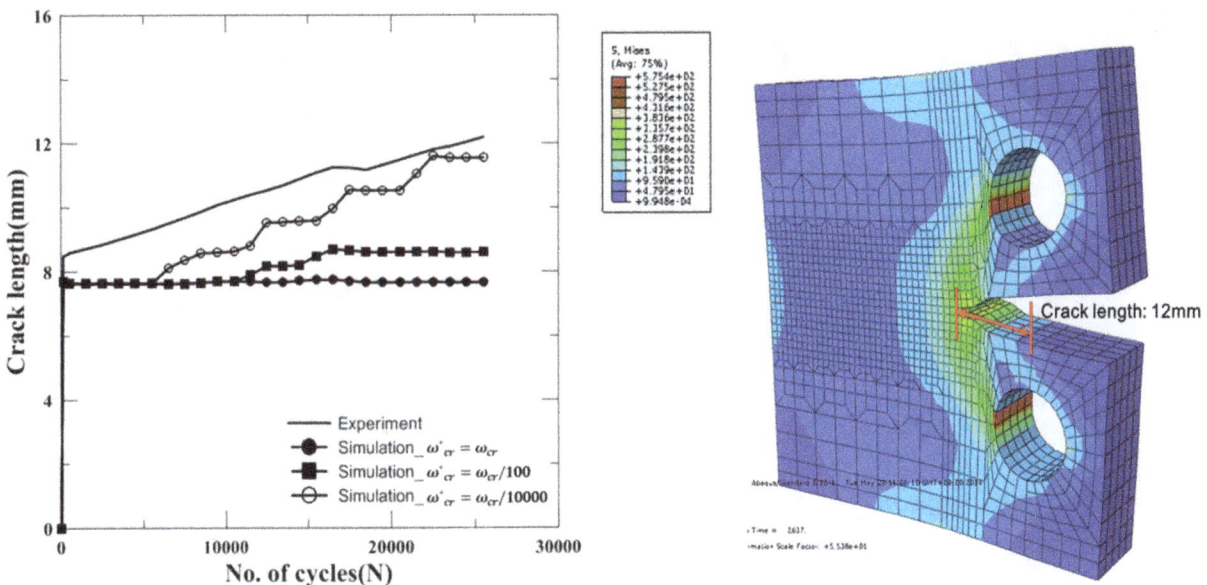

Figure 11. Comparison of relationships between the crack length and the number of cycles obtained from experimental fatigue precrack test and simulation according to varied critical damage (**left**) and the effective stress contour after precrack simulation (**right**).

4.3. Main Test Simulation

The developed ABAQUS UMAT, including the jump-in-cycles procedure, is adopted for computational FCGR tests of 7% nickel steel at room and cryogenic temperatures. Rearranged critical damage is also adopted in order to consider calculate the fatigue damage resulting from the fatigue precrack simulations. The FE model of CT specimen for the main test is the same as model for the fatigue precrack simulation. The loading conditions of the main tests at room and cryogenic temperatures are the same. As shown in Figure 10, the red-outlined sections for the upper and lower holes were similarly controlled using MPC in the experimental tests. The lower hole was restrained in the x-, y-, and z-directions, and the upper hole was restrained in the x- and z-directions. In the y-direction, cyclic loading was applied which has 18 kN and 0 kN of maximum and minimum load levels, respectively, and the load frequency of cyclic loading is 10 Hz. The crack length value increases during the simulation are measured according to intuition, in contrast with the compliance method adopted for the experimental estimation.

Representative simulation results are shown in Figure 12 for comparison with the experimental results. Figure 12 shows a comparison of the relationships between the crack lengths and number of cycles obtained from the experimental results and the numerical estimations. While computational evaluation is terminated when the crack length is approximately 30 mm, it can be observed that the crack growth rate under fatigue loading was sufficiently simulated numerically.

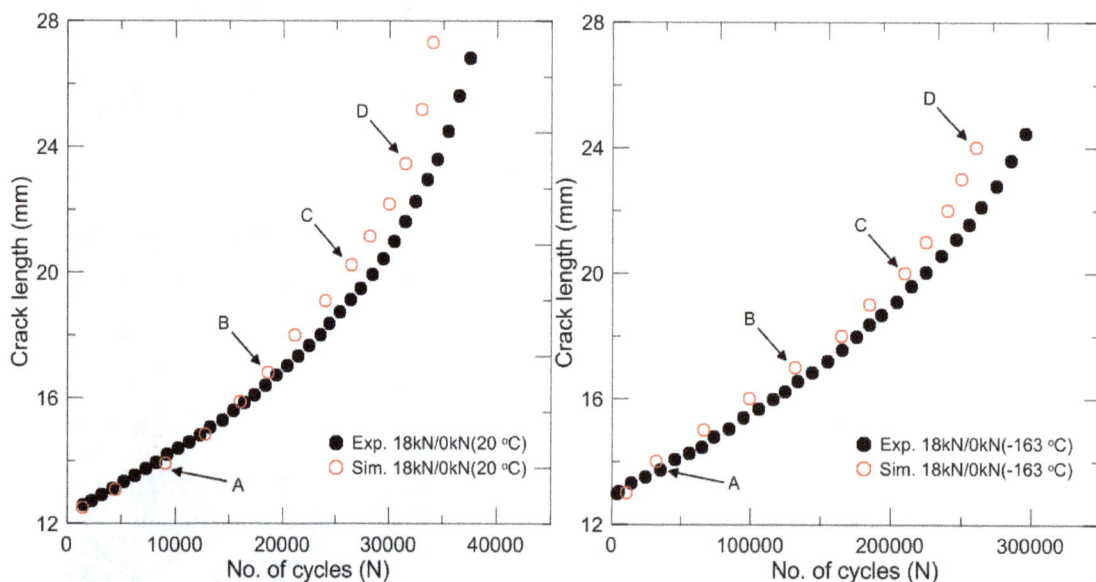

Figure 12. Relationships between the crack length and the number of cycles at room (**left**) and cryogenic (**right**) temperatures.

Figure 13 shows the crack propagations of the CT specimen corresponding to each point in Figure 12 at both temperatures. It can be observed that the crack propagation along the crack tip was sufficiently simulated numerically.

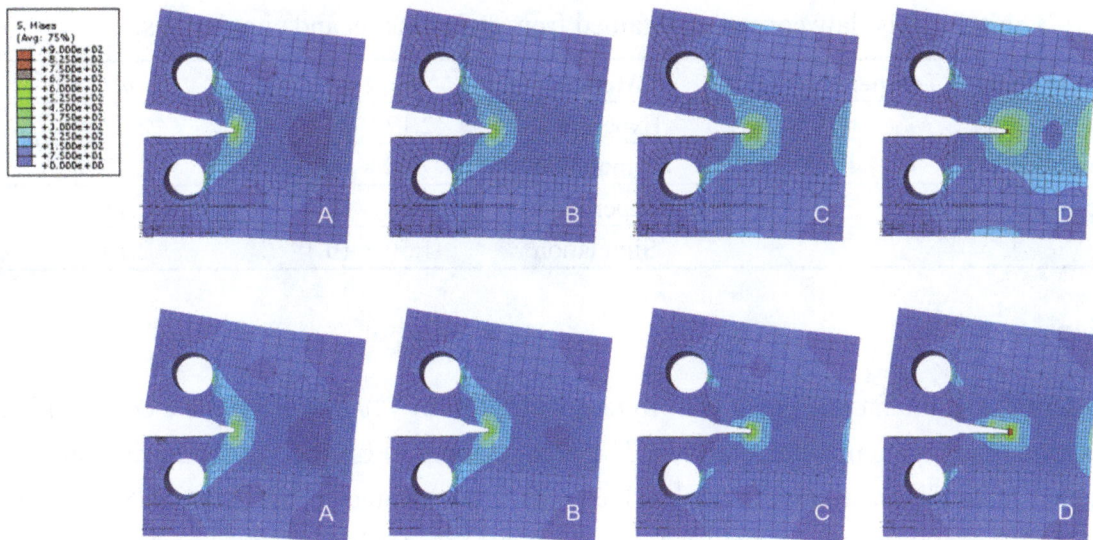

Figure 13. Crack propagations of specimen at room (**upper**) and cryogenic (**bottom**) temperatures.

Figure 14 shows a comparison of the relationship between da/dN and ΔK obtained from the experimental and the numerical estimation. The constants of Paris' law, C and m, carried out from each test are summarized in Table 5. In this study, a systematic approach involving the following steps was used to perform a computational fatigue crack growth rate test: (1) selection of proper constitutive and damage models; (2) comparison of the experimental results with those of typical material testing (e.g., uniaxial tension); and (3) development of a user subroutine and its application to an actual case. For simulation of fatigue crack growth, the procedure for reducing the calculation time (the jump-in-cycles procedure) and the numerical approach (rearrangement of critical damage) are adopted and are implemented to user subroutine. If the adopted constitutive and damage models sufficiently describe the material characteristics and the suitable procedure and rearrangement are employed, this approach can maximize the effectiveness of a user-friendly environment in the commercial FEA code.

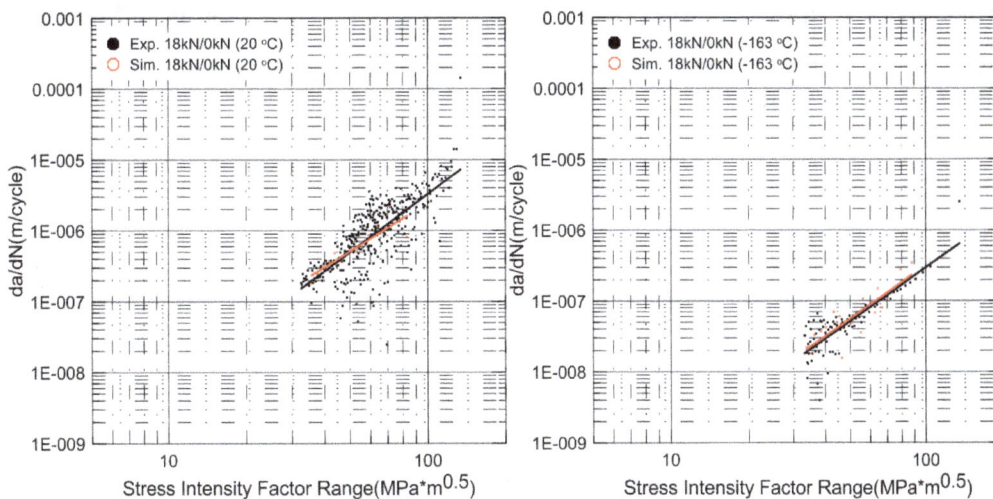

Figure 14. Relationship between da/dN and ΔK at room (**left**) and cryogenic (**right**) temperatures.

Table 5. Paris' law constants obtained from experiments and simulations.

Specimen Type	Temperature (°C)	Method	C	m
Compact Tension	20	Experiment	2.17×10^{-11}	2.57
		Simulation	7.05×10^{-11}	2.27
	−163	Experiment	1.61×10^{-12}	2.71
		Simulation	1.50×10^{-11}	2.10

5. Conclusions

In the present study, the fatigue crack growth rates (FCGRs) of 7% nickel steel are computationally evaluated at room and cryogenic temperatures. The computational FCGR test was carried out using a damage-coupled material constitutive model, which was implemented to the ABAQUS user-defined subroutine, UMAT, for application in the FEA environments. To perform the fatigue damage analysis, the jump-in-cycles procedure and a numerical approach of rearranging critical damage are employed and implemented to the ABAQUS UMAT. The results obtained by the numerical simulation were compared with the experimental results. Some key aspects of the results are highlighted below:

- A set of values for Paris' law constants of 7% nickel steel, which has been recently applied in LNG carrier storage systems, were obtained at room and cryogenic (−163 °C) temperatures. For a compact tension (CT) specimen under tension-type (maximum load: 18 kN, minimum load: 0 kN, stress ratio: 0) cyclic loading, the Paris' law constants C and m were 2.17×10^{-11} and 2.57 at room temperature and 1.61×10^{-12} and 2.71 at cryogenic temperature, respectively.
- It was observed that typical results of the experimental FCGR tests (for Paris' law constants and the relationship between the crack length and the number of cycles) could be obtained numerically. A comparison with the experimental results revealed that the correct use of a damage-coupled material constitutive model and approaches for fatigue damage analysis produces satisfactory results.
- A computational approach incorporating a material constitutive model, the jump-in-cycles procedure, and the rearrangement of critical damage was proposed. The proposed procedure offers useful fatigue crack growth solutions to a designer, particularly when experimental measurement is extremely difficult or unavailable.

Acknowledgments

This work was supported by a National Research Foundation of Korea (NRF) grant funded by the Korean government (MSIP) through GCRC-SOP (No. 2011-0030013). This research was also supported by the Basic Science Research Program funded by the Ministry of Education (No. 2013R1A1A2A10011206) through the National Research Foundation of Korea (NRF).

Author Contributions

Seul-Kee Kim is the first author of the paper and performed the FCGR simulations. Chi-Seung Lee and Jeong-Hyeon Kim aided in the description of the material constitutive model and implementation into FEA. Myung-Hyun Kim aided in the performance and analysis of the experimental FCGR tests. Byeong-Jae Noh and Toshyuki Matsumoto aided in the review of the paper and provided critical comments. Jae-Myung Lee is the corresponding author of the paper, and conceived the paper and supervised the theoretical and numerical processes described in the paper.

Nomenclature

a: Crack length

α: The ratio between crack length and width of specimen

B: Thickness of compact tension type specimen

b: Material parameter that control the characteristics of damage

C: Constants of Paris's law

$C_1 \sim C_3$: Material constants in multiaxial stress function

$C_a \sim C_f$: Constants determined from the measuring location as shown in Figure 4

D_0: Assumed maximum plastic strain rate

D_{ijkl}: Elastic stiffness tensor

ΔK: Stress intensity factor range

ΔN_{JICint}: Jumped blocks of number of cycles during jump-in-cycles procedure

$\Delta \omega_{\text{JICint}}$: Accumulated damage during jump-in-cycles procedure

E: Elastic modulus

$\dot{\varepsilon}_{ij}$: Total strain rate tensor

$\dot{\varepsilon}_{ij}^e$: Elastic strain rate tensor

$\dot{\varepsilon}_{ij}^p$: Plastic strain rate tensor

G_{IC}: Strain energy release rate

h: Material parameter that control the characteristics of damage

I_1^+: First stress invariant

J_2: Second stress invariant

$K_{\max}$: Maximum stress intensity factor

λ: Plastic multiplier

m: Constants of Paris's law

m_h: Rate of isotropic hardening

N: Number of cycle

n: Material parameter that controls the rate sensitivity

N_S: Stabilized cycle of damage accumulation

ω: Damage parameter

ω_{cr}: Critical damage

ω^*_{cr}: Rearranged critical damage

P: Force

$\dot{Q}$: Multiaxial stress function

r: Material constants in multiaxial stress function

S_{ij}: Deviatoric stress tensor

σ_{eff}: Effective stress

σ_{ij}: Stress tensor

σ_{max}^+ : Maximum tensile principal stress

t: Time

v: Displacement between the measurement points, as shown in Figure 4

W: Width of compact tension type specimen

W_P: Plastic work

Z: Total hardening variable

Z_0: Initial isotropic hardening variable

Z_1: Saturated isotropic hardening variable

Conflicts of Interest

The authors declare no conflict of interest.

References

1. Anderson, T.L. *Fracture Mechanics*, 3rd ed.; Taylor & Francis: New York, NY, USA, 2005.

2. Paris, P.; Gomez, M.; Anderson, W.A. A rational analytic theory of fatigue. *Trend Eng.* **1961**, *13*, 9–14.

3. Elices, M.; Guinea, G.V.; Gomez, J.; Planas, J. The cohesive zone model: Advantages, limitations and challenges. *Eng. Fract. Mech.* **2002**, *69*, 137–163.

4. De-Andres, A.; Perez, J.L.; Ortiz, M. Elastoplastic finite element analysis of three-dimensional fatigue crack growth in aluminum shafts subjected to axial loading. *Int. J. Solids Struct.* **1999**, *36*, 2231–2258.

5. Yang, B.; Mall, S.; Ravi-Chandar, K. A cohesive zone model for fatigue crack growth in quasibrittle materials. *Int. J. Solids Struct.* **2001**, *38*, 3927–3944.

6. Roe, K.L.; Siegmund, T. An irreversible cohesive zone model for interface fatigue crack growth simulation. *Eng. Fract. Mech.* **2003**, *70*, 209–232.

7. Bouvard, J.L.; Chaboche, J.L.; Feyel, F.; Gallerneau, F. A cohesive zone model for fatigue and creep-fatigue crack growth in single crystal superalloys. *Int. J. Fatigue* **2009**, *31*, 868–879.

8. Ural, A.; Krishnan, V.R.; Papoulia, K.D. A cohesive zone model for fatigue crack growth allowing for crack retardation. *Int. J. Solids Struct.* **2009**, *46*, 2453–2462.

9. Moes, N.; Dolbow, J.; Belytschko, T. A finite element method for crack growth without remeshing. *Int. J. Numer. Methods Eng.* **1999**, *46*, 131–150.

10. Stolarska, M.; Chopp, D.L.; Moes, N.; Belytschko, T. Modelling crack growth by level sets in the extended finite element method. *Int. J. Numer. Methods Eng.* **2001**, *51*, 943–960.

11. Sukumar, N.; Chopp, D.L.; Moran, B. Extended finite element method and fast marching method for three-dimensional fatigue crack propagation. *Eng. Fract. Mech.* **2003**, *70*, 29–48.

12. Shi, J.; Chopp, D.; Lua, J.; Sukumar, N.; Belytschko, T. Abaqus implementation of extended finite element method using a level set representation for three-dimensional fatigue crack growth and life predictions. *Eng. Fract. Mech.* **2010**, *77*, 2840–2863.

13. Lemaitre, J. A continuous damage mechanics model for ductile fracture. *J. Eng. Mater. Technol. Trans. ASME* **1985**, *107*, 83–89.

14. Lemaitre, J. Micro-mechanics of crack initiation. *Int. J. Fract.* **1990**, *42*, 87–99.

15. Chaboche, J.L. Continuum damage mechanics: Part II–Damage growth, crack initiation, and crack growth. *J. Appl. Mech.* **1988**, *55*, 65–72.

16. Chow, C.L. A damage mechanics model of fatigue crack initiation in notched plates. *Theor. Appl. Fract. Mech.* **1991**, *16*, 123–133.

17. Lee, C.S.; Kim, M.H.; Lee, J.M.; Mahendran, M. Computational study on the fatigue behavior of welded structures. *Int. J. Damage Mech.* **2011**, *20*, 423–463.

18. Bodner, S.R. *Unified Plasticity for Engineering Applications*, 1st ed.; Kluwer Academic and Plenum Publishers: New York, NY, USA, 2002.

19. Lemaitre, J.; Desmorat, R. *Engineering Damage Mechanics*, 1st ed.; Springer: Berlin, Germany, 2005.

20. Murakami, S. *Continuum Damage Mechanics*, 1st ed.; Springer: Berlin, Germany, 2012.

21. Lemaitre, J. *A Course on Damage Mechanics*, 2nd ed.; Springer: Berlin, Germany, 1992.

22. Gurson, A.L. Continuum theory of ductile rupture by void nucleation and growth: Part I–Yield criteria and flow rules for porous ductile media. *J. Eng. Mater. Technol. Trans. ASME* **1977**, *99*, 2–15.

23. Tvergaard, V.; Needleman, A. Analysis of the cup-cone fracture in a round tensile bar. *Acta Metall.* **1984**, *32*, 157–169.

24. Chaboche, J.L. Constitutive equations for cyclic plasticity and cyclic viscoplasticity. *Int. J. Plast.* **1989**, *5*, 247–302.

25. Lemaitre, J.; Chaboche, J.L. *Mechanics of Solid Materials*, 1st ed.; Cambridge University Press: Cambridge, UK, 1990.

26. Ohno, N.; Wang, J.D. Kinematic hardening rules for simulation of ratchetting behavior. *Eur. J. Mech. A-Solids* **1994**, *13*, 519–531.

27. Hamon, F.; Henaff, G.; Halm, D.; Gueguen, M.; Billaudeau, T. A damage model for fatigue crack propagation from moderate to high ΔK levels. *Fatigue Fract. Eng. Mater. Struct.* **2011**, *35*, 160–172.

28. Evans, H.E. *Mechanism of Creep Fracture*, 1st ed.; Elsevier: Amsterdam, The Netherlands, 1984.

29. Cocks, A.C.F.; Leckie, F.A. Creep constitutive equations for damaged materials. *Adv. Appl. Mech.* **1987**, *25*, 239–294.

30. Cadek, J. *Creep in Metallic Materials*, 1st ed.; Elsevier: Amsterdam, The Netherlands, 1988.

31. Mazars, J. A description of micro- and macro-scale damage of concrete structures. *Eng. Fract. Mech.* **1986**, *25*, 729–737.

32. Ju, J.W. On energy-based coupled elastoplastic damage theories: Constitutive modeling and computational aspects. *Int. J. Solids Struct.* **1989**, *25*, 803–833.

33. Mazars, J.; Pijaudier-Cabot, G. Continuum damage theory—Application to concrete. *J. Eng. Mech.* **1989**, *115*, 345–365.

34. Murakami, S.; Kamiya, K. Constitutive and damage evolution equations of elastic-brittle materials based on irreversible thermodynamics. *Int. J. Mech. Sci.* **1997**, *39*, 473–486.

35. Kim, S.K.; Lee, C.S.; Kim, J.H.; Kim, M.H.; Lee, J.M. Computational evaluation of resistance of fracture capacity for SUS304L of liquefied natural gas insulation system under cryogenic temperatures using ABAQUS user-defined material subroutine. *Mater. Des.* **2013**, *50*, 522–532.

36. Lee, C.S.; Yoo, B.M.; Kim, M.H.; Lee, J.M. Viscoplastic damage model for austenitic stainless steel and its application to the crack propagation problem at cryogenic temperatures. *Int. J. Damage Mech.* **2013**, *22*, 95–115.

37. ASTM E647–13ae1. Standard test method for measurement of fatigue crack growth rates. In *Annual Book of ASTM Standards*; ASTM International: West Conshohocken, PA, USA, 2013; Volume 3.01.

38. Kim, J.H.; Lee, C.S.; Kim, M.H.; Lee, J.M. Prestrain-dependent viscoplastic damage model for austenitic stainless steel and implementation to ABAQUS user-defined material subroutine. *Comput. Mater. Sci.* **2013**, *67*, 273–281.

39. Marshall, C.W.; Hehemann, R.F.; Troiano, A.R. The characteristics of 9% nickel low carbon steels. *Trans. ASM* **1962**, *55*, 135.

40. Syn, C.K.; Jin, S.H.; Morris, J.W. Cryogenic fracture toughness of 9Ni steel enhanced through grain refinement. *Metall. Mater. Trans. A-Phys. Metall. Mater. Sci.* **1976**, *7A*, 1827–1832.

41. Lee, J.M. A Study of Computational Mechanics of Thermal Damage Problem for Structural Members. Ph.D. Dissertation, University of Tokyo, Tokyo, Japan, 1999. (In Japanese)

42. Benallal, A.; Billardon, R.; Lemaitre, J. Continuum damage mechanics and local approach to fracture: Numerical procedures. *Comput. Meth. Appl. Mech. Eng.* **1991**, *92*, 141–155.

43. Desmorat, R.; Kane, A.; Seyedi, M.; Sermage, J.P. Two scale damage model and related numerical issues for thermos-mechanical High Cycle Fatigue. *Eur. J. Mech. A-Solids* **2007**, *26*, 909–935.

44. Doudard, C.; Calloch, S.; Cugy, P.; Galtier, A.; Hild, F. A probabilistic two-scale model for high-cycle fatigue life prediction. *Fatigue Fract. Eng. Mater. Struct.* **2005**, *28*, 279–288.

45. Lemaitre, J.; Sermage, J.P.; Desmorat, R. A two scale damage concept applied to fatigue. *Int. J. Fract.* **1999**, *97*, 67–81.

46. Hayhurst, D.R.; Leckie, F.A. Constitutive equations for creep rupture. *Acta Metall.* **1977**, *25*, 1059–1070.

7

An *in situ* Study of NiTi Powder Sintering Using Neutron Diffraction

Gang Chen [1,2,*]**, Klaus-Dieter Liss** [3,4] **and Peng Cao** [1,*]

[1] Department of Chemical and Materials Engineering, the University of Auckland, Private Bag 92019, Auckland 1142, New Zealand

[2] State Key Laboratory of Porous Metal Materials, Northwest Institute for Nonferrous Metal Research, Xi'an 710016, Shaanxi, China

[3] Australian Nuclear Science and Technology Organisation, New Illawarra Road, Lucas Heights, NSW 2234, Australia; E-Mail: kdl@ansto.gov.au

[4] Quantum Beam Science Directorate, Japan Atomic Energy Agency, 2-4 Shirakata-Shirane Tokai-mura, Naka-gun, Ibaraki-ken 319-1195, Japan

* Authors to whom correspondence should be addressed;
E-Mails: mychgcsu@163.com (G.C.); p.cao@auckland.ac.nz (P.C.);

Academic Editor: Hugo F. Lopez

Abstract: This study investigates phase transformation and mechanical properties of porous NiTi alloys using two different powder compacts (*i.e.*, Ni/Ti and Ni/TiH$_2$) by a conventional press-and-sinter means. The compacted powder mixtures were sintered in vacuum at a final temperature of 1373 K. The phase evolution was performed by *in situ* neutron diffraction upon sintering and cooling. The predominant phase identified in all the produced porous NiTi alloys after being sintered at 1373 K is B2 NiTi phase with the presence of other minor phases. It is found that dehydrogenation of TiH$_2$ significantly affects the sintering behavior and resultant microstructure. In comparison to the Ni/Ti compact, dehydrogenation occurring in the Ni/TiH$_2$ compact leads to less densification, yet higher chemical homogenization, after high temperature sintering but not in the case of low temperature sintering. Moreover, there is a direct evidence of the eutectoid decomposition of NiTi at *ca.* 847 and 823 K for Ni/Ti and Ni/TiH$_2$, respectively, during furnace cooling. The static and cyclic stress-strain behaviors of the porous NiTi alloys made from the Ni/Ti and Ni/TiH$_2$ compacts were also investigated. As compared with the Ni/Ti sintered samples, the samplessintered from the Ni/TiH$_2$ compact exhibited a much higher porosity, a higher close-to-total porosity, a larger pore size and lower tensile and compressive fracture strength.

Keywords: NiTi; powder sintering; dehydrogenation; neutron diffraction

---

1. Introduction

NiTi alloys have excellent properties including unique shape memory effect (SME), superelasticity, good biocompatibility and great energy absorption, which have been attracting attention from multiple areas such as medical devices, energy absorbers, actuators and mechanical couplings [1,2]. Powder metallurgy (PM) is a simple, energy-saving and widely used route to produce NiTi alloys [3]. Additionally, powder sintering is an effective technique to produce various porous structures, which are beneficial to bone tissue ingrowth and also provide an effective way of reducing stiffness of the implant [4].

Elemental powder sintering to fabricate porous NiTi alloys has been tremendously successful recently [4–10]. Interestingly, TiH$_2$ powder was frequently used in NiTi powder sintering in previous studies [4,10–17] due to its cleansing effect of dehydrogenation, which lowers oxygen content and potentially promotes chemical homogenization and densification [18,19]. There is no doubt that the use of TiH$_2$ favors final phase homogenization after high temperature sintering in the previous reports [4,10–17]. However, our most recent results [10,17,20,21] and the report from Robertson and Schaffer [14] disclosed a discouraging densification and a much larger porosity when using TiH$_2$ powder. As such, the use of such powder cannot guarantee densification promotion in all NiTi studies, although it does show densification in some other alloys, e.g., pure Ti, Ti-6Al-4V, Ti-5Al-2.5Fe and TiAl [19,22–27]. This might be caused by other factors simultaneously affecting the sintering process and thus the densification. These factors include TiH$_2$ particle size in Refs. [11,12,28,29] and the binders used in the reports [4,16]. Our recent results [17,20,21] also pointed out that it is the dehydrogenation of TiH$_2$ powder that increased the porosity of sample and then hindered its densification, when compared with that using similar particle size of Ti powder.

The process of TiH$_2$ dehydrogenation has been studied for many years [17,19,20,25,27,30–36]. However, most of the studies are conducted in either argon or air atmosphere [15,19,32,33,35]. With respect to the atmosphere, the dehydrogenation usually takes place in the temperature range from 523 to 973 K (250 to 700 °C), which possibly causes the concern of TiH$_2$ oxidation. On the other hand, some studies, e.g., Refs. [31,34], were performed in vacuum, effectively avoiding the oxidation issue. In spite of this, the diffraction instrument used is laboratory low-intensity X-ray diffraction systems [34], which normally require several minutes to one hour to achieve a complete scan for phase analysis and the achieved data is normally semi-accurate. Such "long"-time scanning properly leads to delayed or missing information. These technical limitations can be tackled with high-energy neutron diffraction under vacuum, which is able to penetrate bulk metals, and this type of diffraction has been successfully employed for *in situ* studies for sintering mechanism and reactions [20,36]. The beam intensities allow information from bulk material to be followed on short time scales (less than 60 s), while undergoing an *in situ* heating/cooling cycle to observe phase transformations. Furthermore, due to the strong incoherent neutron scattering from hydrogen, neutron diffraction can also track the development of hydrogen concentration during dehydrogenation [20].

Since dehydrogenation of TiH$_2$ involving in the reaction procedure of powder sintering, this reactive process is thought to be more intricate and different from the case of Ni/Ti blend. To the best of our knowledge, no report has elaborated the reactive sintering mechanism using Ni/TiH$_2$ blend involving dehydrogenation of TiH$_2$ and the mechanism investigation of TiH$_2$ decomposition under vacuum. Bearing in mind, it is of great importance to investigate the combination of dehydrogenation of TiH$_2$ and newly born Ti and Ni sintering hereafter and the comparative study of mechanical properties of as-fabricated NiTi alloys using Ni/Ti and Ni/TiH$_2$ powder blends. In this study, it is the first time to observe and study the combined phase transformation processes of dehydrogenation of TiH$_2$ and the subsequent reactions between new-born Ti and Ni particles using *in situ* neutron diffraction under vacuum as a comparison of the Ni/Ti blend. Further, the systematic mechanical comparison was investigated in terms of pore size, porosity, pore shape and pore size distribution. Therefore, this study is an additional and supplemental report to our recent results in Refs. [17,20].

2. Experimental Section

The mean particle size of Ti, TiH$_2$ and Ni raw powders used in this study was 32.2, 24.6 and 16.4 μm, respectively. Powder mixtures of Ni/Ti and Ni/TiH$_2$ were gently mixed in a ball mill for 10 h. Both powder mixtures had a nominal composition of 51 at.% Ni and 49 at.% Ti.

After mixing, powder mixtures were pressed into cylindrical discs of 12 mm diameter with three heights (*i.e.*, 4, 10 and 20 mm for microstructural characterization, neutron diffraction measurement and compression test, respectively) and tensile testing bars (15 mm in gauge length and 2 mm in thickness) in a single-action steel die under 250 MPa pressure. Stearic acid lubricant was slightly applied to the compaction die wall. Subsequently, the 4- and 20-mm-thick green compacts and tensile bars were sintered in a vacuum furnace at 3×10^{-3} Pa, while the 10-mm-thick green compacts were sintered in a high temperature vacuum furnace (5×10^{-4} Pa) equipped on the WOMBAT for *in situ* neutron diffraction measurements. The WOMBAT is a high-intensity diffractometer at the Australian Nuclear Science and Technology Organization (ANSTO), which uses monochromatic neutrons and is equipped with a two-dimensional area detector [37]. The basic technical information of WOMBAT is detailed in Refs. [20,36]. The sintering profile with a heating rate of 5 K/min will be shown in Section 3.2. The heating process was designed into two stages where the first stage is for dehydrogenation of TiH$_2$ powders, while the second one is to perform final sintering at a temperature of 1373 K (1100 °C) for 2 h, followed by furnace cooling.

A free Rietveld program *MAUD* was chosen to analyze the full powder-diffraction pattern using the Rietveld method, which is to obtain quantitative values of the phase fractions throughout the *in situ* experiments [20]. To determine the phase fractions, each 1-D diffraction pattern was subsequently fed into the Rietveld analysis as a function of time. The analysis was began with a well-fitted analysis file in *MAUD*, which was then used for recursive fitting of the following data files. The batch running was repeated several times with different starting values and constraints to start the iterating process until there was a consistently good fitting throughout the entire run.

Open porosity and sintered density were measured by the Archimedes method as specified in the ASTM B962-08 standard. Pore size distribution analysis was conducted using a pore-size distribution analyzer (GaoQ PDSA-20) using the bubble-point method as per the ASTM F316-03 standard [38]. Microstructures of the as-sintered compacts were observed using an environmental scanning electron microscope (ESEM, FEI Quanta 200F, FEI, Houston, TX, USA) equipped with an energy dispersive X-ray spectrometer (EDX, Oxford Instruments, Oxfordshire, UK). Phase constituents were determined using X-ray diffraction (XRD, Bruker D2 Phaser, Bruker, Karlsruhe, Germany). Differential scanning calorimetry (DSC, Netzsch 404 F3, Netzsch, Selb, Germany) was used to determine the various reactions of compacts during sintering with a heating rate of 5 K/min under flowing argon gas.

The tensile properties of the as-sintered NiTi tensile bars were measured on an Instron 3367 universal machine with a cross-head speed of 0.5 mm/min at ambient temperature. The tensile bars were tensioned approaching to its fracture strength. The compressive properties of the 20-mm-thick samples after 1373 K sintering were measured on an MTS 810 universal machine with a load rate of 0.6 kN/s at room temperature. An alignment cage ensured the parallelism of all samples during testing. The ends of compression cylindrical samples (machined into 10.5-mm diameter and 15-mm height) were polished and smoothed using sand papers, and finally the ends were greased before compression tests. Cyclic experiments were performed to study possible deformation and superelasticity. The cylindrical samples were first compressed until a significant deflection of the linear elastic deformation portion on the stress-strain curve was obtained or the stress level approached to its fracture strength. After that they were unloaded to zero stress and the subsequent cycle followed.

3. Results

3.1. Microstructure

Differential scanning calorimetry (DSC) measurements were conducted to investigate the phase evolution for each compact. Figure 1 shows the DSC curves of the Ni/Ti, Ni/TiH$_2$ and pure TiH$_2$ compacts after 250 MPa compaction with a heating rate of 5 K/min. According to Figure 1, a broad exothermic peak can be seen at *ca.* 1036 K for the Ni/Ti compact, which is followed by an endothermic peak developing with an onset temperature at 1143 K. With increasing temperature, this is immediately followed by an apparent exothermic peak at around 1240 K. The final peak is an endothermic peak whose temperature is 1417 K. As discussed in Ref. [20], the four peaks correspond to formation of intermetallic phases (e.g., NiTi, Ni$_3$Ti and NiTi$_2$, *etc.*), eutectic reaction to generate liquid Ti-rich phase, combustion reaction between molten Ti-rich and Ni-rich phases, and another eutectic reaction between NiTi and Ni$_3$Ti phases, respectively. In contrast, the dehydrogenation of TiH$_2$ is a thermally endothermic process [39]. Therefore, the first two endothermic peaks for Ni/TiH$_2$ and TiH$_2$ compacts correspond to the dehydrogenation, which ranges from ~630 to 920 K. However, the following peaks for the Ni/TiH$_2$ are less manifest as compared with the Ni/Ti compact. This is due to the fact that the dehydrogenation peaks may overlap with the following reaction peaks [20].

The X-ray diffraction (XRD) results are presented in Figure 2 for both compacts sintered at 1373 K. It can be seen that the main sintered phase is austenitic B2 NiTi in both cases, with the existence of martensitic B19', secondary NiTi$_2$, Ni$_3$Ti and Ni$_4$Ti$_3$. The existence of these phases in the as-sintered samples is further confirmed in the ESEM micrographs and EDX analysis (Figure 3). It should be noted that the amount of Ni_4Ti_3 phase is too little to be detected by EDX. The needle-like structural phase is determined to be Ni_3Ti in both samples (Figure 3b,d), which is due to the eutectoid reaction of $NiTi \rightarrow NiTi_2 + Ni_3Ti$ during cooling [20]. However, it is interesting to observe that the amount of secondary phases of the Ni/TiH_2 sintered sample is less compared than that of the Ni/Ti sintered based on the XRD (Figure 2) and energy dispersive X-ray (EDX) results (Figure 3). This means that the final chemical homogeneity of the Ni/TiH_2 sintered is higher than that of the Ni/Ti sintered sample.

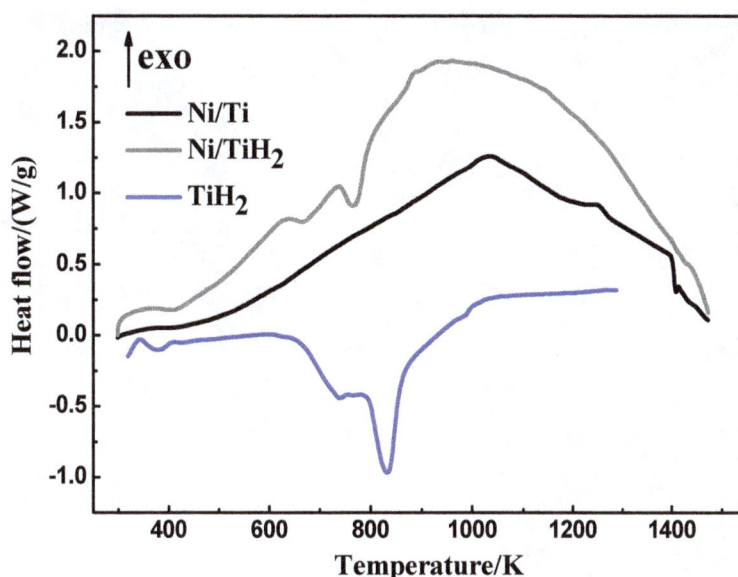

Figure 1. Differential Scanning Calorimetry (DSC) curves of Ni/Ti, Ni/TiH_2 and TiH_2 compacts with a heating rate of 5 K/min.

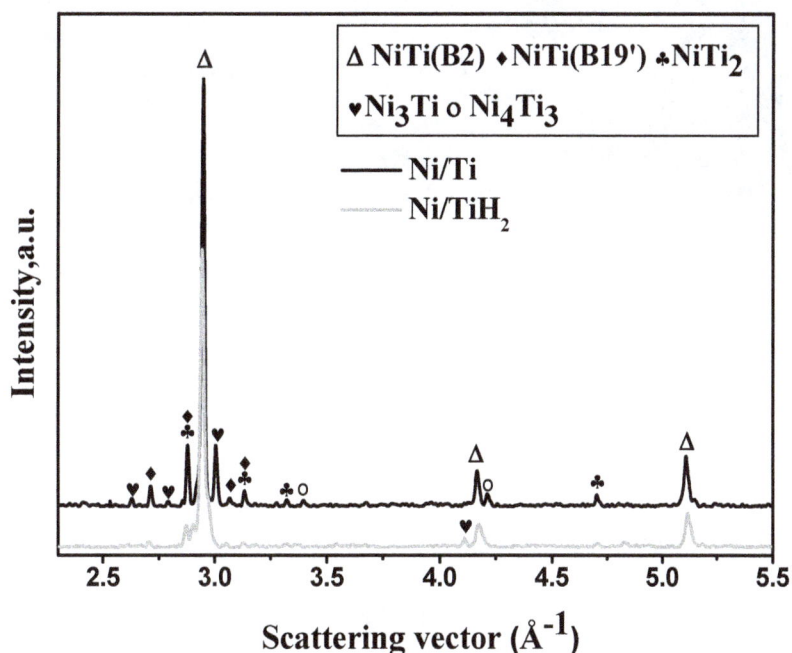

Figure 2. X-ray Diffraction (XRD) patterns of the samples after being sintered at 1373 K.

Table 1 summarizes the basic data of both sintered compacts from 4-mm-thick green samples. It can be figured that the dimension exhibits shrinkage for both sintered samples in terms of either radial or axial direction. Moreover, the shrinkage of the Ni/Ti is larger than that of the Ni/TiH$_2$ after sintering, with the concomitant higher density for former case. In addition to the shrinkage and density, the open porosity and close-to-total porosity ratio are significantly different from each other. For instance, the close-to-total porosity ratio of the Ni/Ti sintered sample is 89.6% $\pm$ 3.4%, while it is only 12.2% $\pm$ 0.8% in the case of Ni/TiH$_2$.

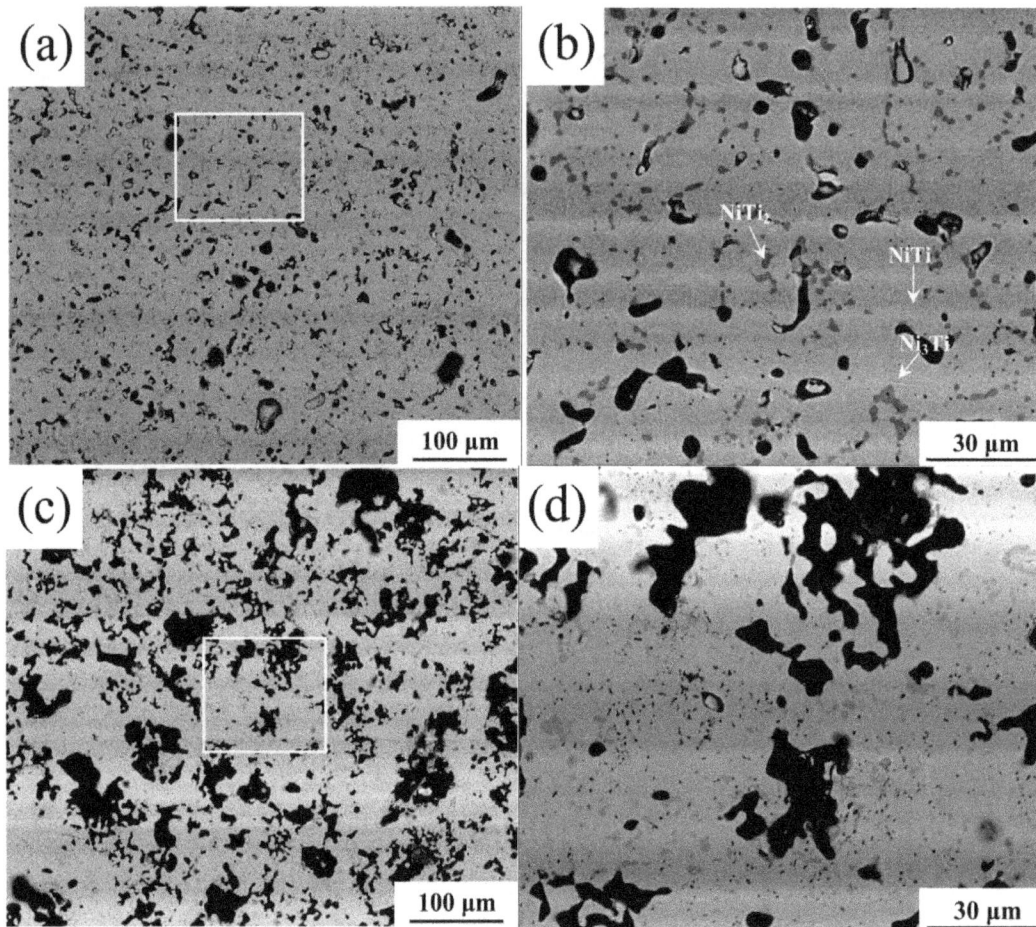

Figure 3. Back-scattered electron images of samples sintered from the Ni/Ti compact at (**a**) 1373 K, (**b**) enlarged square area in (**a**); sintered from the Ni/TiH$_2$ compact at (**c**) 1373 K, (**d**) enlarged square area in (**c**).

Table 1. Characteristics of the 1373 K sintered porous NiTi samples.

Sample	Shrinkage/%		Density/g·cm^{-3}	Open porosity/%	Close-to-total porosity ratio/%
	axial	radial			
Ni/Ti	10.47 ± 1.23	6.49 ± 0.62	5.81 ± 0.11	1.0 ± 0.1	89.6 ± 3.4
Ni/TiH$_2$	5.93 ± 0.49	4.21 ± 0.37	4.47 ± 0.07	26.9 ± 2.9	12.2 ± 0.8

3.2. In situ Neutron Diffraction

Figure 4 presents the neutron diffraction patterns of the Ni/Ti and Ni/TiH$_2$ compacts collected as a function of time in the 2D plot. The intensity is displayed by the grey scale values as a function of scattering vector Q (Q = $4\pi/\lambda \cdot \sin\theta$) on the abscissa and time on the ordinate. It is focused on the dehydrogenation process and its effect on the phase transformation of the Ni/TiH_2 compact as compared with the Ni/Ti compact.

From Figure 4a, it can be seen when the temperature approaches *ca.* 840 K, the intensities of intermetallic phases (*i.e.*, B2 NiTi, Ni_3Ti, $NiTi_2$ and Ni_4Ti_3) start to establish as a result of the intensity decrease of elemental Ni and Ti in the Ni/Ti compacts. Afterwards, the intensities of elemental Ni and Ti gradually decrease and it is almost nil at about 1076 K, while these intermetallic phases largely increase. Until the temperature increases to 1163 K, the peaks of some secondary phases (Ni_3Ti and Ni_4Ti_3) almost disappear while the NiTi and $NiTi_2$ phases still remain with temperature increase even when holding at 1373 K. Additionally, it is interesting to note that the intensities of previously disappeared Ni_3Ti and Ni_4Ti_3 phases re-emerge when the furnace was cooled to *ca.* 847 K. This phenomenon has also been discussed in our recent reports [20,36]. It is due to the eutectoid reaction (NiTi → $NiTi_2$ + Ni_3Ti) taking place at *ca.* 903 K during furnace cooling [20,36,40–42]. Additionally, it is obvious that the peaks are significantly shifted in position, which is attributed to thermal expansion of crystal lattice when the temperature is relatively high [43]. Moreover, the Mo peaks come from the Mo wires holding the samples in the instrument.

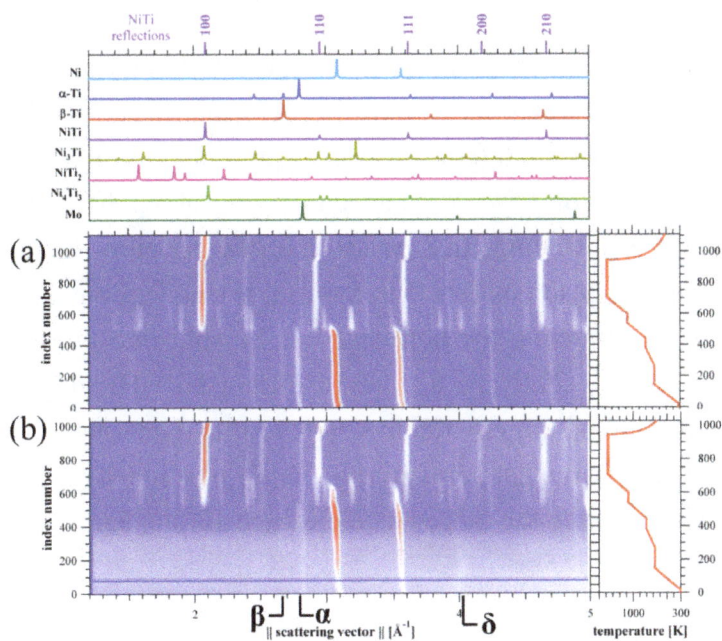

Figure 4. Neutron diffraction patterns as a function of time while temperature is ramped 1373 K from (**a**) Ni/Ti and (**b**) Ni/TiH_2.

In contrast, several differences can be seen between the Ni/TiH_2 compact (Figure 4b) and Ni/Ti compact (Figure 4a) in the heating and cooling process. First, when involving TiH_2 sintering, the initial background is much more significant compared to the Ni/Ti case (Figure 2a). Second, the temperature to establish intensities of intermetallic phases (*i.e.*, B2 NiTi, $NiTi_2$ and Ni_3Ti) is nearly 100 K higher than the Ni/Ti (Figure 4b *cf.* Figure 4a). Third, there is no Ni_4Ti_3 phase formed during sintering in the Ni/TiH_2 case and the intensities of secondary phases are weaker as compared to the Ni/Ti sample. The initial pattern background is caused by the strong incoherent neutron scattering from hydrogen atoms in TiH_2. Then, it gradually decreases with the temperature till ~923 K when it is thought the dehydrogenation of δ-Ti(H) is almost complete. It is noteworthy that both α-Ti(H) and β-Ti(H) phases appeared during decomposition of TiH_2 below 780 K, which is consistent with the recent study by Jiménez *et al.* [33]. Several intermetallic phases (*i.e.*, B2 NiTi, Ni_3Ti and $NiTi_2$) start to form when the temperature reaches ~975 K concomitant with the intensity decrease of elemental Ni and Ti. After this, the intensities of these phases continue to increase until the temperature rises to ~1350 K when the peaks of Ni, Ti and Ni_3Ti phases disappear. There only exist B2 NiTi and minor $NiTi_2$ phases when holding at 1373 K. It is similar with the case of Ni/Ti compact that the intensity of Ni_3Ti phase starts to re-establish when it was cooled to ~823 K.

With a particular focus on the dehydrogenation process of TiH_2, it can be seen from Figure 4b that the starting constituent includes δ-Ti(H) phase, and with increasing temperature another two hydrogen-containing solid solutions, *i.e.*, α-Ti(H) and β-Ti(H) phases, establish their intensities. The α-Ti(H) and β-Ti(H) phase has hcp and bcc structure, respectively, and hydrogen atoms sit randomly on the tetrahedral sites of both phases [44]. When the temperature approaches *ca.* 695 K, the intensity of δ-Ti(H) phase completely vanishes. Afterwards, the β-Ti(H) and α-Ti(H) phases totally transfer to α-Ti phase at ~780 K.

3.3. Pore-Size Distribution

The use of bubble-point method is to measure the pore-size distribution of both green and sintered samples, which can determine the pore-throat size in the pore tunnel as specified in the American Society of Testing Materials (ASTM) F316-03 standard. As presented in Figure 5a, most pores of the green Ni/Ti compact are in the range of 2.5~7.5 μm accounting for about 80% and only few pores are larger than 15.0 μm or smaller than 2.0 μm. In contrast, the original pore size in the green Ni/TiH_2 compact (Figure 5b), which mostly positions less than 5.0 μm, is smaller compared to the green Ni/Ti compact. However, after 1373 K sintering the pore size can be split into two main ranges for each sample, which are 2.5~20.0 and 4.0~20.0 μm for the Ni/Ti and Ni/TiH_2, respectively. This means pore-size distribution is broader and pores become larger after sintering. Such phenomenon is significantly obvious in the Ni/TiH_2 case that pores are previously positioned below 5.0 μm as shown in Figure 5b, while most of them enlarge to the range between 5 and 10 μm, accounting *ca.* 50% porosity, after sintering.

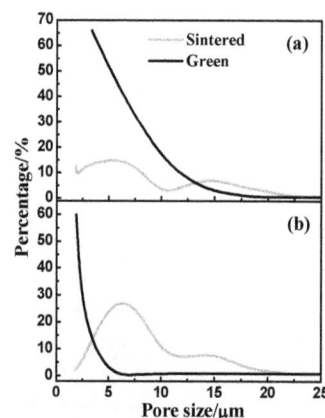

Figure 5. Pore-size distribution of green and 1373 K sintered samples from (**a**) Ni/Ti and (**b**) Ni/TiH_2.

3.4. Mechanical Properties

3.4.1. Static Tensile Test

Figure 6 displays typical stress-strain curves of the NiTi bars being sintered at 1373 K from both compacts. However, both sintered samples exhibited typical brittle fracture behaviors. As presented in Table 2, the fracture tensile strength of the Ni/Ti sintered sample (549.4 ± 9.6 MPa) is much higher than that of the Ni/TiH$_2$ (160.2 ± 7.3 MPa). Accordingly, the fracture strain of the former sample (4.6% ± 0.2%), which is expectable for porous NiTi, is much higher compared to the later sample (0.9% ± 0.1%). Nevertheless, both Ni/Ti sintered bars demonstrated quasi-linear elastic deformation behavior. In contrast, the Young's modulus of both samples is quite similar but significantly lower than that of the wrought NiTi alloys (~70 GPa) [45].

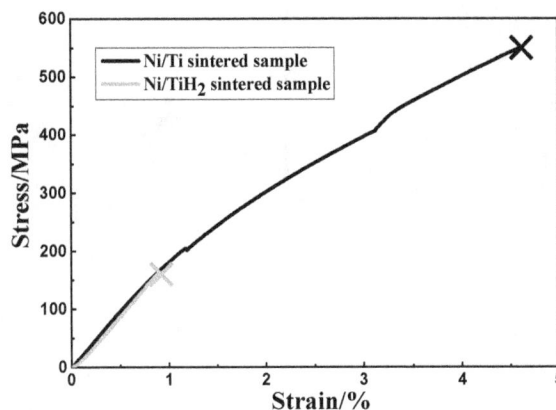

Figure 6. Tensile stress-strain curves for the 1373 K sintered NiTi parts made from Ni/Ti and Ni/TiH$_2$.

Table 2. Static tensile properties of the as-sintered NiTi alloys.

Sample	Fracture tensile strength/MPa	Fracture strain/%	Young's modulus/GPa
Ni/Ti	549.4 ± 9.6	4.6 ± 0.2	18.9 ± 1.1
Ni/TiH$_2$	160.2 ± 7.3	0.9 ± 0.1	18.0 ± 0.9

3.4.2. Cyclic Compressive Test

To investigate the porosity effect on the compressive properties, a total of five cycles was applied to each sintered sample. The cyclic compressive samples were compressed to 500, 800 and 1200 MPa, respectively, for the 1373 K-sintered samples from the Ni/Ti compact and then completely unloaded. In the Ni/TiH$_2$ case, the compressive load changes to 300, 500 and 800 MPa, respectively, since the tensile strength of the Ni/TiH$_2$ sintered sample is much lower than the Ni/Ti sintered sample (Figure 6). Figure 7 shows the strain curves as a function of time for the compressive cycles. The Ni/TiH$_2$ sintered sample failed during the third cycle with a fracture strain of 7.14% under 800 MPa stress (Figure 7b). By contrast, the Ni/Ti sintered sample could withstand all the five cycles under both 500 and 800 stresses only, except the 1200 MPa load, where the sample collapsed at the third cycle (Figure 7a).

There can be seen several interesting aspects of superelasticity originating from these curves. First, the residual strain increases with the compressive stress. On the other hand, it is noteworthy that the residual strain of the Ni/Ti sintered sample is less compared with the case of Ni/TiH$_2$ under the identical compressive load. For instance, the residual strain is 0.75% for the Ni/Ti sintered sample while it is 1.96% in the latter case. Additionally, the maximum strain slightly rises with the cycle number obviously for the higher compressive stress.

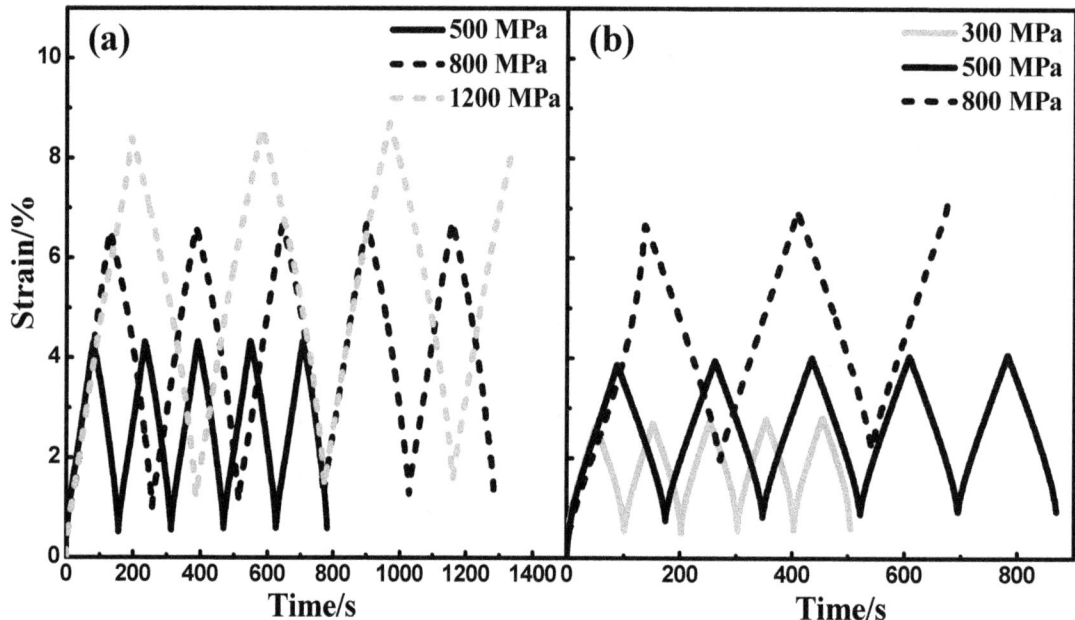

Figure 7. Compressive load-unload-recovery cycles under different compressive stresses for the samples after sintering at 1373 K (**a**) the Ni/Ti compact and (**b**) the Ni/TiH$_2$ compact. A total of five cycles was applied to each sample.

4. Discussion

4.1. Microstructural Evolution

4.1.1. Dehydrogenation process

Several *in situ*/*ex situ* studies have been focused on the thermal decomposition of TiH$_2$ [4–35,46,47]. However, it should be noted that the *ex-situ* XRD and TEM investigations may suffer from instant information loss in terms of the phase transformation during the heating process [19,27,30,46,47]. Additionally, although *in situ* high temperature XRD and X-ray synchrotron/neutron diffraction techniques were applied, their results may still be of concern. First, some experiments were conducted in argon atmosphere [32,33,35], which possibly causes the oxidation problem and may mislead the result. Moreover, other reports using vacuum atmosphere may result in instant information loss or delay due to the fact that XRD scanning required a long time (usually several minutes to one hour to achieve a complete scan) [31,34]. To our best knowledge, it is the first time using the neutron diffraction technique to *in situ* investigate the dehydrogenation process of TiH$_2$. This means could not only solve the long-time scanning problem (needed below 60 s), but also involve vacuum furnace to effectively avoid the oxidation issue.

According to the Ti-H phase diagram (Figure 8), titanium hydride appears as δ, β and α-phase has *ca.* 50~66.7 at.%, 0~50 at.% and 0~8.5 at.% of hydrogen content [48,49], respectively. In our case, the initial titanium hydride phase includes δ phase as shown in Figure 4b. Based on this phase diagram and the neutron diffraction pattern in Figure 4b, it can be concluded that its dehydrogenation could take place as follows: $δ → δ + α → δ + β + α → β + α → α$. This finding is with great agreement with the report in Ref. [33]. Attributed to the strong incoherent neutron scattering from hydrogen atoms, there is an obvious background during the initial heating. In spite of this, the hydrogen release progresses with the temperature and time concomitant with the background slash. The dehydrogenation temperature range in this study occurs between 573 and 1073 K as presented in our DSC curves (Figure 1) and neutron diffraction pattern (Figure 4b), which is consistent with previous studies [15,30,32,33,50]. As a result, the background evolution is consistent with the process of dehydrogenation during heating and finally almost disappears at about 780 K, Figure 4b.

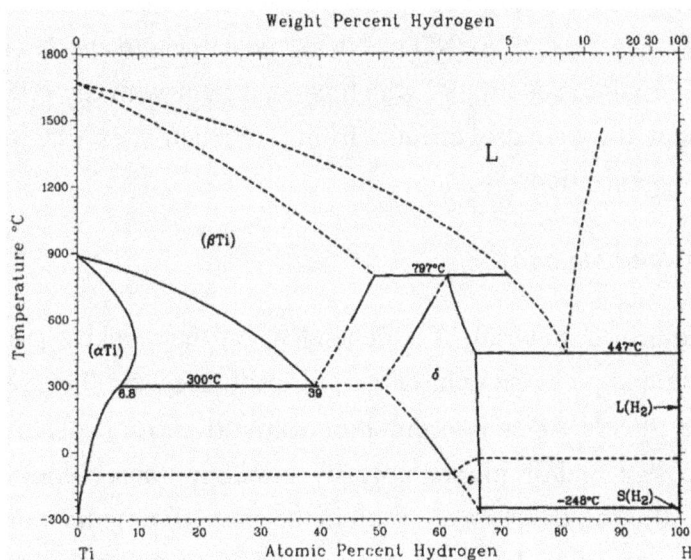

Figure 8. Ti-H binary phase diagram redrawn from Ref. [48].

4.1.2. Pore and Phase Evolution

As discussed in our previous reports [4,10,17,20] together with other studies [5,14,51–56], the pores present in the final sintered samples can be originated from the following four sources: (1) original pores in the green compact, (2) Kirkendall pores formed due to the different diffusion rates between Ni and Ti or newly born Ti elements, (3) pores occurred by the following phase transformation or alloying and (4) large pores caused by liquid phase sintering (LPS). It has been proved in Refs. [17,20] that dehydrogenation in the Ni/TiH$_2$ compact causes porosity increase during sintering and then the diffusion distance between Ni and new-born Ti particles enlarges, which is thought to delay sequential alloying and increase pore size and porosity in the Ni/TiH$_2$ sample. In contrast, LPS has two opposite effects on densification. On the one hand, it would favor densification since it promotes diffusion due to the presence of liquid [57,58]. On the other hand, however, it could give rise to swelling because of pores leaving behind [5]. We recall the microstructure images (Figure 3), density and porosity data (Table 1), and pore-size distribution (Figure 5), it seems the combination of the two factors, which are dehydrogenation and LPS, leads to the fact that the density of Ni/Ti sintered sample is much higher compared to the case of Ni/TiH$_2$ although the relative density of both green compacts is similar (*i.e.*, it is 73.0% and 71.2% for the Ni/Ti and Ni/TiH$_2$ compact, respectively).

The Rietveld quantitative analysis from the neutron diffraction data, shown in Figure 9, further supports the discussion above. Figure 9 displays the weight fraction of various intermetallic phases for both compacts during sintering and furnace cooling. It can be confirmed that the whole sintering process of the Ni/TiH$_2$ compact below 1373 K is postponed compared to the Ni/Ti compact (Figure 9b *cf.* Figure 9a). Nevertheless, at the final holding stage at 1373 K, the amount of B2 NiTi phase is slightly lower for the Ni/TiH$_2$ compact (94.3 wt.%) than that in the Ni/Ti compact (96.2 wt.%). However, such situation occurs oppositely after furnace cooling, because the final B2 phase amount of the Ni/TiH$_2$ compact after cooling (87.3 wt.%) is higher as compared to the Ni/Ti compact (81.3 wt.%). This observation has been reported in our recent result [20] that there is a eutectoid reaction NiTi → Ni$_3$Ti + NiTi$_2$ happened at around 903 K, which means the B2 NiTi phase decomposed into Ni$_3$Ti and NiTi$_2$ phases during cooling and thus gives rise to the phase amount change accordingly. All the amount of secondary phases such as NiTi$_2$, Ni$_3$Ti and Ni$_4$Ti$_3$ in the Ni/Ti sintered sample is higher than that in the Ni/TiH$_2$ sintered sample, which is consistent with the XRD results (Figure 2). This can further confirm that the dehydrogenation from TiH$_2$ activates titanium surface and thus enhances final chemical homogenization.

4.2. Fracture, Superelasticity and Modulus

With regard to the strength of a material, it is dependent on the weakest portion in the material. Normally, porosity, pore size and pore shape have a significant effect on the strength of porous NiTi alloys. For instance, a more severe stress concentration may arise from a sharp edge of the pores. Furthermore, a larger pore size and/or higher porosity result in more reduction in the effective load-carrying cross section [10]. These factors all result in the strength drop the porous NiTi alloys [59]. Recalling the fracture tensile strength and Young's modulus (Figures 6 and 7, Table 2), these values of the Ni/Ti sintered sample are higher compared with the Ni/TiH$_2$ sintered sample.

On the one hand, from a fracture mechanics point of view, the material fails when the stress intensity factor K ($= Y\sigma\sqrt{\pi a}$) reaches its fracture toughness [60]. In this respect, the "a" represents the pore size and pore-size distribution, while the "Y" is a collective parameter of pore shape and orientation in a porous material. In this study, the mean pore size of the Ni/Ti sintered sample is significantly smaller than did in the Ni/TiH$_2$ case, Figures 3 and 5. However, the ESEM micrographs (Figure 3) show that the pore shape is similar in both samples. This implies that the average "Y" value is analogous in both cases, while the "a" value gives rise to a higher stress intensity factor K for the Ni/TiH$_2$ sintered sample. As such, the Ni/Ti sintered sample demonstrated a higher fracture stress, as compared to the case of Ni/TiH$_2$. Alternatively, this means the use of TiH$_2$ powder leads to lower fracture strength caused by larger pore size and lower densification (Table 1) although it shows higher chemical homogenization (Figure 9).

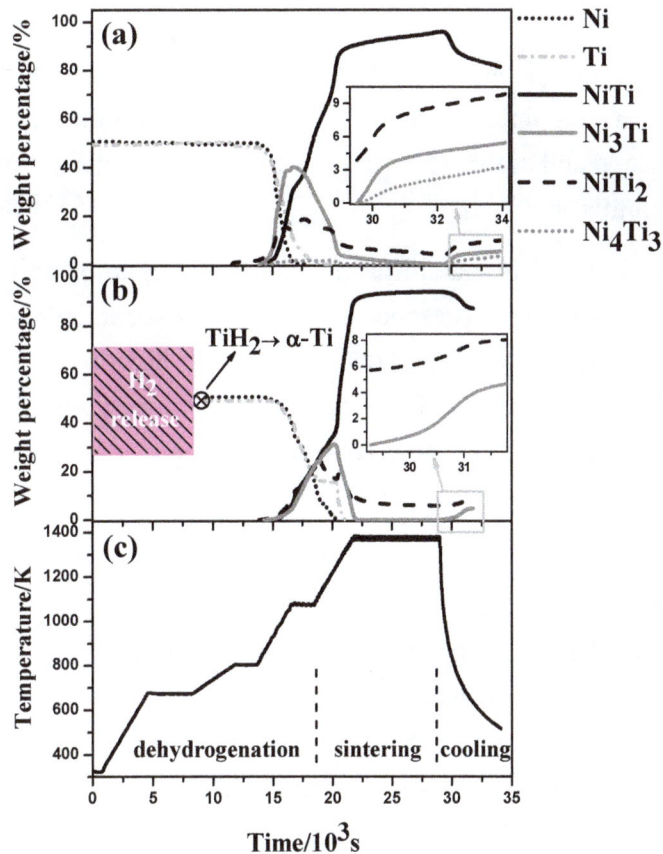

Figure 9. Weight fractions of the detected phases as a function of time (temperature) during *in situ* scan as achieved by Rietveld refinement analysis upon heating and cooling (**a**) the Ni/Ti compact, (**b**) the Ni/TiH$_2$ compact, and (**c**) heating and cooling profile as a function of time.

On the other hand, compressive tests show the typical superelasticity properties of sintered NiTi alloys, which are attributed to the stress-induced martensitic transformation [2]. With increasing the cycle number, the accumulated residual strain increases and then levels off to a constant value (Figure 7). This phenomenon has been discussed regarding to the general shape memory "training process" [20,61]. The Young's modulus of the Ni/Ti sintered sample is greater than that of the Ni/TiH$_2$ sintered sample (Table 2). First, as shown in Table 1 the close-to-total porosity ratio is 89.6% ± 3.4% and 12.2% ± 0.8% for the Ni/Ti and Ni/TiH$_2$ sintered compacts, respectively. Normally, higher ratio of close-to-total porosity would give rise to higher elastic modulus [60,62]. Second, the higher density of the Ni/Ti sintered sample would result in higher elastic modulus than did the Ni/TiH$_2$ sintered sample after 1373 K sintering as shown in Table 1. Additionally, it should be noted that the final phases present in the sintered compacts also affect the elastic modulus. Recalling Figure 9 that the Ni/TiH$_2$ sintered compact contains 8.0 wt.% NiTi$_2$ phase while the Ni/Ti sintered sample has 9.8 wt.% NiTi$_2$. More amount of NiTi$_2$ phase also causes higher elastic modulus for the Ni/Ti sintered sample [20].

5. Summary

In this report, porous NiTi alloys from Ni/Ti and Ni/TiH₂ powder compacts were produced by introducing a conventional press-and-sinter method. The microstructure and mechanical properties of sintered samples were investigated and compared with involving the use of TiH_2 powder. The following conclusions can be drawn from this study.

(1) B2 NiTi phase is the dominant phase identified in both samples after being sintered at 1373 K holding for two hours together with the presence of some minor secondary phases.
(2) Dehydrogenation from TiH_2 leads to a lower density, a much higher porosity, a larger pore size but higher final chemical homogenization after sintering as compared with the Ni/Ti compact.
(3) The use of TiH_2 powder causes lower fracture strength and lower elastic modulus compared with the Ni/Ti sintered sample.

Acknowledgments

We acknowledge the financial support from Ministry of Business Innovation and Employment (MBIE), New Zealand. Gang Chen thanks the China Scholarship Council (CSC) for providing him a doctoral scholarship. We also acknowledge the support of the Bragg Institute, Australian Nuclear Science and Technology Organization (ANSTO), in providing the neutron research facilities used in this work. The authors would like to thank Australian Institute of Nuclear Science and Engineering (AINSE) Ltd for providing financial assistance (award No. P2716) to enable work on WOMBAT to be conducted. The authors also appreciate the funding from Shaanxi Science and Technology Co-ordination and Innovation Project (2014KTZB01-02-04).

Conflicts of Interest

The authors declare no conflict of interest.

References

1. Duering, T.W.; Pelton, A.R. *Materials Properties Handbook: Titanium Alloys*; ASM International, the Materials Information Society: Materials Park, OH, USA, 1994.
2. Yamauchi, K.; Ohkata, I.; Tsuchiya, K.; Miyazaki, S. *Shape Memory and Superelastic Alloys: Technologies and Applications*; Woodhead Publishing: Cambridge, UK, 2011; p. 390.
3. Elahinia, M.H.; Hashemi, M.; Tabesh, M.; Bhaduri, S.B. Manufacturing and processing of NiTi implants: A review. *Prog. Mater. Sci.* **2012**, *57*, 911–946.
4. Chen, G.; Cao, P.; Wen, G.; Edmonds, N.; Li, Y. Using an agar-based binder to produce porous NiTi alloys by metal injection moulding. *Intermetallics* **2013**, *37*, 92–99.
5. Whitney, M.; Corbin, S.F.; Gorbet, R.B. Investigation of the mechanisms of reactive sintering and combustion synthesis of NiTi using differential scanning calorimetry and microstructural analysis. *Acta Mater.* **2008**, *56*, 559–570.
6. Sadrnezhaad, S.K.; Hosseini, S.A. Fabrication of porous NiTi-shape memory alloy objects by partially hydrided titanium powder for biomedical applications. *Mater. Des.* **2009**, *30*, 4483–4487.

7. Tosun, G.; Ozler, L.; Kaya, M.; Orhan, N. A study on microstructure and porosity of NiTi alloy implants produced by SHS. *J. Alloys Compd.* **2009**, *487*, 605–611.

8. Whitney, M.; Corbin, S.F.; Gorbet, R.B. Investigation of the influence of Ni powder size on microstructural evolution and the thermal explosion combustion synthesis of NiTi. *Intermetallics* **2009**, *17*, 894–906.

9. Liu, X.; Wu, S.; Yeung, K.W.K.; Xu, Z.S.; Chung, C.Y.; Chu, P. Superelastic porous NiTi with adjustable porosities synthesized by powder metallurgical method. *J. Mater. Eng. Perform.* **2012**, *21*, 2553–2558.

10. Chen, G.; Cao, P.; Edmonds, N. Porous NiTi alloys produced by press-and-sinter from Ni/Ti and Ni/TiH2 mixtures. *Mater. Sci. Eng. A* **2013**, *582*, 117–125.

11. Li, B.-Y.; Rong, L.-J.; Li, Y.-Y. Stress–strain behavior of porous Ni-Ti shape memory intermetallics synthesized from powder sintering. *Intermetallics* **2000**, *8*, 643–646.

12. Li, B.-Y.; Rong, L.-J.; Li, Y.-Y. The influence of addition of TiH2 in elemental powder sintering porous Ni-Ti alloys. *Mater. Sci. Eng. A* **2000**, *281*, 169–175.

13. Bertheville, B.; Neudenberger, M.; Bidaux, J.E. Powder sintering and shape-memory behaviour of NiTi compacts synthesized from Ni and TiH2. *Mater. Sci. Eng. A* **2004**, *384*, 143–150.

14. Robertson, I.M.; Schaffer, G.B. Swelling during sintering of titanium alloys based on titanium hydride powder. *Powder Metall.* **2010**, *53*, 27–33.

15. Wu, S.; Liu, X.; Yeung, K.W.K.; Hu, T.; Xu, Z.; Chung, J.C.Y.; Chu, P.K. Hydrogen release from titanium hydride in foaming of orthopedic NiTi scaffolds. *Acta Biomater.* **2011**, *7*, 1387–1397.

16. Chen, G.; Wen, G.A.; Cao, P.; Edmonds, N.; Li, Y.M. Processing and characterisation of porous NiTi alloy produced by metal injection moulding. *Powder Injection Moulding Int.* **2012**, *6*, 83–88.

17. Chen, G.; Cao, P. NiTi powder sintering from TiH2 powder: An *in situ* investigation. *Metall. Mater. Trans. A* **2013**, 1–4.

18. Wang, H.; Fang, Z.Z.; Sun, P. A critical review of mechanical properties of powder metallurgy titanium. *Int. J. Powder Metall.* **2010**, *46*, 45–57.

19. Wang, H.T.; Lefler, M.; Fang, Z.Z.; Lei, T.; Fang, S.M.; Zhang, J.M.; Zhao, Q. Titanium and titanium alloy via sintering of TiH2. *Key Eng. Mater.* **2010**, *436*, 157–163.

20. Chen, G.; Liss, K.-D.; Cao, P. *In situ* observation and neutron diffraction of NiTi powder sintering. *Acta Mater.* **2014**, *67*, 32–44.

21. Chen, G. Powder Metallurgical Titanium Alloys (TiNi and Ti-6Al-4V): Injection Moulding, Press-and-Sinter, and Hot Pressing. Ph.D. Thesis, The University of Auckland, Auckland, New Zealand, 2014.

22. Azevedo, C.R.F.; Rodrigues, D.; Beneduce Neto, F. Ti-Al-V powder metallurgy (PM) via the hydrogenation–dehydrogenation (HDH) process. *J. Alloys Compd.* **2003**, *353*, 217–227.

23. Robertson, I.M.; Schaffer, G.B. Comparison of sintering of titanium and titanium hydride powders. *Powder Metall.* **2010**, *53*, 12–19.

24. Ivasishin, O.M.; Eylon, D.; Bondarchuk, V.I.; Savvakin, D.G. Diffusion during powder metallurgy synthesis of titanium alloys. *Defect Diffus. Forum* **2008**, *277*, 177–185.

25. Zhang, J.M.; Yi, J.H.; Gan, G.Y.; Yan, J.K.; Du, J.H.; Liu, Y.C. Research on dehydrogenation and sintering process of titanium hydride for manufacture titanium and titanium alloy. *Adv. Mater. Res.* **2013**, *616–618*, 1823–1829.

26. Ivasishin, O.M.; Savvakin, D.G.; Froes, F.; Mokson, V.C.; Bondareva, K.A. Synthesis of alloy Ti-6Al-4V with low residual porosity by a powder metallurgy method. *Powder Metall. Metal Ceram.* **2002**, *41*, 382–390.

27. Bhosle, V.; Baburaj, E.G.; Miranova, M.; Salama, K. Dehydrogenation of nanocrystalline TiH_2 and consequent consolidation to form dense Ti. *Metall. Mater. Trans. A* **2003**, *34*, 2793–2799.

28. Li, B.Y.; Rong, L.J.; Li, Y.Y. Porous NiTi alloy prepared from elemental powder sintering. *J. Mater. Res.* **1998**, *13*, 2847–2851.

29. Li, B.-Y.; Rong, L.-J.; Li, Y.-Y.; Gjunter, V.E. An investigation of the synthesis of Ti-50 at. Pct Ni alloys through combustion synthesis and conventional powder sintering. *Metall. Mater. Trans. A* **2000**, *31*, 1867–1871.

30. Bhosle, V.; Baburaj, E.G.; Miranova, M.; Salama, K. Dehydrogenation of TiH_2. *Mater. Sci. Eng. A* **2003**, *356*, 190–199.

31. Sandim, H.R.Z.; Morante, B.V.; Suzuki, P.A. Kinetics of thermal decomposition of titanium hydride powder using *in situ* high-temperature X-ray diffraction (HTXRD). *Mater. Res.* **2005**, *8*, 293–297.

32. Liu, H.; He, P.; Feng, J.C.; Cao, J. Kinetic study on nonisothermal dehydrogenation of TiH_2 powders. *Int. J. Hydrog. Energy* **2009**, *34*, 3018–3025.

33. Jiménez, C.; Garcia-Moreno, F.; Pfretzschner, B.; Klaus, M.; Wollgarten, M.; Zizak, I.; Schumacher, G.; Tovar, M.; Banhart, J. Decomposition of TiH_2 studied *in situ* by synchrotron X-ray and neutron diffraction. *Acta Mater.* **2011**, *59*, 6318–6330.

34. Farhana, H.N.; Wang, Y.; Noor, M.M.; Chan, S.I. Static X-ray scans on the titanium hydride (TiH_2) powder during dehydrogenation. *Adv. Mater. Res.* **2013**, *795*, 124–127.

35. Jiménez, C.; Garcia-Moreno, F.; Pfretzschner, B.; Kamm, P.H.; Neu, T.R.; Klaus, M.; Genzel, C.; Hilger, A.; Manke, I.; Banhart, J. Metal foaming studied *in situ* by energy dispersive X-ray diffraction of synchrotron radiation, X-ray radioscopy, and optical expandometry. *Adv. Eng. Mater.* **2013**, *15*, 141–148.

36. Chen, G.; Liss, K.-D.; Cao, P. *In situ* observation of phase transformation of powder sintering from Ni/TiH_2 using neutron diffraction. In *TMS 2014 Supplemental Proceedings*; John Wiley & Sons, Inc.: Hoboken, NJ, USA, 2014; pp. 967–973.

37. Studer, A.J.; Hagen, M.E.; Noakes, T.J. Wombat: The high-intensity powder diffractometer at the opal reactor. *Phys. B Condens. Matter* **2006**, *385–386*, 1013–1015.

38. Yu, J.; Hu, X.; Huang, Y. A modification of the bubble-point method to determine the pore-mouth size distribution of porous materials. *Sep. Purif. Technol.* **2010**, *70*, 314–319.

39. Viswanathan, B.; Murthy, S.S.; Sastri, M.V.C. *Metal Hydrides: Fundamentals and Applications*, 1st ed.; Springer: Berlin, Germany, 1999; p. 189.

40. Duwez, P.; Taylor, J.L. The structure of intermediate phases in alloys of titanium with iron, cobalt, and nickel. *Trans. AIME* **1950**, *188*, 1173–1176.

41. Poole, D.M.; Hume-Rothery, W. The equilibrium diagram of the system nickel-titanium. *J. Inst. Metals* **1954**, *83*, 473–480.

42. Gupta, S.P.; Mukherjee, K.; Johnson, A.A. Diffusion controlled solid state transformation in the near-equiatomic Ti-Ni alloys. *Mater. Sci. Eng.* **1973**, *11*, 283–297.

43. Liss, K.-D.; Bartels, A.; Schreyer, A.; Clemens, H. High-energy X-rays: A tool for advanced bulk investigations in materials science and physics. *Textures Microstruct.* **2003**, *35*, 219–252.

44. Predel, B. H-Ti (Hydrogen-Titanium). In *Ga-Gd-Hf-Zr*; Madelung, O., Ed.; Springer: Berlin/Heidelberg, Germany, 1996; Volume 5f, pp. 1–2.

45. Greiner, C.; Oppenheimer, S.M.; Dunand, D.C. High strength, low stiffness, porous NiTi with superelastic properties. *Acta Biomater.* **2005**, *1*, 705–716.

46. Mandrino, D.; Paulin, I.; Škapin, S.D. Scanning electron microscopy, X-ray diffraction and thermal analysis study of the TiH2 foaming agent. *Mater. Charact.* **2012**, *72*, 87–93.

47. Paulin, I.; Donik, Č.; Mandrino, D.; Vončina, M.; Jenko, M. Surface characterization of titanium hydride powder. *Vacuum* **2012**, *86*, 608–613.

48. Okamoto, H. H-Ti (Hydrogen-Titanium). *J. Phase Equilib. Diffus.* **2011**, *32*, 174–175.

49. Fukai, Y. *The Metal-Hydrogen System, Basic Bulk Properties*; 2nd ed.; Springer: Berlin/Heidelberg, Germany, 2005; p. 497.

50. Igharo, M.; Wood, I.V. Compaction and sintering phenomena in titanium-nickel shape memory alloys. *Powder Metall.* **1985**, *28*, 131–139.

51. Biswas, A. Porous NiTi by thermal explosion mode of SHS: Processing, mechanism and generation of single phase microstructure. *Acta Mater.* **2005**, *53*, 1415–1425.

52. Otsuka, K.; Ren, X. Physical metallurgy of Ti-Ni-based shape memory alloys. *Prog. Mater. Sci.* **2005**, *50*, 511–678.

53. Laeng, J.; Xiu, Z.; Xu, X.; Sun, X.; Ru, H.; Liu, Y. Phase formation of Ni–Ti via solid state reaction. *Phys. Scr.* **2007**, *2007*, 250.

54. Bansiddhi, A.; Dunand, D.C. Shape-memory NiTi foams produced by replication of NaCl space-holders. *Acta Biomater.* **2008**, *4*, 1996–2007.

55. Li, H.; Yuan, B.; Gao, Y.; Chung, C.Y.; Zhu, M. High-porosity NiTi superelastic alloys fabricated by low-pressure sintering using titanium hydride as pore-forming agent. *J. Mater. Sci.* **2009**, *44*, 875–881.

56. Wen, C.E.; Xiong, J.Y.; Li, Y.C.; Hodgson, P.D. Porous shape memory alloy scaffolds for biomedical applications: A review. *Phys. Scr.* **2010**, *2010*, 014070.

57. German, R.M. *Powder Metallurgy Science*; Metal Powder Industries Federation: Princeton, NJ, USA, 1998.

58. German, R.; Suri, P.; Park, S. Review: Liquid phase sintering. *J. Mater. Sci.* **2009**, *44*, 1–39.

59. Ashby, M.F.; Evans, A.; Fleck, N.A.; Gibson, L.J.; Hutchinson, J.W.; Wadley, H. *Metal Foams: A Design Guide*; Butterworth-Heinemann: Boston, MA, USA, 2000.

60. Anderson, T.L. *Fracture Mechanics Fundamentals and Applications*, 3rd ed.; CRC Press: Boca Raton, FL, USA, 2005.

61. Nemat-Nasser, S.; Guo, W.-G. Superelastic and cyclic response of NiTi SMA at various strain rates and temperatures. *Mech. Mater.* **2006**, *38*, 463–474.

62. Gibson, L.J.; Ashby, M.F. *Cellular Solids: Structure and Properties*, 2nd ed.; Cambridge University Press: Cambridge, UK, 1999.

8

"High-Throughput" Evaluation of Polymer-Supported Triazolic Appendages for Metallic Cations Extraction

Riadh Slimi and Christian Girard *

Unité de Technologies Chimique et Biologiques pour la Santé (UTCBS), UMR 8258 CNRS, U 1022 Inserm, Ecole Nationale Supérieure de Chimie de Paris (Chimie ParisTech)/PSL Research University, 11 rue Pierre et Marie Curie, 75005 Paris, France; E-Mail: riadhslimi82@yahoo.fr

* Author to whom correspondence should be addressed; E-Mail: christian.girard@chimie-paristech.fr

Academic Editor: Hugo F. Lopez

Abstract: The aim of this work was to find and use a low-cost high-throughput method for a quick primary evaluation of several metal extraction by substituted piperazines appendages as chelatants grafted onto Merrifield polymer using click-chemistry by the copper (I)-catalyzed Huisgen's reaction (CuAAC) The polymers were tested for their efficiency to remove various metal ions from neutral aqueous solutions (13 cations studied: Li^+, Na^+, K^+, Mn^{2+}, Fe^{3+}, Co^{2+}, Ni^{2+}, Cu^{2+}, Cd^{2+}, Ba^{2+}, Ce^{3+}, Hg^+ and Pb^{2+}) using the simple conductimetric measurement method. The polymers were found to extract all metals with low efficiencies ≤40%), except for Fe^{3+} and Hg^+, and sometimes Pb^{2+}. Some polymers exhibited a selectively for K^+, Cd^{2+} and Ba^{2+}, with good efficiencies. The values obtained here using less polymer, and a faster method, are in fair correspondence (average difference ±16%) with another published evaluation by atomic absorption spectroscopy (AAS).

Keywords: polymer functionalization; click chemistry; CuAAC; complexants; metallic cations; complexation; depollution; catalysis

1. Introduction

Water pollution by metallic ions and other pollutants is becoming an increasing concern nowadays. This modification of the water, in all his reservoirs, is mainly due to Human activities with uncontrolled rejects of such pollutants. The pollution has a strong impact onto the global ecosystem as well as drinkable water sources. There is thus a strong need for methods to analyze traces and to remove the pollutants from water. Usual methods for removing metallic salts from water range from distillation to the use of engineered materials such as zeolites, polymers, membranes, *etc*. The long time known ion-exchange resins can be used for this purpose [1–3]. Usually, the polymers are engineered is such a way that their nature can be hydrophilic [4–7] or phobic [8,9], to meet the requirements for their use [10–14]. Many polymers have been designed in order to include chelatants to fix metal ions to be used in applications such as purification, depollution or catalysis [15–20].

Due to our interest into metal chelation, supported catalysis and Huisgen's reaction, we became interested into the preparation of polymers based on this approach [21–23]. We thus started to use the "click-chemistry" concept for polymer functionalization and especially copper (I)-catalyzed Huisgen's cycloaddition ("copper (I)-catalyzed azide/alkyne cycloaddition" or CuAAC) [24–29]. The use of CuAAC has the advantage to give a quick access of controlled substitutions onto the polymer by the use of its azided version and alkynes with various substituents [30–35]. This CuAAC is linking the azided polymer and the substituents bearing alkyne by forming the 1,4-triazole linkage. All the introduced substituents and the triazole can be implicated into chelation through a "triazole design", or a "pendant design", or both parts implicated into an "integrated design". The chelation can be a mono- or multi-dentate mode due to the vicinity of other chelating entities and the flexible structure of the polymer chains. (Figure 1) [34,35].

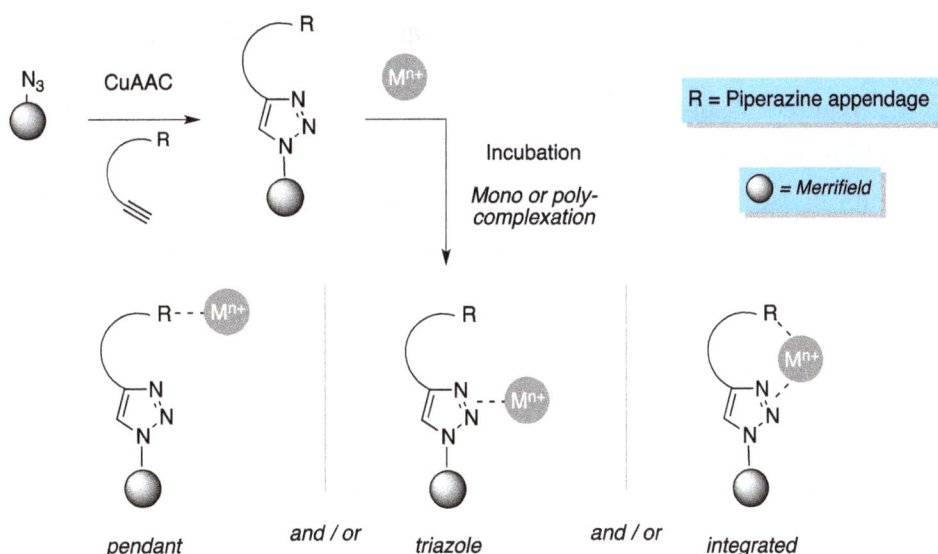

Figure 1. Functionalization of azido-Merrifield polymer by complexing appendages using copper (I)-catalyzed azide/alkyne cycloaddition (CuAAC) for a 1,2,3-triazole linkages and possible chelation modes.

The goal of this work was to try to find a faster method, using less polymer, for the evaluation of several metal cations complexation evaluation. We present in this communication the use of the less sensitive conductimetric method for the study of piperazine-triazole-substituted poly(styrenes). The polymers were tested for their ability to extract metal cations salts (Li$^+$, Na$^+$, K$^+$, Mn^{2+}, Fe^{3+}, Co^{2+}, Ni^{2+}, Cu^{2+}, Cd^{2+}, Ba^{2+}, Ce^{3+}, Hg$^+$ and Pb^{2+}) from neutral aqueous solutions. The results were found to be fair enough to be used for a primary evaluation at a "high-throughput" level when compared to our previous atomic absorption spectroscopy measurements (within ±16% average difference) [36].

2. Experimental Section

2.1. N-Substituted Piperazine Propargylcarbamates and Polymers

Poly(azidomethylstyrene) was prepared from Merrifield polymer as already reported. N'-propargylcarbamates of N-substituted piperazine, and the corresponding polymers containing triazole-linked piperazines preparations were described in a another publication as depicted in Figure 2 [36,37]. Typical procedures are indicated below.

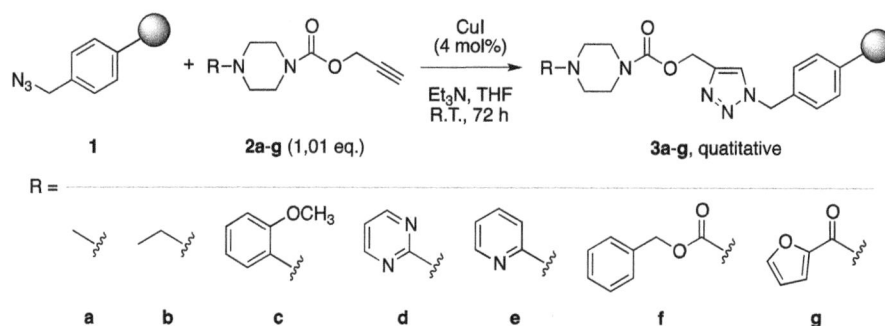

Figure 2. Synthesis of the polymers by CuAAC procedure.

2.1.1. General Procedure for the Synthesis of Propargylcarbamates Derivatives of Piperazines 2a–2g

To a solution of the required N-substituted piperazine **1a** to **1g** (12.0 mmol) in acetonitrile (45 mL) was added Na$_2$CO$_3$ (1.27 g, 12.0 mmol, 1 eq.). Propargyl chloroformate (1.42 g, 1.17 mL, 12.0 mmol, 1 eq.) was then added dropwise. The reaction mixture was stirred for 48 h at room temperature and then filtered and evaporated under vacuum. The resulting carbamates **2a–2g** were sufficiently pure to be used without further purification.

2.1.2. General procedure for the Synthesis of Polymers 3a–3g

Coupling reactions onto the polymer using CuAAC were conducted accordingly to the general procedure indicated below in round bottom flasks equipped with a reflux condenser. To a suspension of 3.00 g of azidomethyl polystyrene **A** (1.82 mmol N$_3$ g^{-1}, 5.46 mmol N$_3$) in THF (60 mL) was added 6.30 mmol (1.15 eq.) of the alkyne (**2a–2g**), 9.00 mL of triethylamine (6.75 g, 66,7 mmol, 12.2 eq.) and 2.40 mg of copper (I) iodide (12.6 μmol, 4 mol%). The suspension was slowly stirred at room temperature 72 h. After this time, the complete disappearance of the IR band of the azide of the polymer **A** (2103 cm^{-1}) was observed. The resulting polymer was filtered on sintered glass and washed sequentially with CH$_2$Cl$_2$, pyridine, and MeOH (60 mL each), the sequential washings being repeated two other times. The resulting polymers **3a–3g** were finally dried overnight in an oven at 50 °C.

2.2. Conductimetric Quick Primary Evaluation of Metals Complexation

2.2.1. Extraction

Neutral aqueous solutions of metal salts were prepared with 50 mg of LiCl (1.179 mmol), NaCl (0.856 mmol), KCl (0.671 mmol), $MnCl_2 \cdot 4H_2O$ (0.253 mmol), $Fe(NO_3)_3 \cdot 9H_2O$ (0.124 mmol), $CoCl_2 \cdot 6H_2O$ (0.210 mmol), $NiCl_2 \cdot 6H_2O$ (0.210 mmol), $CuCl_2 \cdot 2H_2O$ (0.293 mmol), $CdCl_2 \cdot H_2O$ (0.248 mmol), $BaCl_2 \cdot 2H_2O$ (0.205 mmol), $CeCl_3 \cdot 7H_2O$ (0.134 mmol), $HgNO_3$ (0.190 mmol) and $Pb(NO_3)_2$ (0.151 mmol) in 1 L of distilled water.

Aliquots of the each polymers (100 mg; **3a**: 0.137 mmol, **3b**: 0.134 mmol, **3c**: 0.121 mmol, **3d**: 0.126 mmol, **3e**: 0.126 mmol, **3f**: 0.117 mmol and **3g**: 0.123 mmol piperazine) were incubated in triplicate with 20 mL (50 mg L^{-1}, equals to 1 mg of salt) of each metal ion solution (23.6 µmol Li^+, 17.1 µmol Na^+, 13.4 µmol K^+, 5.05 µmol Mn^{2+}, 2.47 µmol Fe^{3+}, 4.20 µmol Co^{2+}, 4.21 µmol Ni^{2+}, 5.87 µmol Cu^{2+}, 4.97 µmol Cd^{2+}, 4.09 µmol Ba^{2+}, 2.68 µmol Ce^{3+}, 3.81 µmol Hg^+ and 3.02 µmol Pb^{2+}) at 25 °C for 24 h. The suspension was then filtrated on filter paper, which was previously washed with distilled water until no difference in conductimetry was observed between the washes and water.

2.2.2. Conductimetric Measurements

Evaluation of the chelated metal was done by conductimetric measurements on a bench Conductivity/TDS/°C Meter CO 3000 L, pHenomenal® by VWR (Paris, France) on the filtrate by comparison with the conduction of the initial solution of the metal salt. The results, average of three experiments, were expressed as percentages of extraction of the metal (Figures 3 (below) and 4, Section 3.2).

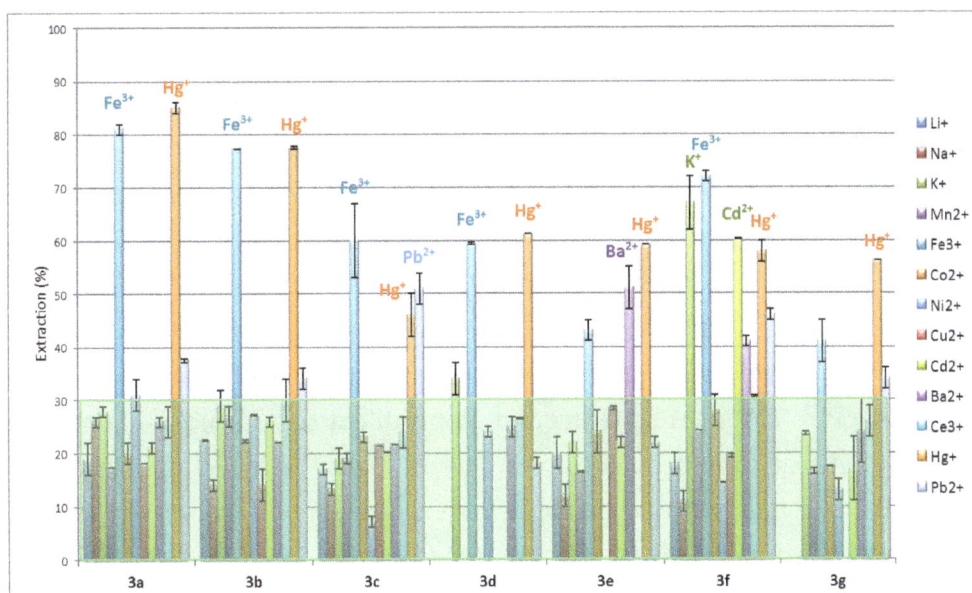

Figure 3. Extraction efficiencies of metal cations by the triazolic piperazine polymers **3a–3g** (cations extracted at levels ≥50% are indicated).

3. Results and Discussion

3.1. Polymers Preparation

The polymers needed for this study have been prepared elsewhere [36]. They were easily accessible, in different substitution motifs, using poly(azidomethylstyrene) (**1**) and various *N*-substituted-piperazine-*N'*-propargylcarbamates (**2a–2g**). A CuAAC procedure afforded a quantitative reaction to form the triazole linkage onto the polymers **3a–3g** (Figure 2, Section 2.1) [36,37].

3.2. Extraction Results

After 24 h incubation on a 100 mg scale of the polymers in 20 mL of 50 mg L^{-1} solutions of the salts (1 mg of salt, see Section 2.2.1 for details), the percentages of extraction for each metal were calculated by conductimetric differences between the initial and final solutions, each experiment having been carried out in triplicate. The results for each polymer as a function of the metallic cations are presented in Figure 3 (Section 2.2.2).

When looking at the whole results, we can observe than most of the cations were poorly extracted, at 30% or below, but exceptions. Since the polymeric structure differs only by the R substituent on the nitrogen (Figure 2, Section 2.1), an analysis has been done to understand the influence of the substituent's nature on the extraction.

In the alkyl series, polymer **3a**, R = Me, was a very good extractant for Fe^{3+} (81% ± 1%) and Hg^+ (85% ± 1%), less for Pb^{2+} (37.5% ± 0.3%), and not very good for other metals (≤30%). When changing R for Et, in polymer **3b**, the same extractive properties were found, only at a little lower level, for Fe^{3+} (77.35% ± 0.05%) and Hg^+ (77.5% ± 0.3%), less for Pb^{2+} (34% ± 2%), the other metals being around or below 30%, as for the previous one.

When entering the aryl series, an obvious decrease in the extraction potentials was observed, alongside some other interesting behaviors. For the uncited cations in the following paragraph, they were extracted ≤30%. For R = 4-methoxyphenyl in **3c**, Fe^{3+} (60% ± 7%) and Hg^+ (46% ± 4%) were less extracted. However, Pb^{2+} (51% ± 3%) was better extracted. By replacing the phenyl group by a nitrogen containing aromatic one, like in the 2-pyrimidyl (**3d**), Fe^{3+} (59.4% ± 0.2%) was extracted at the same level, lower than the alkyl family (**3a**, **3b**). Hg^+ (61.25% ± 0.05%) was a little less extracted than by *N*-alkyl substituted polymers **3a** and **3b** but better than the 4-methoxyphenyl polymer **3c**. In the case of polymer **3d**, Pb^{2+} (18% ± 1%) extraction dropped, but K^+ (34% ± 3%) uptake increased. It is interesting to point out that some metals were not at all extracted by **3d**: Li^+, Na^+, Mn^{2+}, Co^{2+}, Cu^{2+} and Cd^{2+}. When the substituent borne by the nitrogen was 2-pyridyl (**3e**), the extraction level dropped for Fe^{3+} (43% ± 2%), but stayed similar for Hg^+ (59.19% ± 0.04%) and Pb^{2+} (22% ± 1%). Interestingly, another metal was extracted, Ba^{2+} (51% ± 4%), while Ni^{2+} and Ce^{3+} were not extracted.

By changing the nature of the substituent by introducing a carbamate function in **3f** (R = C(O)OCH₂Ph), the polymer became less selective. This polymer (**3f**) still extracted Fe^{3+} (72% ± 1%), but at the same level as the *N*-alkyl substituted ones (**3a**, **3b**), it extracted Pb^{2+} (46% ± 1%) at the level of the 4-methoxyphenyl polymer **3c**, and stayed in the same range as the previous one for Hg^+ (58% ± 2%) and Ba^{2+} (41% ± 1%). Furthermore, the polymer **3f** was a good extractant for K^+ (67% ± 5%) and Cd^{2+} (60.3% ± 0.1%).

Finally, when the piperazine nitrogen onto the polymer was an amide of 2-furoic acid (**3g**), The extraction behavior came back to the triade Fe^{3+} (41% ± 4%), Hg^+ (56.23% ± 0.04%), and Pb^{2+} (34% ± 2%). The levels were similar as the ones of **3e**. Once again some cations were not extracted like Li^+, Na^+, and Cu^{2+}.

From this first analysis, based only on the difference of the R group, it seems that the presence of an alkyl substituent (**3a**, R = Me, **3b**, R = Et) is giving the best extraction levels for Fe^{3+} and Hg^+, probably due to the increased electronegativity of the amine. Inductive and steric hindrance effects can explain the differences between **3a** and **3b**. The introduction of a 2-methoxyphenyl group on the nitrogen (**3c**) changes it to an aniline, less basic, which is extracting less Fe^{3+} and Hg^+. However, the presence of the methoxy group in *ortho* position seems to help in Pb^{2+} extraction. It is possible that this oxygenated group is implicated into the chelation of this metal. Replacement of this aromatic by a 2-pyrimidyl (**3d**) and 2-pyridyl (**3e**), less and more basic respectively when compared one to the other, still gives polymers capable of extracting Fe^{3+} and Hg^+, with similar levels as **3c**. Special features of these polymers are higher extraction of K^+ for **3d** and Ba^{2+} for **3e**, as well as exclusion of some cations: Li^+, Na^+, Mn^{2+}, Co^{2+}, Cu^{2+} and Cd^{2+} for **3d**, and Ni^{2+} and Ce^{3+} for **3e**.

The electronegativities and chelating capabilities of these amino-R groups are however difficult to put in relation with their extracting properties. The most puzzling effect is the presence of a benzylcarbamate onto the nitrogen of the polymer (**3f**). This group is totally modifying the properties of the polymer. In this case, more metallic ions are extracted, with the classical Fe^{3+} and Hg^+. This includes K^+, Cd^{2+}, Ba^{2+}, and Pb^{2+}. This may suggest another chelation mode introduced by the presence of the carbamate. Finally, the presence of a derivative of furoic acid as an amide on the nitrogen (**3g**) do not seems to helps since extraction levels are going down with extraction of the usual Fe^{3+}, Hg^+, and Pb^{2+}, and exclusion of Li^+, Na^+, Cu^{2+}.

The electronegativities of the substituted nitrogen of the piperazine can in part explain some of the relative extraction efficiencies. However, it is difficult to draw a clear conclusion. We have also tried to rationalize the interactions between the polymers and the metal ions based on their electropositivities, ionic radii and water solvatation. Once again, no clear link can be drawn about the extraction efficiencies based on the metal cation properties. The only difference that can explain, once again in part, the preference of the polymers for Fe^{3+} and Hg^+, and in some cases Pb^{2+}, is the counter-ion of the salt used for the study. All metallic salts were chlorides except for Fe^{3+}, Hg^+, and Pb^{2+}, which were used as their nitrates.

The final analysis we have tried to make is to try to find out the chelation mode from the piperazine-triazole in relation with the metal cations. For the best extraction results, over 40%, it was not possible to make a clear discrimination between the three modes (pendant, triazole and integrated). The ratios piperazine-triazole:metal cation varied from 13:1 (**3f** and K^+, 67% ± 5%) to 118:1 (**3e** and Fe^{3+}, 43% ± 2%), since the chelating moieties were in large excess. This cannot gives a clear hit on the chelation mode, which can be of polydentate type or simply a statistical repartition on the chelation sites, without knowledge of the chelation type.

All the results and analyses cannot clearly identify the discrete complexation behavior of the metal by the polymers at the solid/liquid interface. However, we were able to obtain better results than with our first series of triazolic polymers based on propargyl amides and propiolic anilides onto poly(styrene) [38].

Figure 4 presents our results by cation absorption to help to find the best extractant. By drawing a cut-off for selection at 40% extraction level, it is clear that none of the polymers is very efficient for the removal of Li^+, Na^+, K^+, Co^{2+}, Ni^{2+}, Cu^{2+} and Ce^{3+}. For K^+, polymer 3f is the best with 3.51 mg K^+ g^{-1}.

- In the case of Fe^{3+}, as said before, all polymers are complexing this ion. The best results are 1.12 mg Fe^{3+} g^{-1} 3a, 1.07 mg Fe^{3+} g^{-1} 3b and 1.00 mg Fe^{3+} g^{-1} 3f; followed by 0.83 mg Fe^{3+} g^{-1} 3c and 0.82 mg Fe^{3+} g^{-1} 3d, while 3e and 3g were borderline.
- Cd^{2+} and Ba^{2+} are more efficiently removed by 3f (3.37 mg Cd^{2+} g^{-1}, 2.31 mg Ba^{2+} g^{-1}) and 3e (2.87 mg Ba^{2+} g^{-1}).
- For Hg^+, as for Fe^{3+}, all polymers can be used. In order, by sorption capacities, are 3a (6.49 mg Hg^+ g^{-1}), 3b (5.92 mg Hg^+ g^{-1}), 3d (4.68 mg Hg^+ g^{-1}), 3e (4.52 mg Hg^+ g^{-1}), 3f (4.43 mg Hg^+ g^{-1}), 3g (4.30 mg Hg^+ g^{-1}) and 3c (3.51 mg Hg^+ g^{-1}).
- Finally, in the case of Pb^{2+}, polymers 3c (3.19 mg Pb^{2+} g^{-1}) and 3f (2.88 mg Pb^{2+} g^{-1}) are the more efficient.

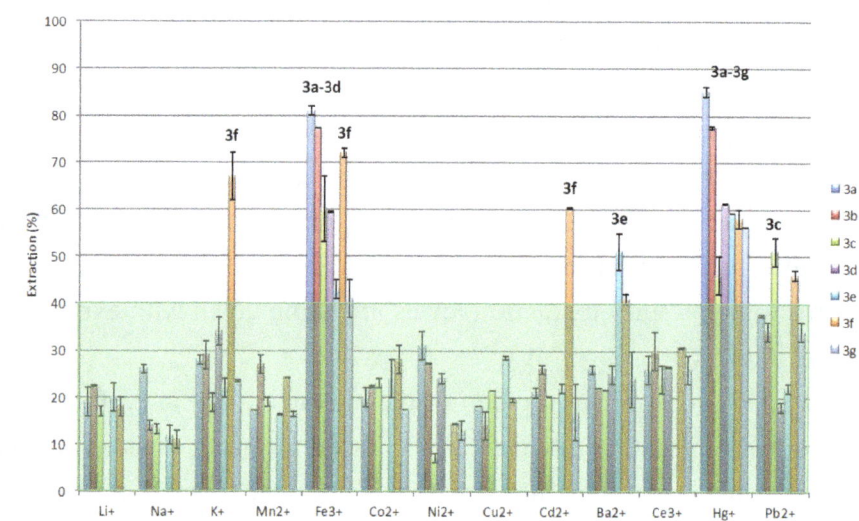

Figure 4. Extraction efficiencies of metal cations by the triazolic piperazine polymers 3a–3g (polymers extracting at levels ≥50% are indicated).

4. Conclusions

In this work, we studied chemical grafting of piperazine chelating units onto commercial poly[styrene] (Merrifield resin) using CuAAC procedure between the azided polymer and selected piperazine-*N*-propargylcarbamates. The synthesized polymers were characterized by FTIR. They were then tested for their efficiency to extract metallic ions from aqueous solution (Li^+, Na^+, K^+, Mn^{2+}, Fe^{3+}, Co^{2+}, Ni^{2+}, Cu^{2+}, Cd^{2+}, Ba^{2+}, Ce^{3+}, Hg^+ and Pb^{2+}). All polymers were found to extract most of the ions at low level (≤40%), with the exception of Fe^{3+}, Hg^+ and Pb^{2+}. Some polymers showed selectivity for K^+, Cd^{2+} and Ba^{2+}. Extraction efficiencies reached up 85%, with the highest sorption capacity at 6.49 mg Hg^+ g^{-1} of polymer (3a).

The conductimetric method used, even having a less precise reputation, was good enough to have a quick evaluation of several cations removal. This method has been selected both for its lower cost in apparatus when compared to AAS and ICP methods. It is also faster and easier to do the measurements in order to speed up the process to find the best and highly selective extractant for a range of engineered polymers.

Even if no clear interpretation can be done with the results for the interfacial chelation process, the good extraction properties encourage us to continue polymers modifications using CuAAC in order to find new polymeric complexants for depollution and catalytic applications. Further studies will be reported in due course.

Acknowledgments

This project was financed by Tunisian 05/UR/12-05 and French UMR 8258 CNRS/U1022 INSERM grants. R. S. is grateful to the Tunisian Ministry of Research and Education for fellowship.

Author Contributions

This is a part of the Ph.D. thesis of R. Slimi under the supervision of C. Girard. Experiments were conducted by R. Slimi. R. Slimi and C. Girard analyzed the data.

Conflicts of Interest

The authors declare no conflict of interest.

References

1. Rifi, E.H.; Leroy, M.J.F.; Brunette, J.P.; Schloesser-Becker, C. Extraction of copper, cadmium and related metals with poly(sodium acrylate–acrylic acid) hydrogels. *Solvent Extr. Ion Exch.* **1994**, *12*, 1003–1119.

2. Hodgkin, J.H.; Eibi, R. Gold extraction with poly(diallylamine) resins. *React. Polym. Ion Exch. Sorb.* **1988**, *9*, 285–291.

3. Rivas, B.L.; Klahenhoff, D.; Perich, I.M. Kinetics of uranium sorption from acidic sulphate solutions onto crosslinked polyethyleneimine based resins. 10. *Polym. Bull.* **1990**, *23*, 219–223.

4. Anspach, W.M.; Marinsky, J.A. Complexing of nickel(II) and cobalt(II) by a polymethacrylic acid gel and its linear polyelectrolyte analog. *J. Phys. Chem.* **1975**, *79*, 433–439.

5. Nichide, H.; Oki, N.; Suchida, E.T. Complexation of poly(acrylic acid)s with uranyl ion. *Eur. Polym. J.* **1982**, *18*, 799–802.

6. Pollack, G.H. Water, energy and life: Fresh views from the water's edge. *Int. J. Des. Nat. Ecodyn.* **2010**, *5*, 27–29.

7. Rifi, E.H.; Rastegar, F.; Brunette, J.P. Uptake of cesium, strontium and europium by a poly(sodium acrylate-acrylic acid) hydrogel. *Talanta* **1995**, *42*, 811–816.

8. Wang, X.; Weiss, R.A. A facile method for preparing sticky, hydrophobic polymer surfaces. *Langmuir* **2012**, *28*, 3298–3305.

9. Loret, J.F.; Brunette, J.P.; Leroy, J.F.M.; Candau, S.G.; Prevost, M. Liquid-lipophilic gel extraction of precious metals. *Solv. Extr. Ion Exch.* **1988**, *6*, 585–603.

10. Nghiem, L.D.; Mornane, P.; Potter, I.D.; Pereira, J.M.; Cattrall, R.W.; Kolev, S.D. Extraction and transport of metal ions and small organic compounds using polymer inclusion membranes (PIMs). *J. Membr. Sci.* **2006**, *281*, 7–41.

11. Peterson, J.; Nghiem, L.D. Selective extraction of cadmium by polymer inclusion membranes containing PVC and Aliquat 336: Role base polymer and extractant. *Int. J. Environ. Technol. Manag.* **2010**, *12*, 359–368.

12. Kebiche Senhadji, O.; Sahi, S.; Kahloul, N.; Tingry, S.; BenAmor, M.; Seta, P. Extraction du Cr(VI) par membrane polymère à inclusion. *Sci. Technol. A* **2008**, *27*, 43–50.

13. Guibaud, G.; Baudu, M.; Dollet, P.; Condat, M.L.; Dagot, C. Role of extracellular polymers in cadmium adsorption by activated sludges. *Environ. Technol.* **1999**, *20*, 1045–1054.

14. Upitis, A.; Peterson, J.; Lukey, C.; Nghiem, L.D. Metallic ion extraction using polymer inclusion membranes (PIMs): optimising physical strength and extraction rate. *Desalin. Water Treat.* **2009**, *6*, 41–47.

15. Zotti, G.; Zecchin, S.; Schiavon, G.; Berlin, A.; Penso, M. Ionochromic and potentiometric properties of the novel polyconjugated polymer from anodic coupling of 5,5'-Bis(3,4-(ethylenedioxy)thien-2-Yl)-2,2'-Bipyridine. *Chem. Mater.* **1999**, *11*, 3342–3351.

16. Brembilla, A.; Cuny, J.; Roizard, D.; Lochon, P. Un nouveau polymère catalyseur bifonctionnel: Le polyvinyl-5 (6)benzimidazoleméthanethiol Synthèse et catalyse de l'hydrolyse de l'acétate de *p*-nitrophényle. *Eur. Polym. J.* **1983**, *19*, 729–735.

17. Rhazi, M.; Desbrières, J.; Tolaimate, A.; Rinaudo, M.; Vottero, P.; Alagui, A.; El Meray, M. Influence of the nature of the metal ions on the complexation with chitosan: Application to the treatment of liquid waste. *Eur. Polym. J.* **2002**, *38*, 1523–1530.

18. Kozlowski, C.; Walkowiak, W. Applicability of liquid membranes in chromium(VI) transport with amines as ion carriers. *J. Membr. Sci.* **2005**, *266*, 143–150.

19. Kozlowski, C.; Apostoluk, W.; Walkowiak, W.; Kita, A. Removal of Cr(VI), Zn(II) and Cd(II) ions by transport across polymer inclusion membranes with basic ion carriers. *Physicochem. Probl. Min. Process.* **2002**, *36*, 115–122.

20. Toy, P.H.; Janda, K.D. Soluble polymer-supported organic synthesis. *Acc. Chem. Res.* **2000**, *33*, 546–554.

21. Girard, C.; Önen, E.; Aufort, M.; Beauvière, S.; Samson, E.; Herscovici, J. Reusable polymer-supported catalyst for the [3+2] Huisgen cycloaddition in automation protocols. *Org. Lett.* **2006**, *8*, 1689–1692.

22. Jlalia, I.; Meganem, F.; Herscovici, J.; Girard, C. "Flash" solvent-free synthesis of triazoles using a supported catalyst. *Molecules* **2009**, *14*, 528–539.

23. Jlalia, I.; Beauvineau, C.; Beauvière, S.; Önen, E.; Aufort, M.; Beauvineau, A.; Khaba, E.; Herscovici, J.; Meganem, F.; Girard, C. Automated synthesis of a 96 product-sized library of triazole derivatives using a solid phase supported copper catalyst. *Molecules* **2010**, *15*, 3087–3120.

24. Li, C.; Finn, M.G. Click chemistry in materials synthesis. II. Acid-swellable crosslinked polymers made by copper-catalyzed azide–alkyne cycloaddition. *J. Polym. Sci. Part A: Polym. Chem.* **2006**, *44*, 5513–5518.

25. Tasdelen, M.A.; Yilmaz, G.; Iskin, B.; Yagci, Y. Photoinduced free radical promoted copper(I)-catalyzed click chemistry for macromolecular syntheses. *Macromolecules* **2011**, *45*, 56–61.

26. Liu, Y.; Díaz, D.D.; Accurso, A.A.; Sharpless, K.B.; Fokin, V.V.; Finn, M.G. Click chemistry in materials synthesis. III. Metal-adhesive polymers from Cu(I)-catalyzed azide–alkyne cycloaddition. *J. Polym. Sci. Part A* **2007**, *45*, 5182–5189.

27. Dag, A.; Durmaz, H.; Demir, E.; Hizal, G.; Tunca, U. Heterograft copolymers via double click reactions using one-pot technique. *J. Polym. Sci. Part A: Polym. Chem.* **2008**, *46*, 6969–6977.

28. Opsteen, J.A.; Brinkhuis, R.P.; Teeuwen, R.L.M.; Lowik, D.W.P.M.; van Hest, J.C.M. "Clickable" polymersomes. *Chem. Commun.* **2007**, 3136–3138.

29. Riva, R.; Schmeits, S.; Jérôme, C.; Jérôme, R.; Lecomte, P. Combination of ring-opening polymerization and "click chemistry": Toward functionalization and grafting of poly(ε-caprolactone). *Macromolecules* **2007**, *40*, 796–803.

30. Löber, S.; Rodriguez-Loaiza, P.; Gmeiner, P. Click linker: Efficient and high-yielding synthesis of a new family of SPOS resins by 1,3-dipolar cycloaddition. *Org. Lett.* **2003**, *5*, 1753–1755.

31. Iskin, B.; Yilmaz, G.; Yagci, Y. ABC type miktoarm star copolymers through combination of controlled polymerization techniques with thiol-ene and azide-alkyne click reactions. *J. Polym. Sci. Part A: Polym. Chem.* **2011**, *49*, 2417–2422.

32. Xue, X.; Zhu, J.; Zhang, Z.; Cheng, Z.; Tu, Y.; Zhu, X. Synthesis and characterization of azobenzene-functionalized poly(styrene)-*b*-poly(vinyl acetate) via the combination of RAFT and "click" chemistry. *Polymer* **2010**, *51*, 3083–3090.

33. Shin, J.-A.; Lim, Y.-G.; Lee, K.-H. Synthesis of polymers including both triazole and tetrazole by click reaction. *Bull. Korean Chem. Soc.* **2011**, *32*, 547–552.

34. Struthers, H.; Mindt, T.L.; Schibli, R. Metal chelating systems synthesized using the copper(I) catalyzed azide-alkyne cycloaddition. *Dalton Trans.* **2010**, *39*, 675–696.

35. Urankar, D.; Pinter, B.; Pevec, A.; De Proft, F.; Turel, I.; Košmrlj, J. Click-triazole N_2 coordination to transition-metal ions is assisted by a pendant pyridine substituent. *Inorg. Chem.* **2010**, *49*, 4820–4829.

36. Slimi, R.; Ben Othman, R.; Sleimi, N.; Ouerghui, A.; Girard, C. Polystyrene-supported triazolic substituted piperazines for metal ions extraction. *React. Funct. Polym.* **2015**, submitted.

37. Arseniyadis, S.; Wagner, A.; Mioskowski, C. A straightforward preparation of amino-polystyrene resin from Merrifield resin. *Tetrahedron Lett.* **2002**, *43*, 9717–9719.

38. Ouerghui, A.; Elamari, H.; Ghammouri, S.; Slimi, R.; Meganem, F.; Girard, C. Polystyrene-supported triazoles for metal ions extraction: Synthesis and evaluation. *React. Funct. Polym.* **2014**, *74*, 37–45.

9

Integral Steel Casting of Full Spade Rudder Trunk Carrier Housing for Supersized Container Vessels through Casting Process Engineering (Sekjin E&T)

Tae Won Kim [1], Chul Kyu Jin [2], Ill Kab Jeong [1], Sang Sub Lim [1], Jea Chul Mun [2], Chung Gil Kang [3,*], Hyung Yoon Seo [4] and Jong Deok Kim [4]

[1] Sekjin E&T Co., Ltd., Sinpyeong-dong 642-19, Saha-Gu, Busan 604-030, Korea;
E-Mails: tacsicom@empal.com (T.W.K.); kaby1031@nate.com (I.K.J.);
sin-zi000@hanmail.net (S.S.L.)

[2] Graduate School of Mechanical and Precision Engineering, Pusan National University,
San 30 Chang Jun-dong, Geum Jung-Gu, Busan 609-735, Korea;
E-Mails: ckjeans82@pusan.ac.kr (C.K.J.); mjch78@nate.com (J.C.M.)

[3] School of Mechanical Engineering, Pusan National University, San 30 Chang Jun-dong,
Geum Jung-Gu, Busan 609-735, Korea

[4] Department of Computer Science and Engineering, Pusan National University, San 30 Chang Jun-dong,
Geum Jung-Gu, Busan 609-735, Korea; E-Mails: tanyak@mobile.re.kr (H.Y.S.);
kimjd@pusan.ac.kr (J.D.K.)

* Author to whom correspondence should be addressed; E-Mail: cgkang@pusan.ac.kr

Academic Editor: Anders E. W. Jarfors

Abstract: In casting steel for offshore construction, integral casted structures are superior to welded structures in terms of preventing fatigue cracks in the stress raisers. In this study, mold design and casting analysis were conducted for integral carrier housing. Casting simulation was used for predicting molten metal flow and solidification during carrier housing casting, as well as the hot spots and porosity of the designed runner, risers, riser laggings, and the chiller. These predictions were used for deriving the final carrier housing casting plan, and a prototype was fabricated accordingly. A chemical composition analysis was conducted using a specimen sampled from a section of the prototype; the analytically obtained chemical composition agreed with the chemical composition of the existing carrier housing. Tensile and Charpy impact tests were conducted for determining the mechanical material properties. Carrier housing product after normalizing (920 °C/4.5 h, air-cooling) has 371 MPa of yield strength, 582 MPa of tensile strength, 33.4% of elongation as well as 64 J (0 °C) of impact energy.

Keywords: carrier housing; casting steel; mold design; casting defects; mechanical properties

---

1. Introduction

Today, resource development activities, which have become active in a greater variety of places on Earth, are increasingly expanding into the deep seas. Accordingly, the demand for steel structures that have properties suitable for offshore construction has been annually increasing. The weldability of steel structures must be excellent to allow for easy assembly and maintenance. Moreover, the material must possess good strength and toughness to be able to withstand strong impacts such as high waves or tsunamis. High tensile alloy steel, which is a basic material in heavy industries, is typically used in steel structures for deep-sea regions. It can be categorized into two main types: Quenched and tempered martensitic steel and high-strength low-alloy steel [1,2]. In the first type, the initial austenite is adjusted for grain size in the austenite region and then quenched, after which transformation hardening materials such as martensite or bainite are adjusted for strength and toughness through tempering; in the tempering process, the alloy composition is adjusted without compromising the weldability [3,4]. On the other hand, the latter type acquires strength through precipitation hardening, as well as grain refinement performed through thermomechanical treatment and using microalloying elements (*i.e.*, niobium (Nb), titanium (Ti), and Vanadium (V)) [5–8].

To ensure that the structural alloy steel has the appropriate fracture toughness for it to withstand impacts and fulfill the requirements of weldability and high strength, a low carbon content in the range of 0.03%–0.15% is generally maintained in the alloy [9,10]. Typically, in the case of mild steel, strength can be improved by increasing the quantity of pearlite inside the structure through an increase of the carbon content. However, this conversely reduces impact toughness. By maintaining a low carbon content between 0.03% and 0.15%, the pearlite quantity can be reduced while improving the strength and impact toughness [11]. Accelerated cooling (AcC) was first introduced in a Japanese hot strip mill in 1982 [12]. Ever since then, ferrite grain refinement has been achieved through AcC. This AcC based ferrite grain refinement process is based on the effects of suppressing ferrite grain growth during the cooling process, which in turn improves toughness and strength. As a result, usage of the AcC process became popular worldwide; it has subsequently developed into a thermomechanical controlling process together with the hot-controlled rolling process established during the 1960s and 1970s [13,14]. It is now used in producing many grades of steel. Although the steel products created according to their purpose by using the aforementioned materials have their own benefits (e.g., low carbon content and appropriate material properties), steel plates thus produced must usually be welded for use when forming large steel structures. Such welded products are increasingly becoming inappropriate for use in areas with brief periods of repeated extreme stress (e.g., waves in deep seas). As a result, products welded from the steel plates mentioned above are gradually being replaced with casted steel products to provide endurance against intense stress. Furthermore, among today's Korean industries, the casted steel industry is an essential materials industry that supports the automotive, ship, and machine tool industries.

Consequently, there is a rapidly increasing need for advanced casting technologies to produce high value-added products of casted steel.

The average defect rate in the casting industry is approximately 10%, of which 70%–80% is due to incorrect casting. Designing an accurate casting method is an extremely difficult problem because the process of casting molten metal of high temperature into a mold for solidification involves complicated factors with regard to both thermal and physical aspects. Until now, most workshops have mainly relied on experience and the application of basic principles. Moreover, verifying the accuracy of such methods is time-consuming and expensive as it involves repeated casting tests. In other words, anticipating the appropriate casting methods in advance is important for their implementation. Against this backdrop, welded casted steel structures are not used for deep-sea structures in order to safeguard against the formation of any fatigue fractures in the stress raisers of structures operated under environmental loads such as great sea depths, low temperatures, and repeated loads. Instead, casted steel structures for deep seas are manufactured as integral structures from low-temperature high-strength casting steel so that the casted steel product possesses sufficient strength and ductility at 0 °C (as per the requirements of the usage environment).

Therefore, in this study, the flow and solidification of molten metal during casting were predicted and studied for developing an integral steel casting of a full spade rudder trunk carrier housing for supersized container vessels. In addition, MAGMA S/W was used for the casting analysis of the carrier housing to design the required mold. Jin and Kang used MAGMA S/W to design mold for a thin plate with 0.8 mm thickness and to predict the casting defects. They showed that simulation results were completely consistent with experimental results [15–17]. The manufacturing process conditions of the carrier housing were derived using a mold structure designed in a computer aided engineering (CAE) environment. The aim of the study was to ultimately save time and achieve a cost reduction by establishing an optimized casting plan with regard to the molten metal filling method, shape and size of the runner, and size and location of the riser, through the use of computer simulations. Tensile and Charpy impact tests were performed for evaluating material mechanical properties of the specimen.

2. Mold Design by Casting Simulation

Figure 1 shows the design of the carrier housing fabricated in this study. The carrier housing was fabricated by casting the structure integrally without any welds to prevent cracking in stress risers due to repeated friction and fatigue when supporting the rudder trunk, which weighs at least 40 tons. The analysis was conducted by finding a material from the MAGMA Database (Giessereitechnologie GmbH, Aachen, Germany) with a composition similar to that of the material used in the research, which was GS22CrNi3_14. Table 1 shows the chemical composition of GS22CrNi3_14. Table 2 lists the conditions employed for the casting analysis, which was carried out by setting the mold temperature to 20 °C, molten metal pouring temperature to 1600 °C, temperature-dependent coefficient as the heat transfer coefficient, and a filling time of 120 s, which is the actual molten metal pouring time.

Since adopting designs of runner, riser and chiller in the actual experiment would require too much cost and time, a casting simulation program was used instead: flow behaviors were analyzed as melt filled the mold. Based on the porosity result after melt was completely solidified, the final design method for casting was selected.

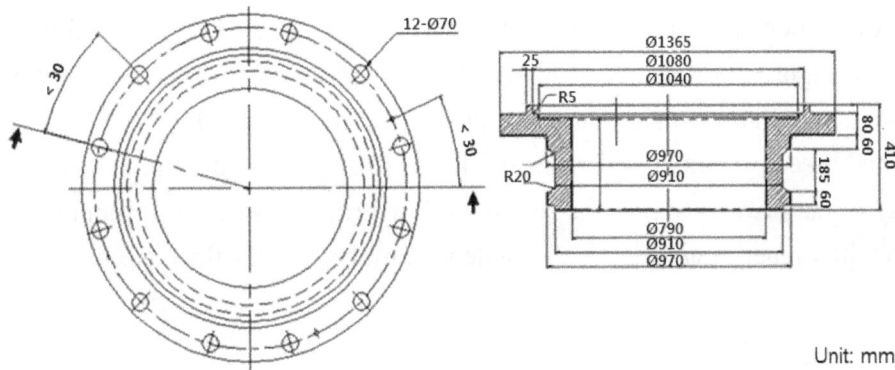

Unit: mm

Figure 1. Design of the carrier housing.

Table 1. Chemical composition of GS22CrNi3_14 (wt %).

C	Si	Mg	Cr	Ni	Fe
0.22	0.30	0.40	1.00	3.00	Bal

Table 2. The conditions employed for the casting simulation.

Parameters	Unit	Values
Solidus temperature	°C	1425
Liquidus temperature	°C	1600
Initial inlet temperature	°C	1600
Initial mold temperature	°C	20
Heat transfer coefficient between material and mold	W/m²K	Temperature dependent
Filling time	sec	120
Number of control volume	EA	3,278,471
Number of metal cells	EA	324,582

2.1. Runner

Two runner system design shapes for building the carrier housing are shown in top view in Figure 2. A casting simulation was conducted using the two shapes to determine which of the two shapes ensures smooth supply and distribution of molten metal with stable flow. The selected shape would be used for prototype fabrication. In both runner systems 1 and 2, the gate and the ingate of the runners were designed to have diameters of 80 mm, and the ingates were placed at two locations for smooth supply and distribution of molten metal to the product. However, in runner system 1 the gate was bent by 90°; therefore, it was expected that turbulence would be generated as the molten metal passed through the gate. This turbulence could result in gas mixing with the molten metal. On the other hand, in the case of runner system 2, the gate was designed as a semicircle for ensuring a more stable flow of molten metal. A comparison of the filling behavior between runner systems 1 and 2 at 5% intervals of the filling rate was shown in bottom view in Figure 2. The filling behavior when molten metal was filled through runner system 1 with the horizontal casting method was observed. As the molten metal passes through the runner and fills the product, there is a severe turbulent aspect in the flow of molten metal from the start of the pouring till up to 10% filling because of the perpendicular bend of the gate, as shown in Figure 2a. Owing to this turbulence, gas is introduced to and trapped inside the product, thereby inducing internal defects. In the case of runner system 2, in which the gate is streamlined as a semicircle to avoid any turbulence, as shown in Figure 2b, it can be observed that the laminar flow aspect of molten metal being filled into the product is more stable than that in the case of runner system 1.

Risers were designed based on runner system 2, which was finally selected for prototype preparation as it improved the stability of the flow as well as the stability of the process of filling of molten metal inside the cavity in comparison with runner system 1. This is evident from the laminar flow aspect shown in Figure 2.

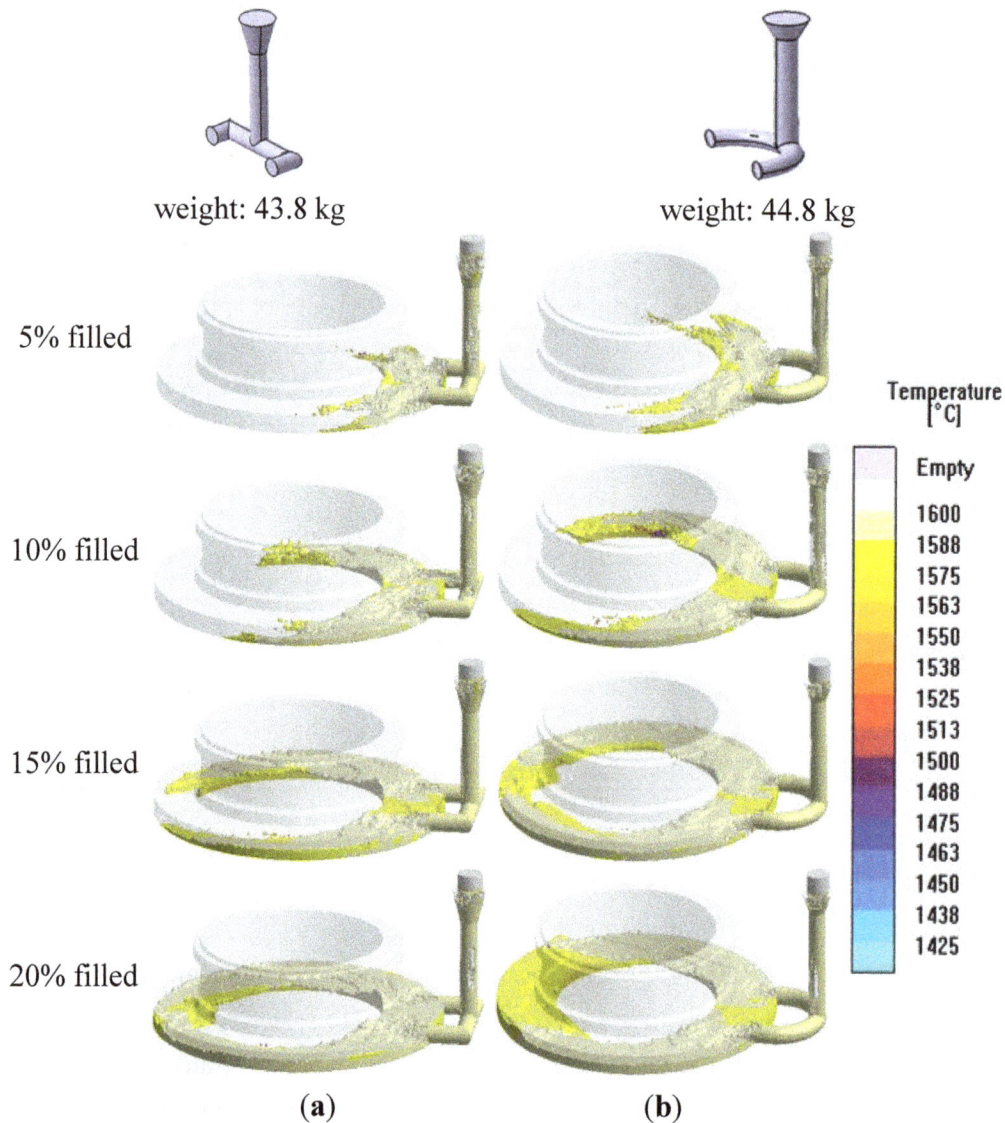

Figure 2. Comparison of the filling behavior between runner systems. (**a**) Runner system 1; (**b**) Runner system 2.

2.2. Riser

Risers are designed to remove the defects due to molten metal porosity resulting from product shrinkage during solidification and the hot spots that are created in the thick areas. Furthermore, riser laggings are designed to be placed around the riser to allow the riser to continuously provide a smooth supply and distribution of molten metal needed in the product cavity. The upper riser was designed as an open-type oval-shaped riser in contact with air, whereas the lower riser was designed as a blind riser with its upper part completely surrounded by the mold and not in contact with air. Figure 3 shows the riser system design for the carrier housing. Table 3 summarizes the dimensions of the risers and the riser laggings for three systems. A riser system with an upper riser size of A200–A240 and a lower riser size of A180–A220 was designed with appropriate riser laggings according to riser sizes. This was done to ensure that the risers are able to supply a sufficient amount of molten metal inside the product cavity.

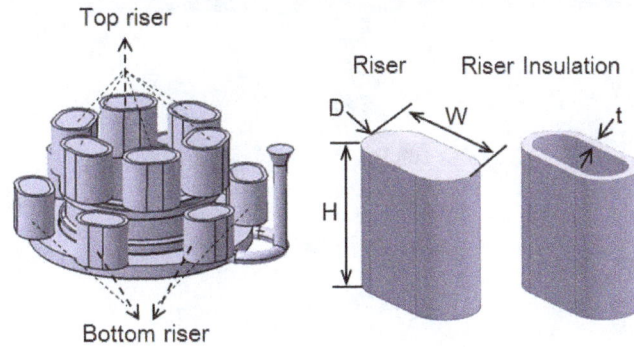

Figure 3. Riser system shape for carrier housing.

Table 3. The basic shapes and size of each riser system.

Parameters	Riser System 1		Riser System 2		Riser System 3	
	Top	Bottom	Top	Bottom	Top	Bottom
Diameter, D (mm)	Ø200	Ø180	Ø200	Ø200	Ø240	Ø220
Width, W (mm)	300	270	300	300	360	330
Height, H (mm)	300	270	300	300	360	330
Thickness, t (mm)	26	25	26	26	28	26
Weight (kg)	192.1		222		339.7	

Figure 4 shows the porosity distribution inside the product. The figures in the graph represent the proportion of solids in the total product volume; this means that the lower figure panel indicates higher porosity. Because products with a solid fraction of 90% are highly likely to exhibit diminished mechanical properties, 90% is set as the minimum value. It was observed that in the case of riser system 1, areas with porosity less than 90% are concentrated locally in the thick upper part and the sidewall of the lower part. This is owing to the filling defects inside the product during casting, and it raises concerns regarding diminished mechanical properties and the occurrence of surface defects in highly porous areas. In the case of riser system 2, the porosity distribution improved by a minimum of 54.7% because the riser dimensions in system 2 were greater than those of riser system 1. However, it was observed that areas with <90% porosity are concentrated locally in the thick upper part and the sidewall of the lower part. Riser system 3 improved porosity distribution by a minimum of 86.5% because the riser dimensions in system 3 were larger than those in riser systems 1 and 2. However, in the case of riser system 3 as well, isolated areas with <90% porosity were observed in the thick upper part and the side wall of the lower part in the product. It is believed that this will lead to the occurrence of defects because any trapped gas will not be released sufficiently through the risers as the molten metal is poured. The riser dimensions selected for prototype fabrication were those of riser system 3; selecting dimensions bigger than riser system 3 would lead to interference among the risers as well as with the product. Additionally, a chiller was designed to remove any porosity present inside the product by allowing for directional solidification within the casting.

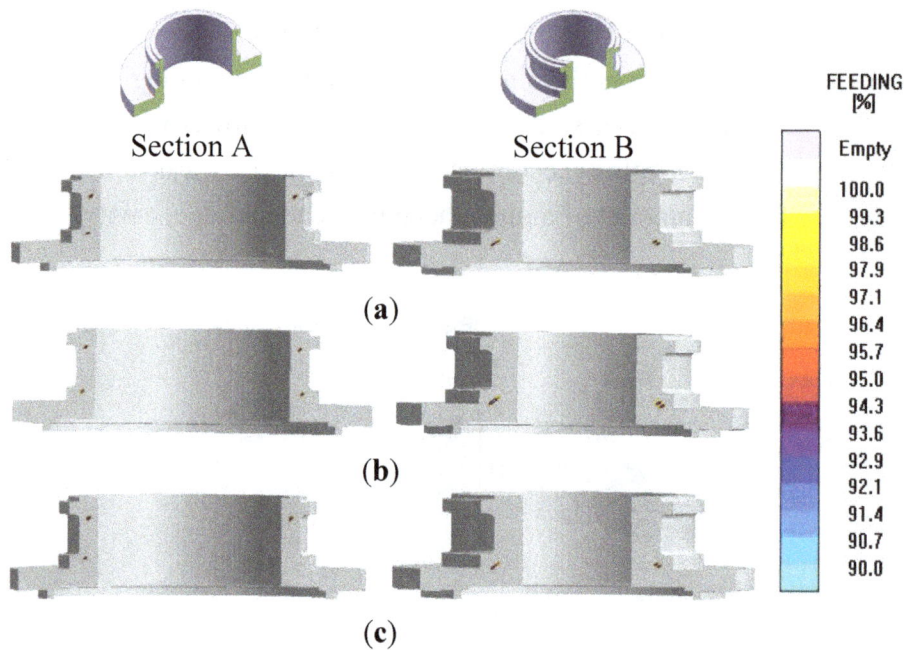

Figure 4. The porosity distribution inside carrier housing according to different riser systems. (**a**) riser system 1; (**b**) riser system 2; (**c**) riser system 3.

2.3. Chiller

By designing a chiller in the thick lower part of the product in the carrier housing where porosity is produced, internal defects such as internal porosity or hot spots can be removed by inducing directional solidification. For determining the appropriate chiller size to achieve directional solidification, the conditions were divided into two types. The designed chiller system for the carrier housing is shown in Figure 5. The chiller system 1 with dimensions of 200 mm × 100 mm × 50 mm and the chiller system 2 with dimensions of 60 mm × 100 mm × 50 mm were designed, respectively. There are 12 chillers each for inducing directional solidification in the cast product.

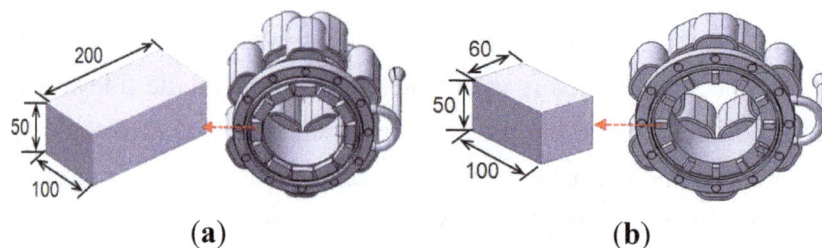

Figure 5. Two designed chiller systems for choosing the optimal system. (**a**) chiller system 1; (**b**) chiller system 2. A chill is an object used to promote solidification in a specific portion of a metal casting mold.

Figure 6 shows the porosity distribution inside the product. It can be observed that porosity is present despite the inclusion of chiller system 1, which is bigger than chiller system 2, yielded excessive chilling effect, which led to rapid solidification in the area where the chiller is attached. This resulted in twisting of the solidification direction, thus generating greater porosity owing to the trapped gas. The chiller system 2 led to the elimination of porosity inside the product through directional solidification because of an adequate chiller effect as shown in Figure 6b.

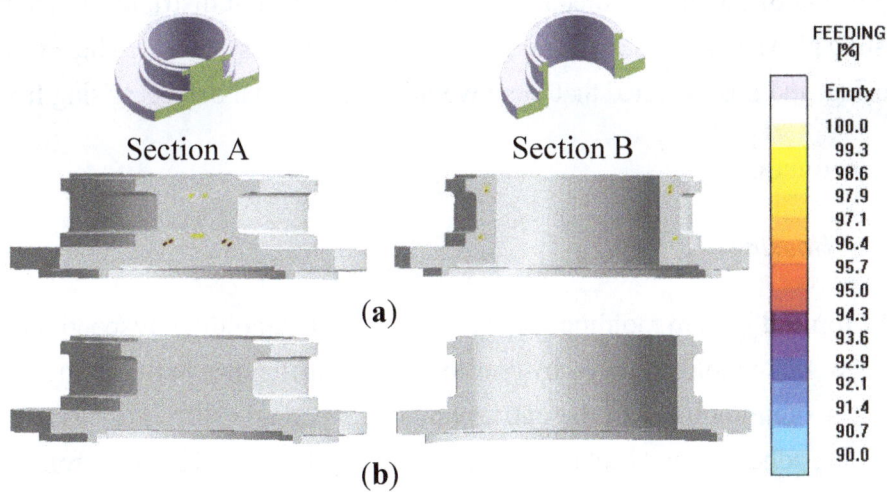

Figure 6. The porosity distribution inside carrier housing according to different chiller systems. (**a**) chiller system 1; (**b**) chiller system 2.

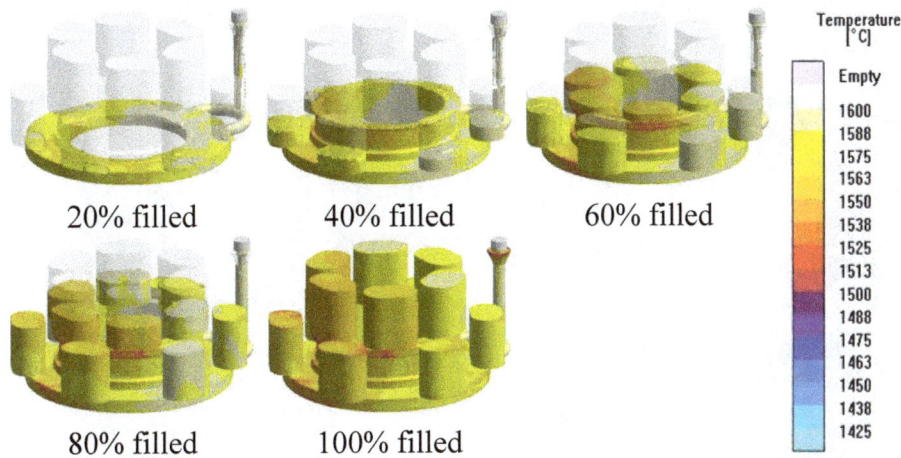

Figure 7. Filling behavior in mold for final casting plan.

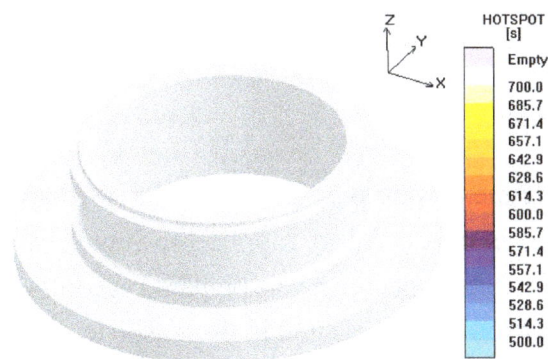

Figure 8. The hot spot distribution inside carrier housing.

Figure 7 shows the filling behavior of the molten metal in the final casting plan (runner system 2, riser system 3 and chiller system 2) determined by performing casting simulations from 20%–100% at an interval of 20%. The filling method was horizontal pouring with runner system 2, which has semi-circular shaped runners. A stable laminar flow aspect can be observed under gentle flow of the molten metal into the product during pouring of the molten metal. Figure 8 shows hot spot distribution inside the product designed by final casting plan (runner system 2, riser system 3 and chiller system 2). There were no hot spots inside the carrier housing, and it is expected that there would be no internal defects arising from hot spots.

3. Experimental Procedures

3.1. Casting of Carrier Housing

Figure 9 shows the top and bottom molding, sand core, assembled molding, respectively. A prototype was fabricated using a wooden mold prepared by considering the difference between the casting analysis results and the actual on-site molding. For the carrier housing, synthetic silica #6 was used considering factors such as ventilation, adhesion, and heat resistance during molding. In addition, chrome sand was used in areas with partial curvatures and in the ingate areas, where there is direct heat transfer. Furthermore, in the case of the core, sand removal was performed easily through the use of ultra black phosphorus disintegrant together with synthetic silica. The mold filled with the molten metal was allowed to stand for approximately 24 h to provide sufficient time for solidification. Thereafter, sand was removed, followed by the disassembly of the runner, risers, and others parts.

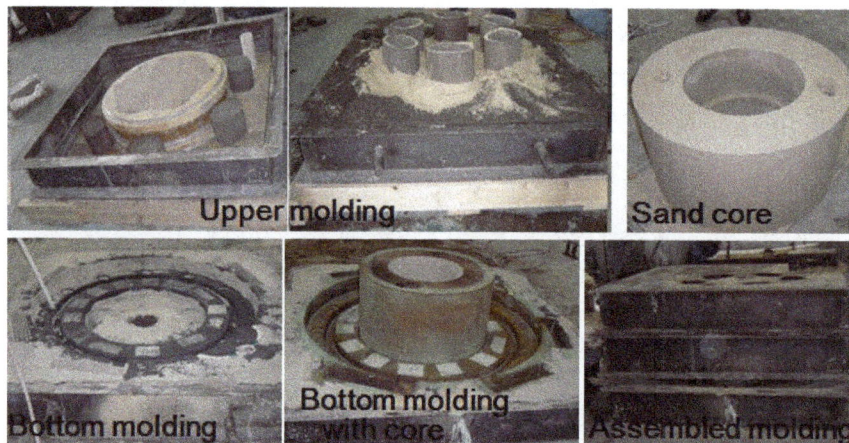

Figure 9. Wooden mold, sand core and assembled mold.

3.2. Mechanical Property Evaluation

The tensile and impact test specimens were prepared using the Y-Block attached to the side of the carrier housing. The tensile test was conducted at room temperature using a hydraulic, servo-type multipurpose tensile tester. The tensile test specimen was prepared according to ASTM A370. The yield strength, tensile strength and elongation were measured in the tensile test. An impact test was conducted using a digital Charpy impact tester, and the specimen was prepared as an A type according to ASTM E23. Figure 10 shows the specimens for tensile test and impact test, respectively.

The tensile and impact test were performed three times according to two heat treatment conditions (Normalizing), respectively. For first normalizing, the specimens for tensile and impact test were heat treated at 920 °C for 3 h, considering specimen thickness, to induce a stable austenitic state. Thereafter, the specimens were air-cooled for obtaining the desired tensile strength and toughness. For second normalizing, the specimens were heat treated 880 °C for 3 h and were air-cooled.

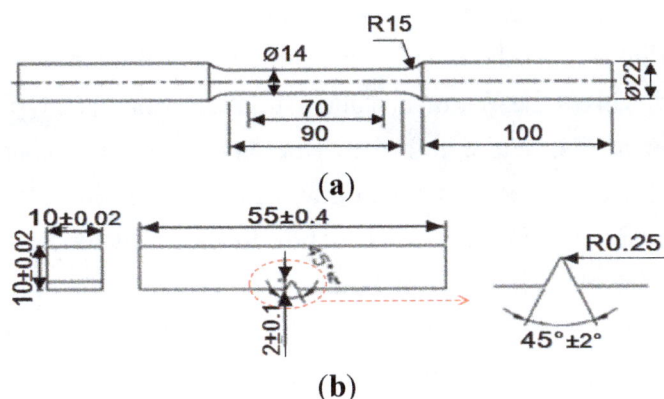

(a)

(b)

Figure 10. The dimension of tensile test specimen and impact test specimen. (**a**) tensile test specimen; (**b**) impact test specimen.

4. Results of Experiment

Figure 11 shows the casted carrier housing attached Y-Block. The casting for the carrier housing went through a sand removal process after it was solidified, and the runners and risers of the castings were then removed. A Y-Block to prepare test specimens for the tensile test and impact test can be seen.

A SPECTROMAXx chemistry analyzer (SPECTRO, Nuremberg, Germany) was used for chemical composition analysis of the specimen sampled from the ladle of the prototype. Table 4 lists the results of the carrier housing's chemical composition. The results compared favorably to the chemical composition of the material used in existing carrier housings. The carbon equivalent (CE) was 0.43%, which satisfies the requirement of 0.45% by becker marine systems.

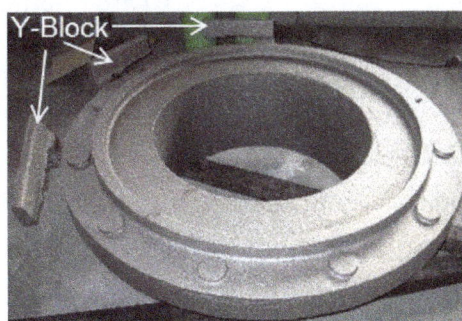

Figure 11. The fabricated carrier housing with Y-block.

Table 4. The chemical composition of the carrier housing (wt %).

Classification	C	Si	Mn	P	S	Cr	Ni	Mo	Cu	V	CE
Spec (Max) [15]	0.22	0.80	1.50	0.040	0.040	0.50	0.50	N/A	N/A	N/A	0.45
Test	0.206	0.46	1.03	0.022	0.0089	0.178	0.189	0.0087	0.075	0.0042	0.43

* Ceq: C + Mn/6 + Si/24 + Ni/40 + Cr/5 + Mo/4 + V/14 (%).

Specimens were collected from the Y-block to observe microstructures of the casted carrier housing. Figure 12 shows the microstructures (at the magnification of 200-fold) of the casted specimen and the normalizing heat treated specimen, respectively. The gray part in the microstructure presents ferrite, while the black part represents pearlite. As can be confirmed from the casted microstructure from Figure 12a, needle-line ferrites are formed. This is the Widmannstätten structure which is a typical in a slowly-cooled thick steel casting. Widmannstätten structure is formed when molten steel reaches transformation point A_3, an α-ferrite and Fe3C and precipitated as a needlelike form in the grain boundary of γ-austenite by the eutectic ($\gamma \rightarrow \alpha$ + Fe3C) reaction [18]. This is a structure formed as α-ferrite or Fe3C is precipitated first along the specific crystalline plane in the grain boundary of austenite, and then pro-eutectoid ferrite and cementite are propagated and replaced. The unstable structure like this Widmannstätten structure has a very weak strength. Therefore, one can realize that casting stress has to be removed while improving the structure. Figure 12b shows the microstructure after normalizing heat treatment. It became a representative type whereby austenite is transformed to ferrite and pearlite during air cooling. During cooling, the microstructure became homogenized while ferrite grains were separated from austenite grain, and pearlite was formed; thus, it became homogenized. Further, ferrite became considerably coarse after the normalization.

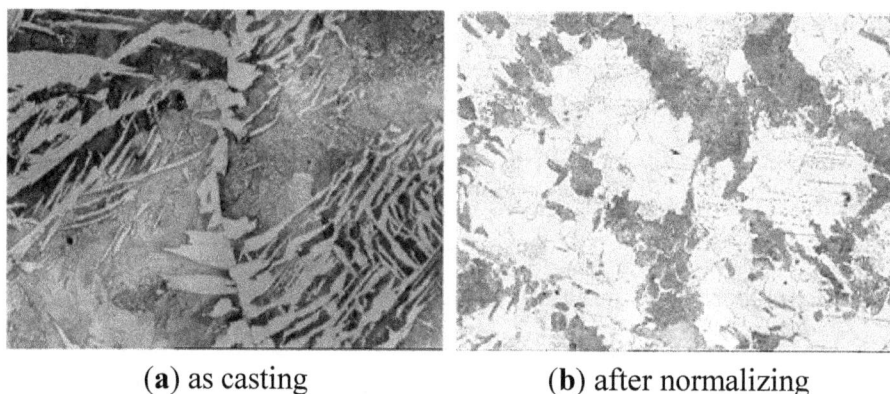

(a) as casting   (b) after normalizing

Figure 12. Microstructures of carrier housing: (a) as casting and (b) after normalizing.

Figure 13 shows the shape and fractured surface of the tensile test specimen and impact test specimen. Fractured tensile specimen had a cup and cone shear fracture. This is a generally occurring ductility fracture, which indicates that a considerable plastic deformation occurred before the fracture was propagated.

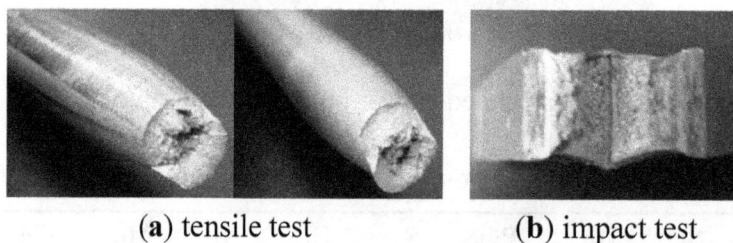

(a) tensile test   (b) impact test

Figure 13. Fracture surface of specimens of tensile test and impact test. (a) tensile test; (b) impact test

The mechanical properties for the casted carrier housing after the normalizing obtained by the tensile test and impact test are listed in Table 5. The air cooling of the specimen after austenizing treatment for the specimen at 920 °C for 4 h and 30 min resulted in the higher yield strength by 1 MPa, higher tensile strength by 26 MPa, and the elongation higher by 5.4% than those of the specimens austenized at 880 °C for 3 h. However, the impact energy by the air cooling after austenizing at 920 °C for 4 h and 30 min in the specimen was lower by 18.93 J (0 °C) that of the specimens with the autenizing treatment at 880 °C for 3 h. This result implies that higher austenizing temperature with a longer holding time is beneficial for increasing the strength and toughness in the steel casting, but the impact strength would be decreased. The results obtained under the both normalizing conditions are in good agreement with the physical properties values proposed by the Becker Marine Systems, Germany [19].

Table 5. The mechanical properties after normalizing heat treatment.

Normalizing Conditions	Yield Strength (MPa)		Tensile Strength (MPa)		Elongation (%)		Impact Energy 0 °C(J)	
Target [19]	210		410		24		27	
880 °C/3 h	370	+6 / −27	556	+14 / −2	28	−2	83	+23
920 °C/4.5 h	371	+3 / −28.4	582	−24	33.4	+2.6 / −5.8	64.07	+7.79

A radiographic test was carried out for the Y-block sample to investigate the internal defect in the specimen during casting. The radiographic test results are presented in Figure 14. From the X-ray photos of specimen, it was confirmed that the void and impurities which causes inferior mechanical properties after the casting were not presented. Therefore, it was judged that a casting process would not play on the mechanical properties of steel castings.

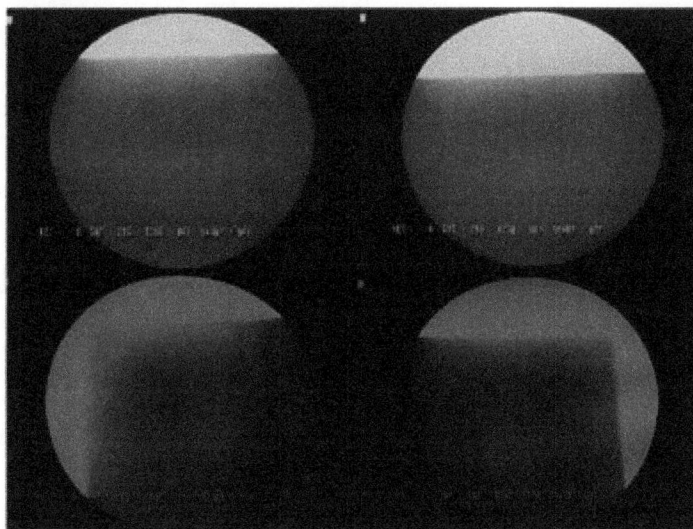

Figure 14. X-ray analysis for specimen.

5. Conclusions

Several casting plans were prepared for designing the optimal mold, and a casting simulation program, was used for casting analysis and deriving the optimal mold design. The results show that the gas generated inside the mold during casting owing to the oxidized layer that enters in the beginning is eliminated. In addition, the casting plan incorporating riser lagging and chiller design for promoting directional solidification in the riser can provide sufficient amounts of molten metal to the product cavity without resulting in any internal or surface defects during casting. A chemical composition analysis was carried out using a specimen sampled from the ladle during carrier housing casting. Furthermore, the Y-Block attached to the carrier housing side during its fabrication was used for preparing the specimen for evaluating the mechanical properties of the carrier housing. A comparison of the values showed that the results are similar to the ones of existing carrier housings. Thus, carrier housings produced by the developed casting process with optimal mold design can suitably replace existing carrier housings.

Acknowledgments

This work was supported by a National Research Foundation of Korea (NRF) grant funded by the Korea government (MSIP) through GCRC-SOP (No. 2011-0030013).

Author Contributions

Chul Kyu Jin and Tae Won Kim designed mold of casting by performing the simulation; Chul Kyu Jin, Ill Kab Jeong, Sang Sub Lim, Jae Chul Mun, Hyung Yoon Seo and Jong Deok Kim conducted experiment and analysis the results; Chung Gil Kang maintained and examined the results of simulation and experiment; All authors have contributed to discussing and revising.

Conflicts of Interest

The authors declare no conflict of interests.

References

1. Horn, R.M.; Ritchie, R.O. Mechanisms of tempered martensite embrittlement in low alloy steels. *Metall. Trans. A* **1978**, *9*, 1039–1053.
2. Korchynsky, M. Cost effectiveness of micro- alloyed steels. International symposium proceedings: Steel for fabricated structures, Cincinnati, OH, USA, July 1999.
3. Krauss, G.; Marder, A.R. The morphology of martensite in iron alloys. *Metall. Trans.* **1971**, *2*, 2343–2357.
4. Bethlehem Steel Corporation. *Modern Steels and Their Properties*, 7th ed.; Bethlehem Steel Co.: Bethlehem, PA, USA, 1972; pp. 27–57.
5. Baird, J.D.; Preston, R.R. *Relationships Between Processing, Structure and Properties in Low Carbon Steels*; Gray, J.M., Ed.; TMS-AIME: Warrendale, PA, USA, 1973; pp. 1–46.
6. Hall, E.O. Deformation and ageing of mild steel: III Discussion of results. *Proc. Phys. Soc. B* **1951**, *64*, 747–753.
7. Petch, N.J. The cleavage strength of polycrystals. *J. Iron Steel Inst.* **1953**, *174*, 25–28.
8. Irvine, K.J. *Strong Tough Structural Steels*; Iron and Steel Institute: London, UK, 1967; pp. 1–7.
9. Irani, J.J.; Burton, D.; Jones, J.D.; Rothwell, A.B. *Strong Tough Structural Steels*; Iron and Steel Institute: London, UK, 1967; p. 110.
10. Graville, B.A. *Cold Cracking in Welds in HSLA Steels*; Welding of HSLA (Microalloyed) Structural Steels: Rome, Italy, 1976; pp. 85–101.

11. Pickering, F.B. *Physical Metallurgy and the Design of Steels*; Applied Science Publishers Ltd.: London, UK, 1978; p. 71.

12. Tsukada, K.; Matsumoto, K.; Hirabe, K.; Takeshige, K. Application of on-line accelerated cooling (OLAC) to steel plate. In Proceedings of the 23rd Mechanical Working and Steel Processing Conference. Pittsburgh Hilton, Pittsburgh, PA, USA, 28–29 October 1981; pp. 347–370.

13. Shina, A.K. *Ferrous Physical Metallurgy*; Butterworth-Heinemann: Oxford, UK, 1989; pp. 608–609.

14. Tamura, I.; Sekine, H.; Tanaka, T.; Ouchi, C. Prediction and control of microstructural change and mechanical properties in hot-rolling. In *Thermomechanical Processing of High Strength Low Alloy Steels*; Butterworth & Co., Ltd.: London, UK, 1988; pp. 202–225.

15. Jin, C.K.; Kang, C.G. Fabrication process analysis and experimental verification for aluminum bipolar plates in fuel cells by vacuum die-casting. *J. Power Sources* **2011**, *196*, 8241–8249.

16. Jin, C.K.; Kang, C.G. Fabrication by vacuum die casting and simulation of aluminum bipolar plates with micro-channels on both sides for proton exchange membrane (PEM) fuel cells. *Int. J. Hydrog. Energy* **2012**, *37*, 1661–1676.

17. Jin, C.K.; Jang, C.H.; Kang, C.G. Vacuum die casting process and simulation for manufacturing 0.8 mm-thick aluminum plate with four maze shapes. *Metals* **2015**, *5*, 192–205.

18. Todorov, R.P.; Khristov, Kh.G. Widmanstatten structure of carbon steels. *Met. Sci. Heat Treat.* **2004**, *46*, 49–53.

19. Becker Marine Systems. Available online: http://www.becker-marine-systems.com (accessed on 29 April 2015).

10

On the Effect of Pouring Temperature on Spheroidal Graphite Cast Iron Solidification

Alex Escobar [1,†]**, Diego Celentano** [2,†,*]**, Marcela Cruchaga** [3,†] **and Bernd Schulz** [1]

[1] Departamento de Ingeniería Metalúrgica, Universidad de Santiago de Chile (USACH), Av. Bernardo O'Higgins 3363, Santiago, Chile; E-Mails: alexesc@gmail.com (A.E.); bernd.schulz@usach.cl (B.S.)

[2] Departamento de Ingeniería Mecánica y Metalúrgica, Centro de Investigación en Nanotecnología y Materiales Avanzados (CIEN-UC), Pontificia Universidad Católica de Chile (PUC), Av. Vicuña Mackenna 4860, Santiago, Chile; E-Mail: dcelentano@ing.puc.cl

[3] Departamento de Ingeniería Mecánica, Universidad de Santiago de Chile (USACH), Av. Bernardo O'Higgins 3363, Santiago, Chile; E-Mail: marcela.cruchaga@usach.cl

[†] These authors contributed equally to this work.

[*] Author to whom correspondence should be addressed; E-Mail: dcelentano@ing.puc.cl

Academic Editor: Hugo F. Lopez

Abstract: This work is focused on the effect of pouring temperature on the thermal-microstructural response of an eutectic spheroidal graphite cast iron (SGCI). To this end, experiments as well as numerical simulations were carried out. Solidification tests in a wedge-like part were cast at two different pouring temperatures. Five specific locations exhibiting distinct cooling rates along the sample were chosen for temperature measurements and metallographic analysis to obtain the number and size of graphite nodules at the end of the process. The numerical simulations were performed using a multinodular-based model. Reasonably good numerical-experimental agreements were obtained for both the cooling curves and the graphite nodule counts.

Keywords: spheroidal graphite cast iron; solidification; microstructure; pouring temperature

1. Introduction

Nodular cast iron, also known as spheroidal graphite cast iron (SGCI) or ductile iron, was firstly developed in 1949 using a Mg-Cu alloy as a spherodizing agent [1]. As a structural material, SGCI has the following properties: versatility, good performance/cost ratio, high corrosion and wear resistance, ductility and high tensile resistance. Studies focusing on this material have been carried out during the last few years due to its extensive use and the growing trend of replacing forged steel because of its manufacturing cost which, in general, is higher than that of SGCI. In particular, the automotive industry has shown great interest and trust in this material by using it in safety components such as steering knuckles and calipers [2]. Another critical application that reflects the good performance and properties of this material is in the storage and transport of nuclear waste [1].

The microstructure of nodular cast iron is responsible for the useful properties of this material and the main reason for the many studies devoted to it since its appearance [3]. By understanding the microstructure, greater control over the physical properties of the final product can be established [4]. One of the most important parameters for microstructure control of SGCI is the cooling rate, *i.e.*, a high cooling rate produces large amounts of carbides in the matrix, causing a significant decrease in ductility and toughness of the material and thus requiring expensive heat treatments for the dissolution of such carbides.

Several models have been proposed to describe the kinetic mechanisms that occur during the liquid-solid transformation of a SGCI with eutectic composition, some of them being based on the uninodular theory [5–14] and others based on the multinodular theory [15–19]. Nevertheless, the so-called multinodular models are nowadays gaining acceptance due to their good prediction capabilities. According to the multinodular theory of nucleation and growth, during the cooling process of a nodular cast iron with eutectic composition, once the eutectic temperature is reached, austenite and graphite nucleate independently in the liquid (austenite growing in a dendritic shape and graphite in a spherical shape) such that at a certain instant (experimentally determined in reference [16]), graphite is enveloped by the thinnest arms of dendritic austenite and subsequent growth proceeds by carbon diffusion through this phase. It should be mentioned that the multinodular-based model developed by Boeri [16] was widely used and validated with the experiments in reference [20].

In the present work, the effect of pouring temperature on the thermal-microstructural response of the solidification of a nodular cast iron with eutectic composition is analyzed both experimentally and numerically. This specific analysis constitutes an extension of another study [20] which emphasized the influence of the cooling rate on the microstructural evolution of the analyzed part. In order to analyze the evolution of the microstructure during the cooling process, two solidification experiences, *i.e.*, two experiences with filling temperatures of 1200 °C and 1250 °C, were carried out in a wedge-like mold. Details of the experimental procedure are presented in Section 2. Section 3 briefly describes the multinodular-based model used in this study. The discussions, comparison and experimental validation of the numerical results provided by this model, which encompass cooling curves and graphite nodule distributions at different positions of the samples, are given in Section 4. Finally, the main concluding remarks are found in Section 5.

2. Experimental Procedure

Two castings, W1 and W2, were carried out in a wedge-like mold in order to analyze the effect of the pouring temperature on the thermal-microstructural response of the solidification of a nodular cast iron with eutectic composition. A schematic view of the analyzed part is shown in Figure 1. The temperature measurements were made using five 0.51 mm type K thermocouples located at the center of the casting and 50 mm from each other along its length. Due to the high temperatures at these points, the wire of the thermocouple was covered with ceramic tiles and fiber glass cases. The temperature-time curves were obtained with a MICRO MAC thermal interface Model Personal Daq55 (pdaqview55) connected to a laptop.

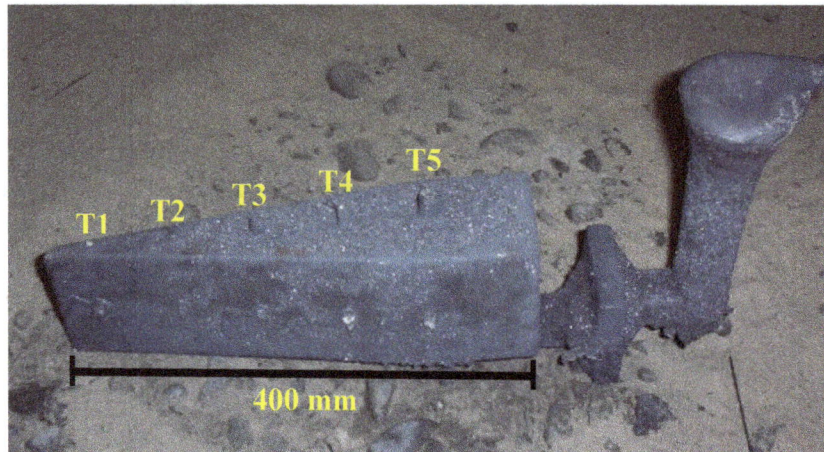

Figure 1. Mold used in the solidification experiences W1 and W2.

2.1. Solidification Experiences

The pouring temperature for the two solidification experiences W1 and W2 was among the standard ones used at the foundry where these experiences were carried out. They led to a maximum recorded liquid temperature at the mold cavity of 1200 ± 5 °C for W1 and 1250 ± 5 °C for W2. The error in this temperature refers to the measurements registered by the five thermocouples at early stages. Thus, these maximum temperatures, which strictly corresponded to the filling temperatures, are considered the initial temperatures owing to the short filling times of the whole part in both tests. The solidification experiences were carried out in an induction furnace operating under standard conditions. Although the total weight of the casting was that of the total furnace capacity, *i.e.*, 560 kg, the total weight of the metal used for the experiences was only 25 kg. The charge for the solidification experiences was as follows: 53.4% scrap, 41.3% returns, 2.0% low sulfur graphite and 3.3% Fe75Si.

The melting process had an extension of 1.5 h, approximately. The carbon equivalent (CE) adjustment was carried out by the addition of FeSi to the molten metal. The composition of the liquid metal was measured with a standard carbon equivalent meter system. Temperature control was carried out by lance immersion in the liquid. The CE, %C and %Si values were obtained approximately at 1450 °C with cup thermal analysis. The desired chemical composition of the liquid was %C: 3.7–3.9 and %Si: 1.8–2.0. The nodularizing treatment was carried out with a FeMg alloy using the sandwich method [21]. In this procedure, the treatment ladle was previously heated with a burner or by pouring molten metal from the furnace into the ladle. The nodularizing and inoculant agents were placed into the ladle cavity, 1% and 0.2%, respectively. Then, steel scrap (2%–3% of metal weight) or plates of steel or nodular cast iron were placed over these agents. After that, the metal was poured into the treatment ladle to carry out the nodularizing and inoculant processes.

The chemical composition of experiences W1 and W2 is shown in Table 1. The equivalent carbon is given by CE = %C + 0.28%Si + 0.30%P + 0.007%Mn + 0.033%Cr.

Table 1. Chemical composition of experiences W1 and W2.

C	Si	Mn	S	P	Ni	Cu	Cr	Mg	CE
3.39	2.67	0.52	0.047	0.053	0.02	0.01	0.08	0.042	4.16

2.2. Metallographic Analysis

In order to measure the size, morphology and nodule distribution as a function of the pouring temperature, a metallographic analysis was carried out. The samples were taken out from the center of the wedge cavity and close to the thermocouple location. The cooling curves were obtained in five points at the center line of the wedge, as already shown in Figure 1. Those points were labeled W11 to W15 and W21 to W25 for solidification experiences W1 and W2, respectively.

The surface nodule count was carried out using the software Image-Pro Plus (Measurement Computing Corporation, Norton, MA, USA). With this software, it was possible to measure the size and shape factor of the graphite nodules. Once these data were obtained, the nodule sizes were grouped into families. The volumetric nodule count was then obtained using the following expression that comes from equating the surface and volume graphite fractions:

$$N_V = \frac{3}{4R} N_A \tag{1}$$

where N_V is the volumetric nodule count (nodules per mm^3), N_A is the surface nodule count (nodules per mm^2) and R is the nodule radius (mm).

The maximum and minimum nodule sizes in the two experiences were 10 μm and 75 μm, respectively. From these lower and upper bounds, five experimental families were considered, each one 13 μm long. The total nodule count was obtained by summing the respective values per family at each point. The family details used to carry out the experimental nodule count are shown in Table 2.

Table 2. Radius range of each family.

Family	Radius range (μm)
1	10–23
2	23–36
3	36–49
4	49–62
5	62–75

2.3. Cooling Curves Characterization

This stage consists in determining the relevant temperatures and times for the many reactions that occur during the solidification process of the nodular cast iron in experiences W1 and W2. The cooling process is carried out in the mold up to 300 °C. In order to carry out the cooling curve characterization, the standard method is applied [22], which consists in determining the first derivative of the cooling curve and then projecting both curves in the same chart. By doing this, it is possible to identify the times at which the slope of the cooling rate curve changes (blue curve) and, then, to obtain the temperatures at such times in the cooling curve (black curve). Figure 2 shows the applied method in a graphical form where the temperature and time of the different reactions that take place during the cooling process can clearly be seen. The temperature T_{EG} corresponds to the bulk eutectic reaction start, T_{EU} is the maximum undercooling temperature, T_{ER} is the maximum temperature reached after the recalescence and T_{ES} is the final solidification temperature.

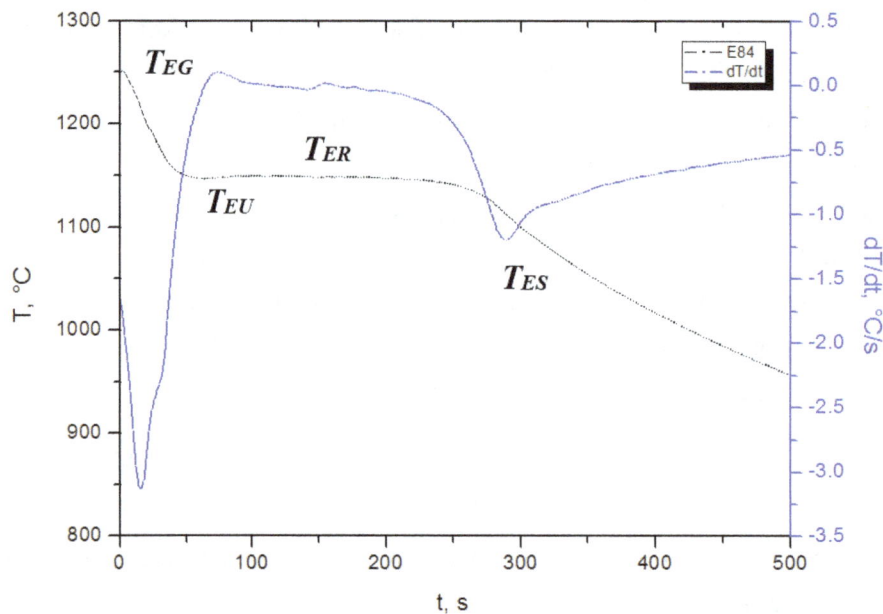

Figure 2. Schematic representation of the procedure applied for the cooling curve characterization.

3. Numerical Modeling

3.1. Thermal Model

In order to take into account the phase changes that occur during the cooling process, the well-known energy equation is considered [17–20]:

$$\rho c \dot{T} + \rho L \dot{f}_{pc} = \nabla \cdot \left(k \nabla T \right) \tag{2}$$

where ρ is the density, c is the specific heat, k is the thermal conductivity, T is the temperature, L is the phase change latent heat and f_{pc} is the phase change function ($0 \leq f_{pc} \leq 1$) that is considered in this context to be the liquid fraction $f_l = 1 - f_s$, where f_s is the liquid fraction provided by the microstructural model described below. A dot over the variable indicates time derivative and ∇ is the gradient operator. The solution of this equation is obtained with initial and boundary conditions and it is carried out through a spatial discretisation using the Finite Element Method (FEM) and a time discretisation using the Finite Differences Method (FDM) [23,24].

3.2. Microstructural Model

The numerical model implemented to simulate the solidification process of an eutectic nodular cast iron poured at two different temperatures in a wedge-like mold is based on the multinodular theory of nucleation and growth [16]. This model assumes that graphite spheres and dendrites of austenite nucleate in the liquid and, after a period of time when the nodule reaches 6 μm in size, the graphite nodules are assumed to be enveloped by the thinner arms of the austenite dendrites. The subsequent growth takes place via carbon diffusion through the austenite envelope. The main aspects of the nucleation and growth laws considered in this approach laws are described below (details of this model can be found in reference [20]).

3.2.1. Graphite Nucleation

The rate of the graphite nucleation density N_{gr} is assumed to follow an exponential law in the form:

$$\dot{N}_{gr} = f_l \, b \, \Delta T \exp\left(-\frac{c}{\Delta T}\right) \tag{3}$$

where $\Delta T = T_E - T$ is the undercooling, with T_E being the equilibrium eutectic temperature, and b and c are the nucleation parameters that are assumed to remain constant for a given composition and liquid treatment. The factor f_l is included to take into account the continuous disappearance of the nucleation sites with liquid consumption. In this stage, it must be clarified that the model assumes that the nucleation stops when the recalescence is achieved and starts again when the temperature falls below the last maximum undercooling until the end of the solidification process.

3.2.2. Graphite Growth

As already mentioned, two different growing stages are considered: at the beginning, the graphite nodules are only in contact with liquid and, afterwards, the graphite particles are enveloped by the thinner dendrites arms of austenite. The respective growth laws are given by:

$$\dot{r}_{gr} = \frac{1}{2r_{gr}} \frac{C^{l/\gamma} - C^{l/gr}}{C^{gr} - C^{l/gr}} \frac{\rho_l}{\rho_{gr}} D_C^l \tag{4}$$

$$\dot{r}_{gr} = \frac{1.91}{r_{gr}} \frac{C^{\gamma/l} - C^{\gamma/gr}}{C^{gr} - C^{\gamma/gr}} \frac{\rho_\gamma}{\rho_{gr}} D_C^\gamma f_l^{2/3} \tag{5}$$

where r_{gr} is the graphite radius and ρ_l, ρ_{gr} and ρ_γ are the liquid, graphite and austenite densities, respectively. Furthermore, D_C^l and D_C^γ are the respective diffusion coefficients of carbon in liquid and austenite while $C^{l/\gamma}$, $C^{\gamma/l}$, $C^{l/gr}$ and $C^{\gamma/gr}$ correspond to the equilibrium concentrations at the temperature T of liquid with austenite, austenite with liquid, liquid with graphite and austenite with graphite, respectively ($C^{gr} = 100\%$ is the carbon concentration in graphite). These parameters are schematically presented in Figure 3. In this figure, T_{AS} correspond to the austenite solidus temperature, T_{AL} is the austenite liquidus temperature, T_{GL} is the graphite liquidus temperature and T_{AG} determines the carbon solubility variation in austenite.

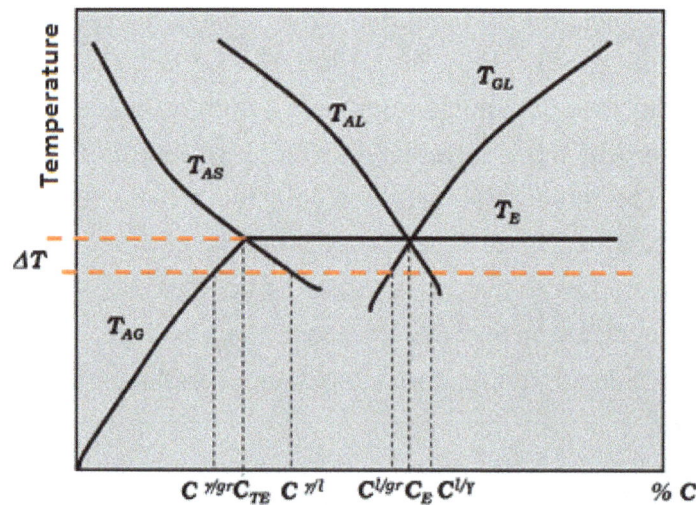

Figure 3. Equilibrium compositions considered in the microstructural model.

3.2.3. Graphite and Austenite Fractions

The graphite fraction is calculated according to:

$$f_{gr} = \sum_{1}^{j} \frac{4}{3} \pi N_{gr_j} r_{gr_j}^3 \tag{6}$$

where j relates to the number of different nodule radii that are present in a volume element due to the non-simultaneity of nucleation.

The austenite fraction is obtained by:

$$f_\gamma = \frac{100\% - C_E}{C_E - C_{TE}} f_{gr} \tag{7}$$

where C_{TE} and C_E are the carbon solubility in austenite at the eutectic temperature and the eutectic carbon content (in %), respectively.

Finally, the solid fraction is equal to:

$$f_s = f_{gr} + f_\gamma \tag{8}$$

4. Results and Discussion

This section firstly presents the experimental results obtained in experiences W1 and W2 described in Section 2. Then, the numerical results obtained with the model summarized in Section 3 are discussed and compared with the corresponding measurements in order to establish its limitations and predictive capabilities.

4.1. Experimental

Figure 4 shows the experimental cooling curves of the five analyzed points in experiences W1 and W2 together with the stable (T_{EG}) and metastable (T_{EM}) eutectic temperatures. These temperatures were obtained using an experimental correlation derived from the works of Chipman [25], Gustafson [26] and Henschel [27], among others; this correlation is identified as CGH. The same temperatures were obtained with the software Thermocalc [28], labeled Th, which does not consider the phosphorus content in its database. When comparing the cooling process of each point in both experiences, it is observed that the higher pouring temperature of W2 produces longer solidification times than those of W1 and, in addition, the temperatures of the different reactions in experience W2 are slightly higher than those of W1. In both experiences, it is also seen that, particularly in the first cooling stage, as the thickness of the part increases, the cooling rate decreases. As a consequence of this, a continuous increase in the solidification time of the five points along the part is observed. As expected, the points located at the thicker part has a longer solidification time than the ones located at the thinner parts, which evidences the effect of wedge thickness on the cooling process. This effect is also influenced by the pouring temperature, i.e., as the pouring temperature increases, the solidification times increases as well.

As mentioned above, the stable and metastable eutectic temperatures are included in Figure 4 with the aim of analyzing and evaluating the cementite nucleation in each point along the part. The different temperatures obtained with both methods (i.e., correlations CGH and Th) can clearly be appreciated. It is observed that, according to the values obtained with the correlation Th in the thinner parts of the wedge (locations W11 and W21), the metastable reaction would take place since the cooling curves of these points intersect the metastable eutectic temperature. On the other hand, it should be noted that this effect is not predicted in the same way with the CGH correlation. Considering this last correlation, it is seen that all the temperature-time curves fall within the range T_{EG}-T_{EM}, which means that, regardless of what is predicted by the Th correlation, all points of the wedge would solidify according to the stable Fe-graphite diagram. However, from the metallographic analysis carried out at the different points, the validity of the prediction given by the Th correlation is confirmed since, as mentioned below, a high fraction of white eutectic was found in locations W11 and W21.

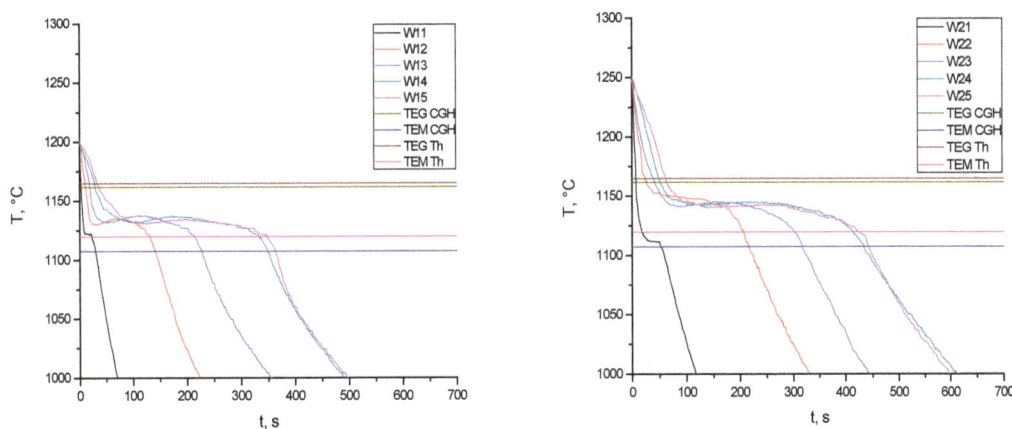

Figure 4. Experimental cooling curves of experiences W1 and W2.

Table 3 summarizes the results obtained from the standard procedure carried out to determine the characteristic times and temperatures of the solidification experiences. The eutectic undercooling $\Delta T_{EG} = T_{EG} - T_{EU}$, the recalescence amplitude $\Delta T_r = T_{ER} - T_{EU}$ and the elapsed time from the eutectic reaction until the solidification ending $\Delta t_{ES} = t(T_{ES}) - t(T_{EG})$ are also included. It can be seen that, as the pouring temperature increases, ΔT_r decreases and Δt_{ES} increases.

Table 3. Experimental cooling curves characterization for experiences W1 and W2.

Point	tT_{EG}, s	T_{EG}, °C	tT_{EU}, s	T_{EU}, °C	tT_{ER}, s	T_{ER}, °C	tT_{ES}, s	T_{ES}, °C	ΔT_r	ΔT_{EG}	Δt_{ES}	dT/dt, °C/s
W11	2.1	1172.8	11.8	1121.9	15.6	1122.5	37.1	1091.3	0.5	50.9	54.1	13.9
W12	3.8	1188.9	31.5	1130.1	77.8	1135.0	150.1	1096.4	5.0	58.8	174.5	3.7
W13	10.8	1181.6	51.4	1131.9	113.9	1137.6	238.0	1094.8	5.7	49.7	241.5	2.1
W14	15.7	1181.9	101.9	1132.1	175.9	1137.2	356.0	1100.8	5.2	49.9	338.5	1.3
W15	20.0	1183.2	127.4	1130.9	198.4	1134.3	372.9	1095.3	3.5	52.3	355.9	1.2
W21	5.9	1180.4	35.0	1111.5	49.8	1111.9	65.8	1087.0	0.4	68.9	59.9	12.1
W22	17.4	1180.6	----	----	----	----	224.6	1098.7	----	----	207.2	3.3
W23	20.2	1190.1	88.4	1141.6	157.0	1145.3	316.3	1108.8	3.7	48.6	296.1	1.7
W24	34.3	1191.4	129.8	1142.5	219.9	1145.5	433.0	1107.9	3.0	48.9	398.6	1.5
W25	50.4	1182.4	141.1	1140.4	244.7	1142.9	438.3	1113.9	2.5	42.0	387.9	1.3

The as-cast microstructures at the different locations in experiences W1 and W2 are shown in Figure 5. The experimental phase fractions for each of those points are summarized in Table 4. In both cases it is observed that, as the pouring temperature increases, the nodule density decreases. This would be due to the effect of the high pouring temperature and the longer time that the molten metal is submitted to this high temperature, producing the growth of the graphite nodules which relates to a lower nodule density. As can be seen in Figure 5, a continuous growth of the graphite nodules is produced from the thinner to the thicker parts of the wedges because of the effect of both cooling rate and pouring temperature on the nucleation and growth time of a graphite particle during the solidification process. As also can be seen in this figure, in the thinner parts of both wedges, random carbides distributed in the structure are observed which, unlike the experimental correlation CGH, is also predicted by Thermocalc. This can be seen in Figure 4, where the cooling curve of points W11 and W21 do not intersect the metastable eutectic temperature curves calculated with this correlation (CGH), thus the white eutectic nucleation is not allowed to succeed. It is important to point out that the presence of carbides is also observed in point W22, a fact that according to Figure 4 would not be plausible. Despite this, it is necessary to take into account the presence of alloying elements in the alloy, mainly manganese and chromium. These elements are strong carbide promoters [29] which, together with the high cooling rate in the thinner sections of the part, could be the reason for the presence of this hard and brittle phase in those zones, even in the 20 mm thick part of wedge W2 (point W21). Towards the thicker parts of the wedges the cementite phase is no longer observed. In these positions, the microstructure is mainly composed by variable quantities of pearlite, ferrite and graphite nodules that depend on the pouring temperature, the cooling rate and the chemical composition of the alloy. As a consequence, when comparing the microstructure of the thinner parts of the wedges to the other locations, there is a change in the graphite fraction, since part of the available carbon forms carbides, producing the stable phase fraction to be lowered; see Table 4. Regarding the phases present in each point, it can be seen that in both cases the carbides fraction is around 5% and part of a ferritic-pearlitic matrix with variables quantities according to the cooling conditions in each point. Towards the thicker parts of the wedges (from W12 upwards for W1 and from W23 upwards for W2), no carbides are observed in the final solidification microstructure of the parts.

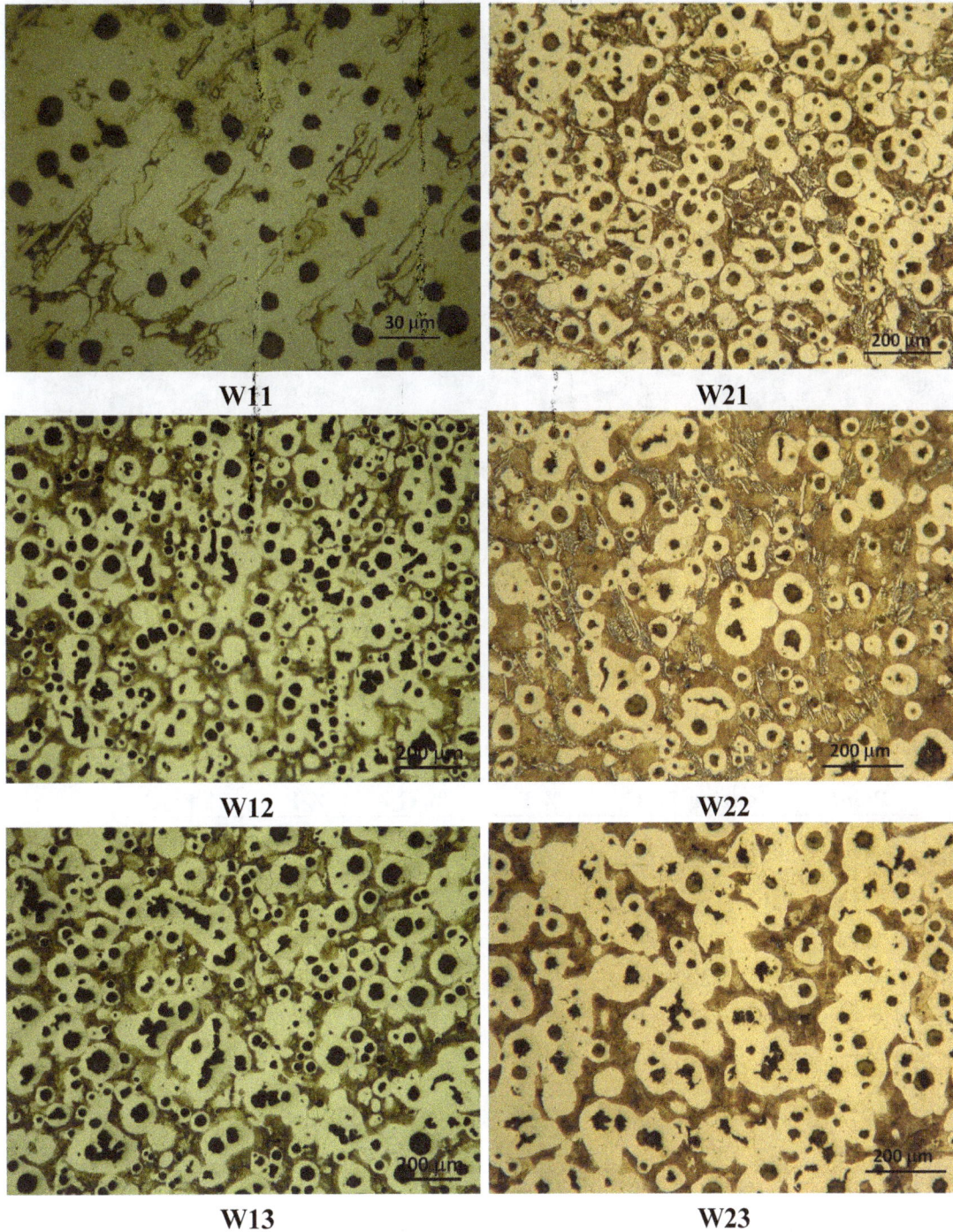

Figure 5. *Cont.*

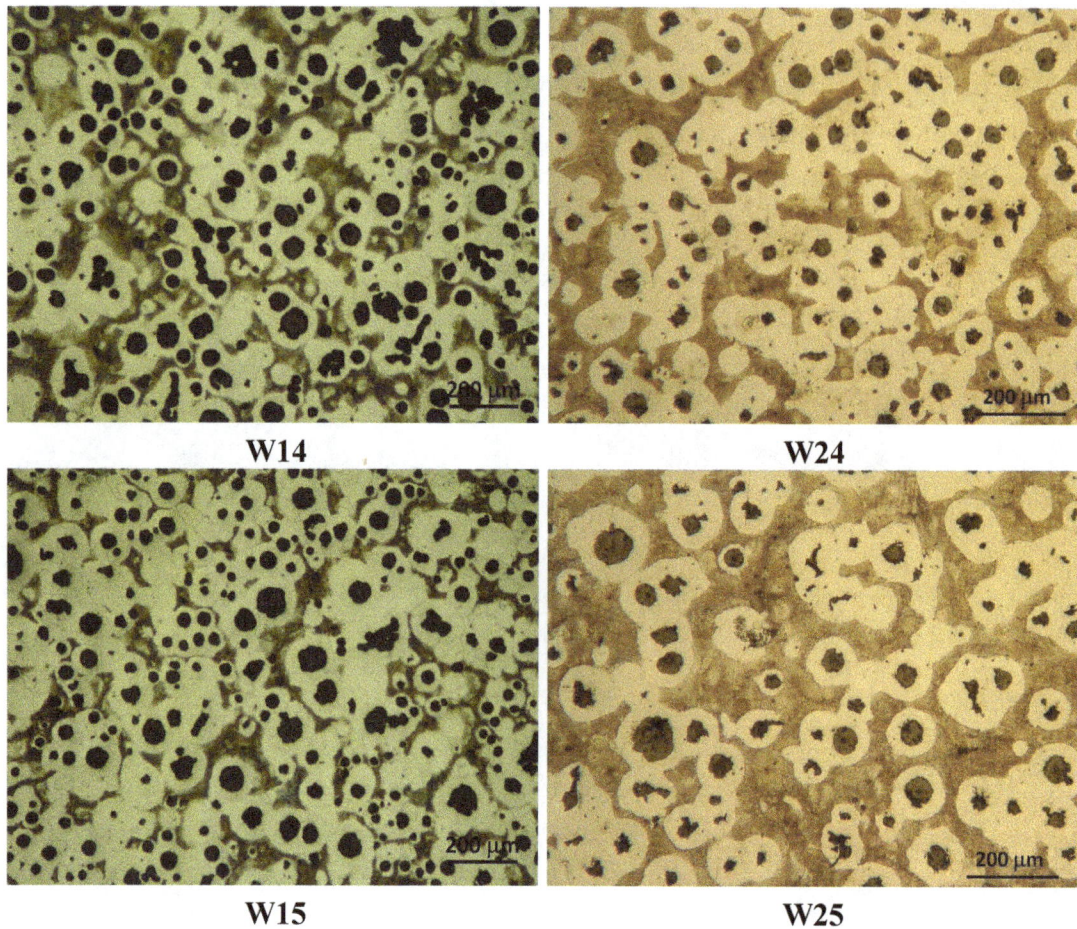

Figure 5. Solidification microstructure at the five points in experiences W1 and W2.

Table 4. Experimental phase fraction at different locations in experiences W1 and W2.

Point	Pouring T, °C	% F	% G	% P	% Fe_3C
W11	1200	76.9	6.9	11.6	4.6
W12	1200	59.9	12.0	28.1	0
W13	1200	59.2	13.2	27.6	0
W14	1200	61.6	12.4	26.0	0
W15	1200	68.6	12.8	18.6	0
W21	1250	42.3	9.4	42.5	5.8
W22	1250	36.9	5.6	51.2	6.3
W23	1250	62.7	7.7	29.6	0
W24	1250	62.8	6.8	30.4	0
W25	1250	53.0	9.4	37.6	0

4.2. Numerical Simulation and Experimental Validation

The nucleation parameters b and c were chosen to be the ones with which the better numerical-experimental adjustment was achieved, *i.e.*, those exhibiting the lower experimental-numerical discrepancy for both the cooling curves and the graphite nodule counts. The parameters that simultaneously minimize this error for experiences W1 and W2 were $b = 4.0 \times 10^{12}$ nuclei/(m °C s) and $c = 340$ °C. The numerical results obtained with these parameters are compared to the experimental measurements presented in the previous section. They are summarized in Table 5, considering the total nodule count at each point together with the numerical-experimental error, computed as:

$$\varepsilon = \frac{\left(experimental\_value - numerical\_value\right)^2}{\left(experimental\_value\right)^2} \qquad (9)$$

It can be seen that, as the pouring temperature increases, the nodule density decreases. This means that bigger nodules are produced because of the effect of a prolonged growth time of these particles in contact with the liquid phase giving rise to a lower quantity of large graphite particles.

Table 5. Error between the average experimental and numerical nodule counts obtained with $b = 4.0 \times 10^{12}$ nuclei/(m °C s) and $c = 340$ °C.

Point	Experimental nodule count N_V (nuclei/m³)	Numerical nodule count N_V (nuclei/m³)	Error
W11	2.47×10^{13}	1.12×10^{13}	0.30
W12	5.45×10^{12}	1.01×10^{13}	0.73
W13	6.53×10^{12}	8.23×10^{12}	0.07
W14	6.75×10^{12}	6.82×10^{12}	0.00
W15	7.62×10^{12}	8.22×10^{12}	0.01
W21	6.28×10^{12}	1.14×10^{13}	0.66
W22	3.47×10^{12}	9.38×10^{12}	2.90
W23	1.46×10^{12}	7.96×10^{12}	19.82
W24	2.00×10^{12}	6.48×10^{12}	5.02
W25	1.23×10^{12}	7.82×10^{12}	28.71

Once the nucleation parameters were determined, Figure 6 plots the experimental and numerical cooling curves for the five points in experiences W1 and W2 together with the experimental cooling rate at the initial stage of cooling. It can be seen that in all the locations, because of the lower pouring temperature, the initial cooling rates are higher in W1 than in W2, which produces the mentioned difference in the graphite nodules size. In both cases, it is observed that in the thinner points of the wedge, the numerical-experimental fit is not as desired. However, the fit improves towards the thicker parts. It is also observed that, as the pouring temperature increases, the solidification time increases as well. In all the locations for experience W1, the computed cooling curves predict a solidification time very similar to the experimental one. It is also notable that in all the locations for experience W1, the calculated cooling curves are located below the experimental ones. The temperature for the eutectic reaction is not well-simulated from point W11 to W13 as in points W14 and W15 where the fit improves slightly. Besides this, although in all the locations the experimental and numerical plateaus differ slightly in temperature, the time extension of such plateaus is very similar, especially in the thick part of the wedge, *i.e.*, points W14 and W15. At the end of the solidification process, the slopes of the experimental and numerical cooling curves are very similar which means that, in general, the numerical-experimental fit is very acceptable. At this point, it can be established that the computed results improve in the thicker parts of the wedge. For experience W2, it can be seen that in all points the calculated cooling curves fall below the experimental ones, with the exception of point W21, where the opposite is seen. In this experience, the best and worse fits were found in points W23 and W22, respectively. For point W23, the numerical and experimental cooling curves are practically superimposed. Regarding characteristic times, the numerical-experimental fit for the eutectic reaction is very acceptable, especially in the thicker parts of the wedge. Besides this, the extensions of the thermal plateaus and the slopes of the calculated and experimental cooling curves at the initial and final stages of the cooling process are very similar, which further improve at the points where the cooling rate is lower. In both experiences, the numerical simulation also predicts longer solidification times for the points located in the thicker part of the sample compared to those located in the thinner part of it.

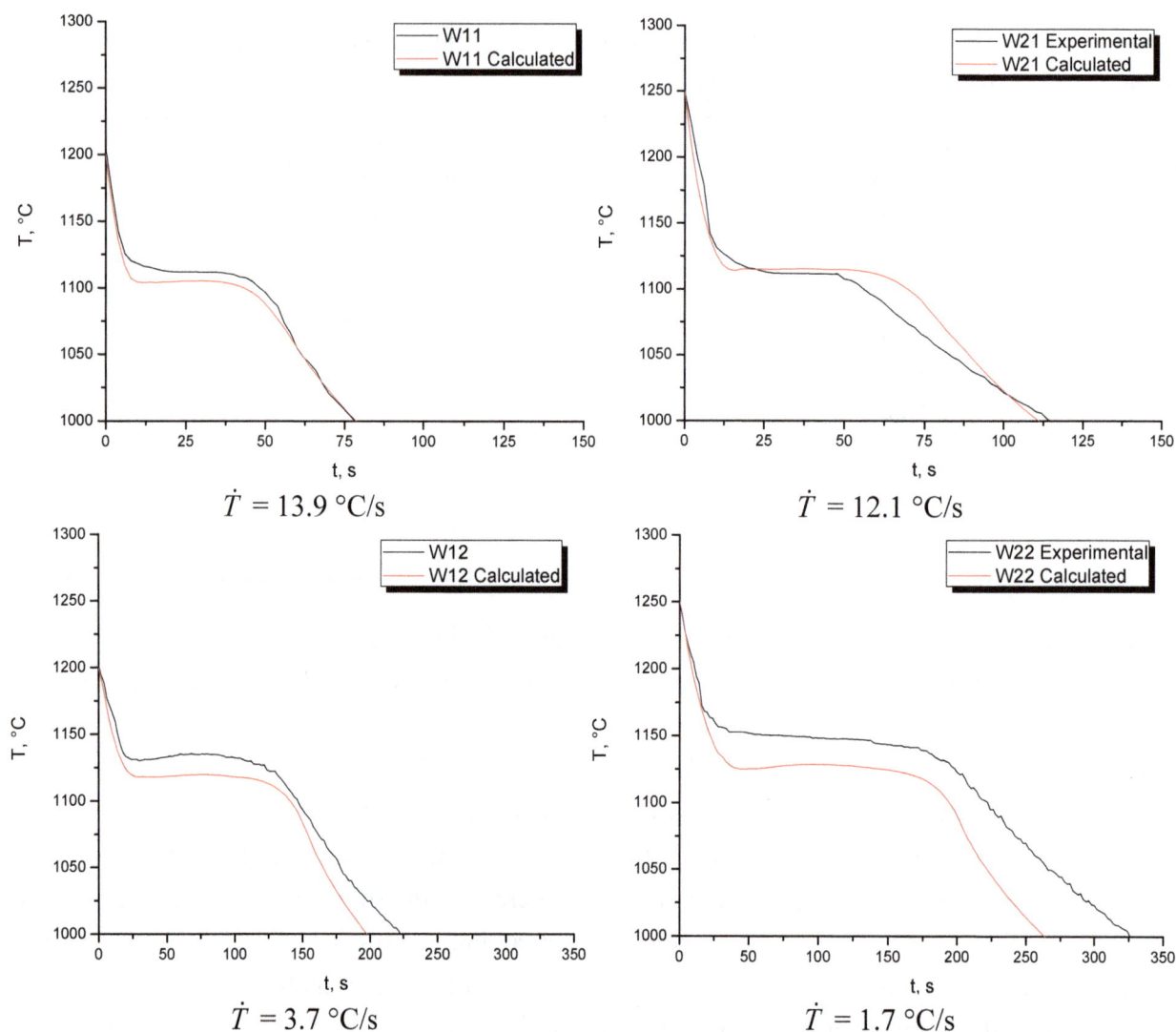

$\dot{T} = 13.9 \,°C/s$ $\dot{T} = 12.1 \,°C/s$

$\dot{T} = 3.7 \,°C/s$ $\dot{T} = 1.7 \,°C/s$

Figure 6. *Cont.*

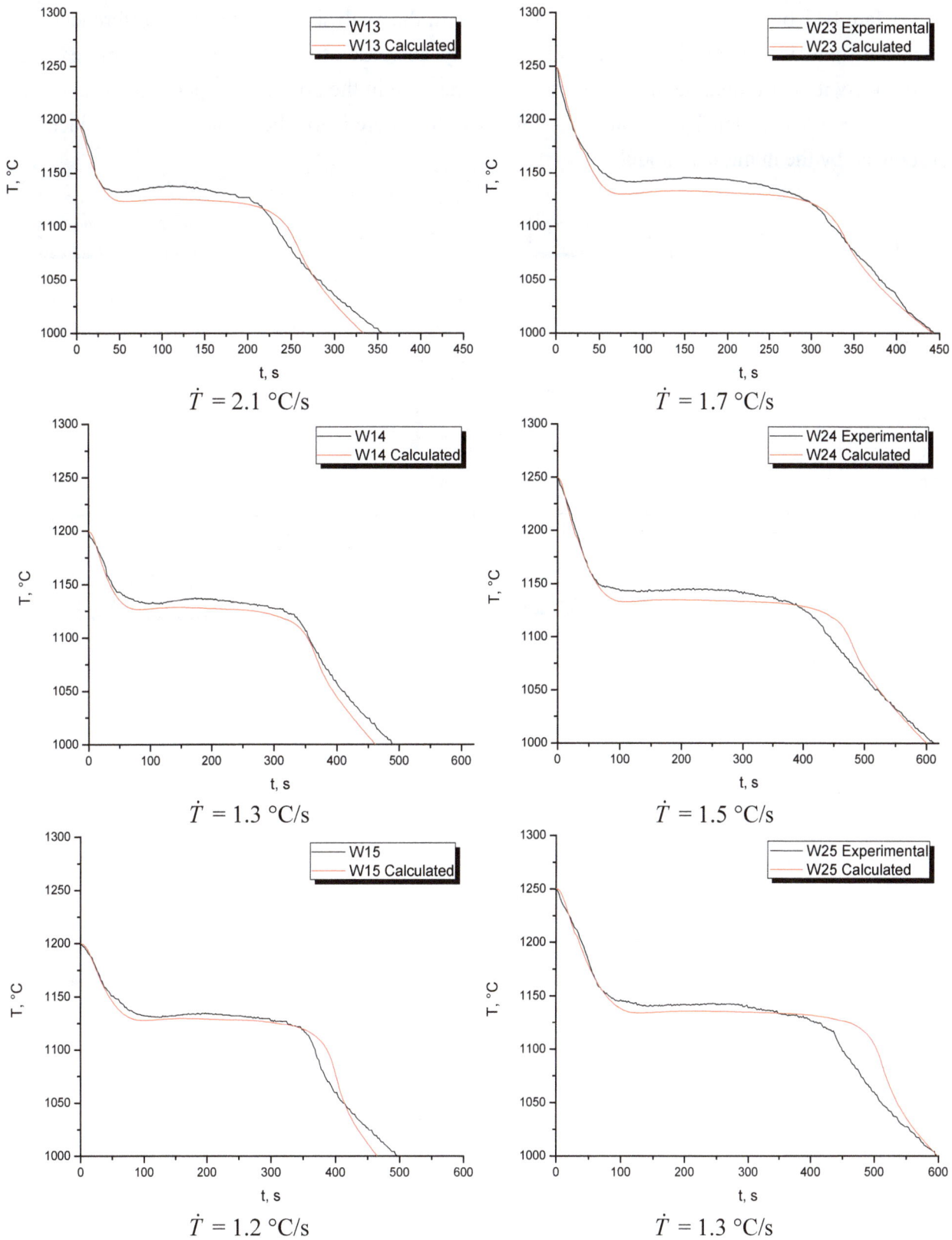

Figure 6. Numerical and experimental cooling curves in experiences W1 and W2.

The experimental and calculated nodule counts for each location in experiences W1 and W2 are shown in Figure 7. Three measurements close to the thermocouples location were made (center of each sample). As can be seen, in both experiences the higher numerical-experimental discrepancies were found in the smaller family, *i.e.*, in the small size nodules. This difference is more significant in W2, not only for family 1 but for the five families considered in the analysis. Despite these differences, in both cases the trend is the same, i.e., the nodule density decreases towards the larger families. As mentioned above, it is also notable in this figure that an increase in the pouring temperature produces a decrease in the nodule density in all the families considered in the analysis. This fact is well-reproduced by the numerical model.

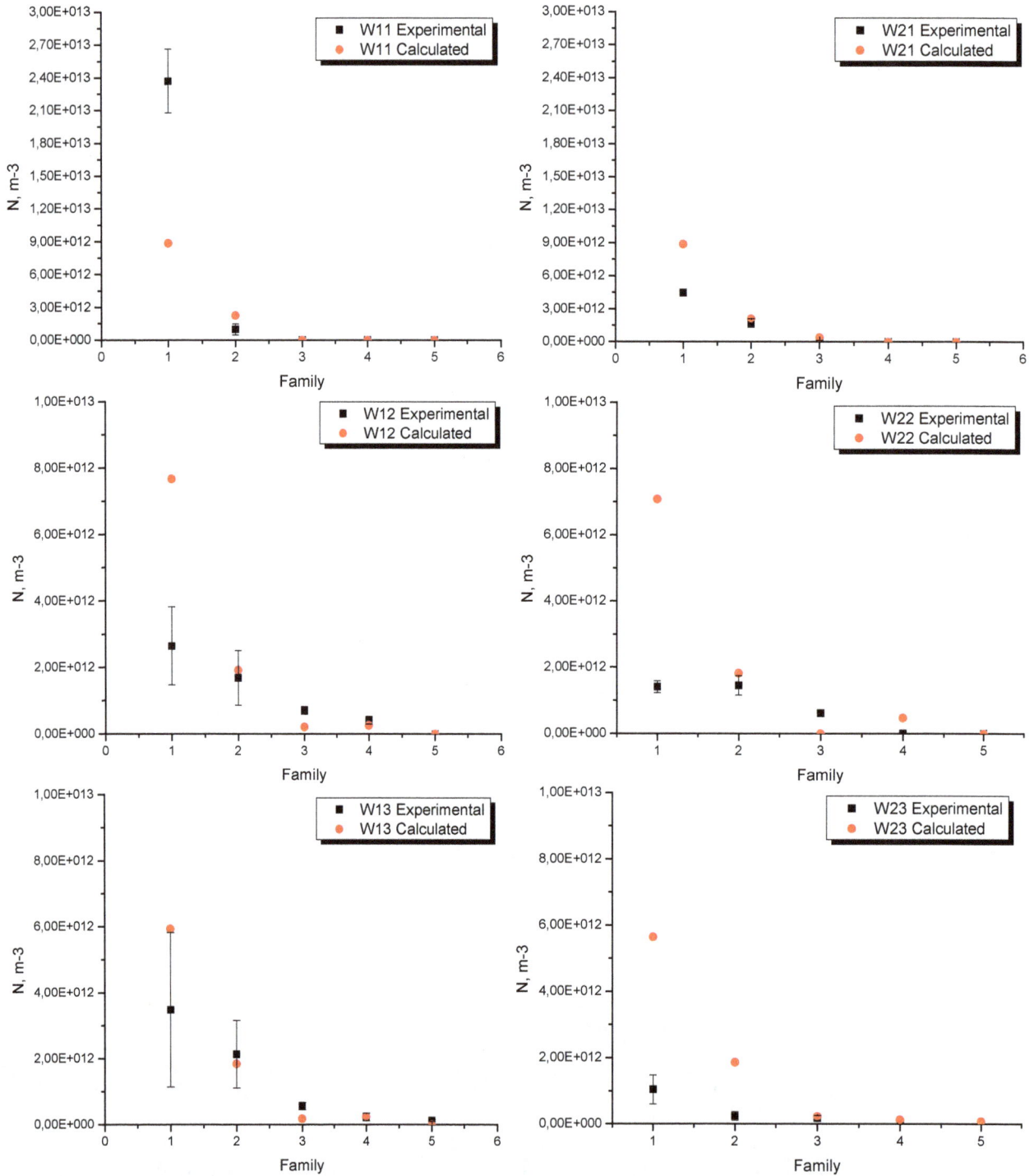

Figure 7. *Cont.*

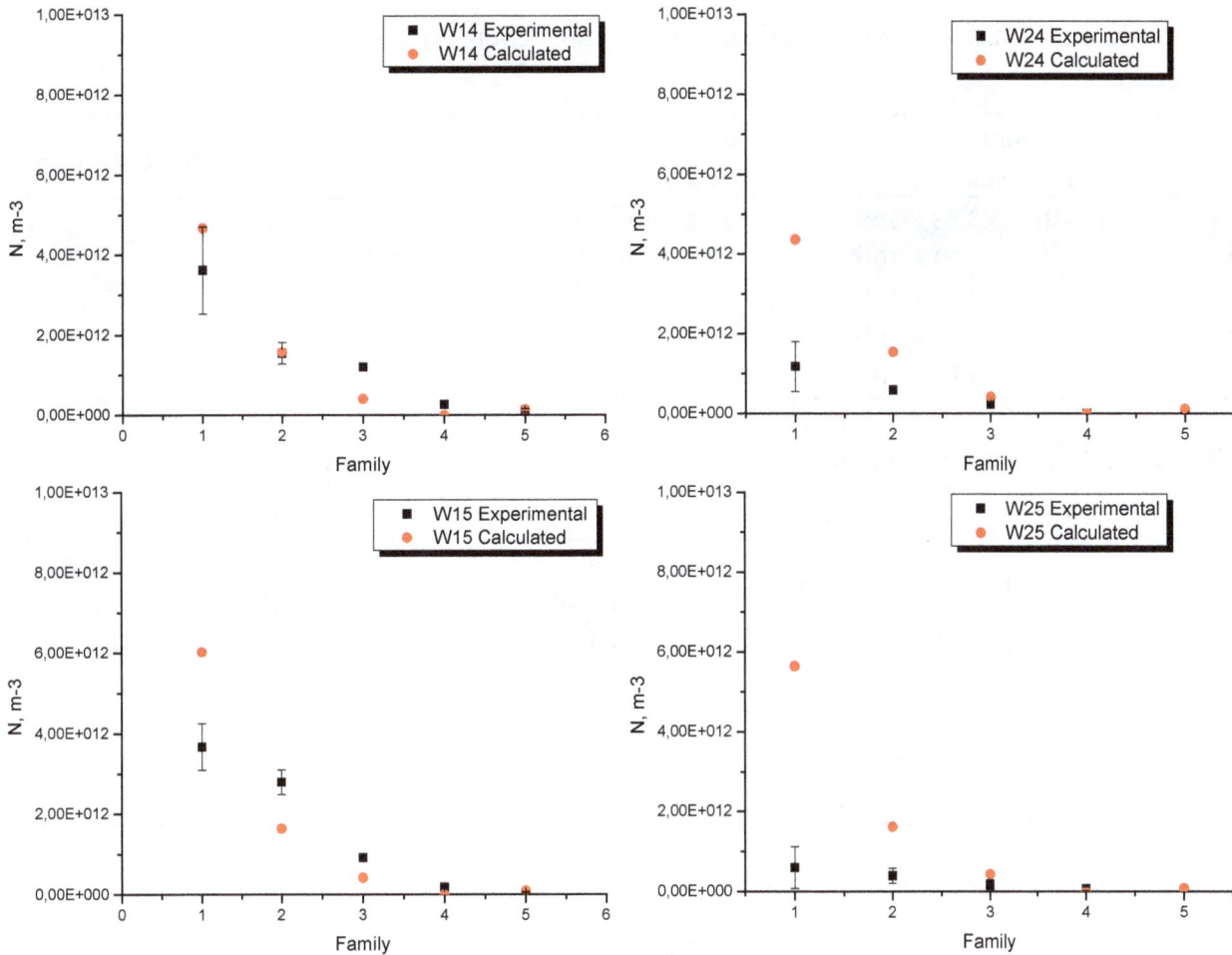

Figure 7. Experimental and numerical nodule count in experiences W1 and W2.

Tables 6 and 7 summarize the experimental and numerical total nodule count and solidification times for experiences W1 and W2. The effect of the pouring temperature on both the microstructure and the solidification times at each location along the part are apparent where, in addition, it is notable that both effects are well-predicted by the numerical model. In particular, it is seen in experience W1 that, considering the extreme values, the nodule count increases with the cooling rate, from the thicker to the thinner part of the wedge. The nodule count and the cooling rate of the points W12 to W14 are similar. Thus, it is possible to affirm that a high cooling rate will produce a high quantity of graphite nodules, which is also well-predicted by the numerical simulation which, in turn, also overestimates the nodule count in W15. In general, the computed solidification time is close to the experimental values, which agree with the graphical presentation in Figure 6. For experience W2, like in W1, as the cooling rate increases, the nodule count increases as well. However, the effect of this parameter is less marked than in W1, where the difference between the tip and the back of the wedge is more significant. Furthermore, in all points for experience W2, the calculated nodule density is higher than the experimental one, even though the calculated solidification times are very similar to the experimental ones.

Table 6. Experimental and numerical total nodule count and solidification times in experience W1.

Point	N_V (nuclei/m³) Experimental	N_V (nuclei/m³) Calculated	$t(T_{ES})(s)$ Experimental	$t(T_{ES})(s)$ Calculated
W11	$2.47 \times 10^{13} \pm 3.43 \times 10^{12}$	1.12×10^{13}	56.2	57.6
W12	$5.45 \times 10^{12} \pm 2.23 \times 10^{12}$	1.01×10^{13}	178.3	156.3
W13	$6.53 \times 10^{12} \pm 3.62 \times 10^{12}$	8.23×10^{12}	252.3	258.6
W14	$6.75 \times 10^{12} \pm 1.50 \times 10^{12}$	6.82×10^{12}	354.2	366.7
W15	$7.62 \times 10^{12} \pm 1.01 \times 10^{12}$	8.22×10^{12}	375.9	402.1

Table 7. Experimental and numerical total nodule count and solidification times in experience W2.

Point	N_V (nuclei/m³) Experimental	N_V (nuclei/m³) Calculated	$t(T_{ES})(s)$ Experimental	$t(T_{ES})(s)$ Calculated
W21	$6.28 \times 10^{12} \pm 3.94 \times 10^{11}$	1.14×10^{13}	65.8	68.9
W22	$3.47 \times 10^{12} \pm 1.56 \times 10^{11}$	9.38×10^{12}	224.6	202.8
W23	$1.46 \times 10^{12} \pm 6.03 \times 10^{11}$	7.96×10^{12}	316.3	341.4
W24	$2.00 \times 10^{12} \pm 7.13 \times 10^{11}$	6.48×10^{12}	433.0	480.7
W25	$1.23 \times 10^{12} \pm 7.26 \times 10^{11}$	7.82×10^{12}	438.3	513.8

Figure 8 shows the numerical and experimental relationship between the total nodule count and the wall thickness for experiences W1 and W2. It is observed that, as the wall thickness increases, the nodule density decreases, which is also well-predicted by the simulation, especially in experience W1. It is both numerically and experimentally seen that the nodule density does not greatly vary along the wedge, which is directly related to the similar cooling conditions of the points located from 20 mm upwards (from W12 and W22). With these results, it is possible to infer that the thermal contribution of the thicker parts plays an important role on these solidification experiences' morphologies.

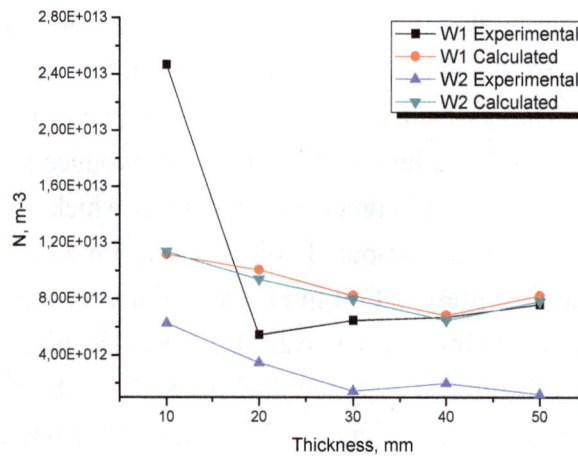

Figure 8. Relationship between numerical and experimental total nodule count and wall thickness for experiences W1 and W2.

5. Conclusions

From the realization of this study the following conclusions can be drawn:

- For each solidification experience, an overall reasonably good numerical-experimental fit was achieved.
- In the thinner parts of experience W1 (points W11 and W12), the model predicts a lower temperature for the eutectic reaction to start and a longer solidification time than the experimental ones.
- For experience W2, the fact mentioned above is clearly notable at point W22, where an undesirable numerical-experimental fit is obtained.
- An increase in the pouring temperature produces larger graphite nodules, which produces a lower nodule density towards the larger families.
- The effect mentioned above, which is mainly caused by the longer time that graphite particles are submitted to this high temperature, is well-predicted by the numerical model.
- Another effect of the high pouring temperature is decreased recalescence and increased solidification time. These facts are well-predicted by the numerical model in all the wedge locations.
- Related to the previous point, as the pouring temperature increases, the cooling curves move to higher temperatures, *i.e.*, the temperature for the eutectic reaction and the maximum temperature reached after recalescence increase.
- From the experimental validation carried out, a good fit is obtained between the experimental and the calculated cooling curves, especially in the thicker sections of the wedges.
- The mentioned above applies for the two solidification experiences where the larger discrepancies were observed at the thinner parts of the wedge. This could be related to the low capability of the model to reproduce extremely high cooling rates.
- It was numerically and experimentally verified that, as the pouring temperature increases, the cooling rate decreases.
- Regarding the nodule count, it can be seen that in both cases, the numerical-experimental fit greatly improves towards the larger families.
- From wedge thicknesses larger than 20 mm, the nodule count is very similar along the wedges. These results are related to the similar cooling rates that occur in these locations.
- Finally, the analysis carried out in this work establishes that is possible to understand and achieve greater control over the pouring temperature of the final microstructure of a nodular cast iron.

Acknowledgments

The authors thank CONICYT (Chilean Council of Research and Technology) for the support provided by Project Fondecyt 1130404.

Author Contributions

Alex Escobar: experimental work, analysis and discussion of results, writing and revision of the manuscript.

Diego Celentano: computational implementation of the model, discussion of results, revision of the manuscript.

Marcela Cruchaga: discussion of results, revision of the manuscript.

Bernd Schulz: discussion of results.

Conflicts of Interest

The authors declare no conflict of interest.

References

1. Sorelmetal. *Ductile Iron Data for Design Engineers*; Rio Tinto Iron & Titanium Inc.: Quebec, QC, Canada, 1990.

2. Ashraf Sheikh, M. Production of carbide-free thin ductile iron castings. *J. Univ. Sci. Technol. Beijing* **2008**, *15*, 552–555.

3. Stefanescu, D.M. *Science and Engineering of Casting Solidification*, 1st ed.; Kluwer Academic/Plenum Publishers: New York, NY, USA, 2002.

4. Pedersen, K.M.; Tiedje, N.S. Graphite nodule count and size distribution in thin-walled ductile cast iron. *Mater. Charact.* **2008**, *59*, 1111–1121.

5. Su, K.C.; Ohnaka, I.; Yamauchi, I.; Fukusako, T. Computer simulation of solidification of nodular cast iron. *MRS Proc.* **1985**, *34*, 181–189.

6. Lesoult, G.; Castro, M.; Lacaze, J. Solidification of spheroidal cast iron-I. Physical modelling. *Acta Mater.* **1998**, *46*, 983–995.

7. Lacaze, J.; Castro, M.; Lesoult, G. Solidification of spheroidal cast iron-II. Numerical simulation. *Acta Mater.* **1998**, *46*, 997–1010.

8. Rappaz, M.; Richoz, J.D.; Thévoz, P. Modelling of solidification of nodular cast iron. In Proceedings of the Euromat 89, Aachen, Germany, 22–24 November 1989; pp. 1–6.

9. Castro, M.; Alexandre, P.; Lacaze, J.; Lesoult, G. Microstructures and solidification kinetics of cast irons: Experimental study and theoretical modeling of equiaxed solidification of S.G and grey cast iron. In Proceedings of the Cast Iron IV Conference, Tokyo, Japan, 4–6 September 1989; p. 433.

10. Fras, E.; Kapturkiewicz, W.; Burbielko, A. Micro-macro modeling of casting solidification controlled by transient diffusion and undercooling. In *Modelling of Casting, Welding and Advanced Solidification Processes VII*; Cross, M., Campbell, J., Eds.; The Minerals, Metals and Material Society: Warrendale, PA, USA, 1995; pp. 679–686.

11. Liu, J.; Elliot, R. Numerical modelling of the solidification of ductile iron. *J. Crystal Growth* **1998**, *191*, 261–267.

12. Fras, E. Computer-aided simulation of the kinetics of solidification of the eutectic ductile cast iron. *MRS Proc.* **1984**, *34*, 191–199.

13. Fredriksson, H.; Svensson, I. Computer simulation of the structure formed during solidification of cast iron. *MRS Proc.* **1985**, *34*, 273–284.

14. Stefanescu, D.M.; Kanetcar, C.S. Computer modeling of the Solidification of Eutectic Alloys: The case of Cast Iron. *Comput. Simul. Microstruct. Evolut.* **1985**, 171–188.

15. Rivera, G.; Boeri, R.; Sikora, J. Influence of the solidification microstructure on the mechanical properties of ductile iron. *Int. J. Cast Met. Res.* **1999**, *11*, 533–538.

16. Boeri, R. The Solidification of Ductile Cast Iron. Ph.D. Thesis, the University of British Columbia, Vancouver, BC, Canada, 1989.

17. Celentano, D.J.; Dardati, P.M.; Godoy, L.A.; Boeri, R. Computational simulation of microstructure evolution during solidification of ductile cast iron. *Int. J. Cast Met. Res.* **2008**, *21*, 416–426.

18. Dardati, P.M.; Celentano, D.J.; Godoy, L.A.; Chiarella, A.A.; Schulz, B.J. Analysis of ductile cast iron solidification: Numerical simulation and experimental validation. *Int. J. Cast Metals Res.* **2009**, *22*, 390–400.

19. Dardati, P.M.; Godoy, L.A.; Celentano, D.J. Microstructural simulation of solidification process of spheroidal-graphite cast iron. *ASME J. Appl. Mech.* **2006**, *73*, 977–983.

20. Escobar, A.; Celentano, D.; Cruchaga, M.; Lacaze, J.; Schulz, B.; Dardati, P.; Parada, A. Experimental and numerical analysis of effect of cooling rate on thermal–microstructural response of spheroidal graphite cast iron solidification. *Int. J. Cast Met. Res.* **2014**, *27*, 176–186.

21. Abdul Halil, F.B. Investigation of Ductile Iron Sandwich Treatment Process Parameter. Bachelor of Mechanical Engineering Thesis, University of Malaysia Pahang, Kuantan, Malaysia, 2009.

22. Elliott, R. *Cast Iron Technology*; Jaico Publishing House: London, UK, 2005.

23. Celentano, D.J. Un Modelo Termomecánico para Problemas de Solidificación de Metales. Ph.D. Thesis, Polytechnical University of Catalunya, Barcelona, Spain, 1994.

24. Celentano, D.J.; Oller, S.; Oñate, E. A coupled thermomechanical model for the solidification of cast metals. *Int. J. Solids Struct.* **1996**, *33*, 647–673.

25. Chipman, J.; Alfred, R.M.; Gottet, L.W.; Small, R.B.; Wilson, D.M.; Thomson, C.N.; Guernsey, D.L.; Fultons, J.C. The solubility in molten iron and in iron-silicon and iron-manganese alloys. *Trans. ASM* **1952**, *44*, 1215–1232.

26. Gustafson, P. A thermodynamic evaluation of the Fe-C system. *Scand. J. Metall.* **1985**, *14*, 259–267.

27. Henschel, C.; Heine, R.W. Some effects of oxygen on the solidification of cast irons. *AFS Cast Metals Res. J.* **1971**, *7*, 93–104.

28. Andersson, J.O.; Helander, T.; Höglund, L.; Shi, P.F.; Sundman, B. Thermocalc and DICTRA, Computational tools for materials science. *Calphad* **2002**, *26*, 273–312.

29. Torres Camacho, G.; Lacaze, J.; Bak, C. Redistribution of alloying elements during graphitisation of mottled spheroidal graphite cast iron. *Int. J. Cast Met. Res.* **2003**, *16*, 173–178.

11

Graphite and Solid Fraction Evolutions during Solidification of Nodular Cast Irons

Amaia Natxiondo [1], Ramón Suárez [1], Jon Sertucha [2] and Pello Larrañaga [2,*]

[1] Veigalan Estudio 2010, Durango E-48200, Spain; E-Mails: anatxiondo@veigalan.com (A.N.); rsuarez@veigalan.com (R.S.)

[2] Engineering, R&D and Metallurgical Processes Department, IK4-Azterlan, Durango E-48200, Spain; E-Mail: jsertucha@azterlan.es

* Author to whom correspondence should be addressed; E-Mail: plarranaga@azterlan.es

Academic Editor: Hugo Lopez

Abstract: Ductile iron casting production is strongly affected by austenite and graphite distribution obtained after the solidification process. At the same time it is accepted that solidification behavior can be considered as hypo-, hyper- or eutectic depending on the chemical composition; there is still some misconception about the growth evolution of graphite nodules and about solid fraction progression. Quenching experiments were performed on two different carbon equivalent compositions using inoculated and non-inoculated thermal analysis standard samples with the aim of freezing the existing phases at different solid fractions for each alloy. As a result of these experiments, it was possible to study the structural features found at different locations of each sample and at different stages of solidification. Additionally nodule evolution during the liquid-solid transformation was also analyzed and discussed regarding the chemical and processing characteristics of the prepared alloys.

Keywords: solidification; ductile irons; quenching experiments; graphite nodules; solid fraction; austenite

1. Introduction

Ductile iron solidification is a complex process in which the mechanisms that occur during the mushy zone development have not yet been fully mastered [1]. A proper understanding of solid fraction evolution, regarding graphite and austenite nucleation and growth, together with the corresponding contraction-expansion phenomena is of special interest in order to establish the most appropriate industrial procedure to manufacture sound castings and to avoid the formation of shrinkage defects [2]. However this task is not easy as there are a wide variety of solidification possibilities for a given chemical composition. Room temperature samples are commonly available for metallographic characterization but they only show minor traces of graphite and austenite developments during solidification.

Attempts to obtain suitable modeling of the solidification process have been a challenge for a while. One of the first approaches to quantitatively explain and to obtain a feasible structure of cast iron solidification was done by Oldfield as early as 1966 [3]. The quenching experiments performed at that time were then adapted and repeatedly used with the aim to freeze the structural changes that occur under specific conditions and to determine their evolution during solidification. In addition to these aspects quenching methods have been also used to validate the mathematical and thermodynamic bases applied to develop the obtained solidification models. In this way several methodologies have been used to quench cast iron samples. The most used one is to keep the samples in a furnace device equipped with a cooling holder below the furnace. Thus samples are dropped at different temperatures into the quenching liquid [4,5]. This methodology has two main advantages: full temperature control of the samples just before quenching is possible and is highly reproducible. Other researchers quenched different samples during the solidification process by immersing them in water [6,7]. This method is a laborious work that requires good coordination efforts and has significant difficulties in reproducing the experiment exactly at the desired solid fraction. More advanced quenching procedures have also been reported in the literature such as the one where portions of melt are extracted using a quartz tube in order to obtain different solid fractions in the same sample [8].

Regarding information obtained from quenching experiments, various goals were achieved by this technique such as the detection of the graphite branching process and its relationship with modifying elements to form nodules [9], the study of the growing process in different graphite particle shapes [10] (mainly in nodules and compacted particles), the determination of chemical element distribution in the solidifying alloys [11] and effective segregation coefficients [8] as primary solid can easily be detected on quenched samples. Other quenching studies showed how inoculation affected the equiaxial solidification evolution [6] or permitted the quantification of the precipitated phases and related them to Fourier Thermal Analysis [4].

Quenching methodologies can be used to deepen phase transformation knowledge. Previous examples of similar topics are the analysis of microstructural evolution during solidification [12] and during the solid transformation of two ductile iron alloys made by Guo and Stefanescu [5] as well as the primary study performed by the authors on solidification in both ductile and gray iron alloys with different carbon equivalent contents and inoculation conditions [7,13]. In the present work, a detailed study of solid fraction, graphite area fraction and nodule count evolutions during solidification of four ductile cast irons with different chemical and inoculation conditions was approached in combination with the thermal characterization of these alloys.

2. Experimental Section

All samples studied in the present work were obtained from two different cast iron alloys whose chemical compositions are shown in Table 1. Carbon content is determined by combustion technique (and is in good agreement with the one obtained by the thermal analysis system, referred to as C_{AT} in Table 2), silicon by gravimetric method and the rest of the elements by Inductively Coupled Plasma (ICP) (Perkin Elmer Optima 5300V, Shelton, CT, USA). The accuracy of each element according to the methodology used is given in brackets for each element in the table.

Table 1. Chemical composition of the prepared cast irons (wt.%).

Alloy	C (0.06)	Si (0.02)	CE * (0.07)	Mn (0.01)	P (0.004)	S (0.001)	Cr (0.002)
A	3.20	2.33	3.98	0.18	<0.015	0.010	0.04
B	3.53	2.66	4.42	0.21	0.017	0.008	0.06
Alloy	Mo (0.002)	Ni (0.007)	Cu (0.005)	Mg (0.004)	Al (0.004)	Ti (0.002)	Sn (0.003)
A	<0.01	0.04	0.03	0.045	0.010	0.016	<0.005
B	<0.01	0.06	0.03	0.047	0.010	0.014	<0.005

* Carbon equivalent calculated as CE = C + Si/3.

Melts were prepared in a 100 kg medium frequency induction furnace (250 Hz, 100 kW) where metallic charges composed of 37% cast iron returns, 27% low alloyed steel scrap, and 36% low alloyed pig iron were introduced. Particular amounts of a commercial graphite (C = 98.9 wt.%, S = 0.03 wt.%) and of FeSi alloy (Si = 74.6 wt.%, Ca = 0.3 wt.%, Al = 0.7 wt.%) were also added to metallic charges so as to approach the designed carbon and silicon contents in the prepared base alloys. After melting, the composition was checked and finally adjusted according to the required contents of these two elements. Then the melt temperature was increased to 1510–1520 °C and its surface was skimmed before being transferred to a 50 kg capacity ladle for nodularization treatment with 0.55 kg (1.1 wt.% of the batch weight) of a FeSiMg alloy (grain size 2–20 mm, Si = 43.54 wt.%, Mg = 5.96 wt.%, Ca = 0.95 wt.%, Al = 0.48 wt.% and RE = 1.08 wt.%) by the sandwich method. The FeSiMg alloy was positioned at the bottom of the ladle and then covered with steel scrap (grain size 5–15 mm) before tapping the melt from the furnace. The treatment temperature was in the range 1470–1495 °C.

Four different series each one composed of five thermal analysis (TA) standard cups with a K-type thermocouple located at the center of them were previously placed close to the nodulization treatment area. These TA cups are 35 mm × 35 mm (average) in horizontal section while their height varies from 37 to 39 mm, giving a geometric modulus in the range of 0.61–0.63 cm (final weight variation from 330–350 g). In two series denoted as A1 and B1, 0.7 g (0.20 wt.% of the sample weight) of a commercial inoculant (grain size 0.2–0.5 mm, Si = 69.9 wt.%, Al = 0.93 wt.%, Ca = 1.38 wt.%, Bi = 0.49 wt.% and RE = 0.37 wt.%) were added to all cups to obtain the inoculated samples. The other two series were denoted as A2 and B2 and contained plain cups for obtaining the non-inoculated samples. All TA cups were connected to the Thermolan system [2] to record the corresponding cooling curves and then to obtain the most relevant thermal parameters from each one.

After completion of the Mg-treatment reaction for each batch, a sample was taken in order to determine its chemical composition (see Table 1 where the contribution of the inoculation carried out in the A1 and B1 series has not been included) and successive samples were then taken by rapidly pouring the five TA cups of the corresponding series. In each of these series four samples were successively quenched by introducing them into a cold water tank (10–15 °C) while one of the samples was retained in the TA cup holder to record the whole cooling curve. All quenched samples were vigorously shaken when introducing them into the water bath in order to promote effective quenching of the samples. After cooling to room temperature all the samples were vertically cut in half and one of the newly obtained entire surfaces was prepared for metallographic examination. The surfaces were divided into twelve areas as shown in Figure 1 to evaluate the effect of the quenching process on them. Optical microscopy observations were carried out in these areas in order to estimate the solid fraction (f_S), graphite area fraction (f_{GA}) and nodule count (N) values using commercial image analysis software.

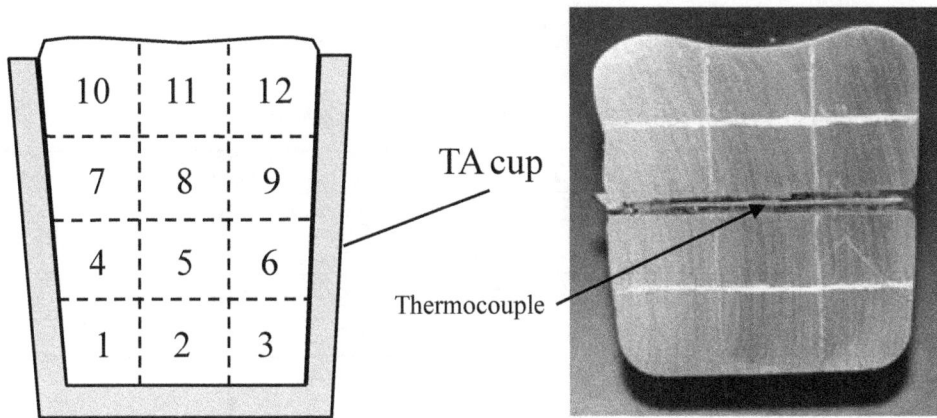

Figure 1. Fields distribution of thermal analysis (TA) samples made for metallographic analyses.

Solid fraction values were determined after etching the surfaces with Nital-5 etchant. The criteria adopted for analyzing all the structures obtained was previously explained in reference [7], *i.e.*, those areas occupied by carbides and martensite were respectively considered as former liquid and austenite regions before quenching effect. Nodule count and f_{GA} values were determined on these surfaces without etching.

3. Results and Discussion

3.1. Thermal Characterization of Cast Irons and Quenching Process

Thermal characterization made on the two cast iron alloys led to the record of the four cooling curves plotted in Figure 2. These records correspond to the non-quenched samples with and without inoculant addition in the TA cup for each composition. The most relevant data obtained from the recorded cooling curves are included in Table 2 according to reference [2]. It can be observed that solidification starts at higher temperatures for the A cast irons than for the B ones, which can be easily recognized by higher liquidus temperature values (T_{liq}). While A alloy shows a clear hypo-eutectic behavior, B alloys, considered as hyper-eutectic by chemical composition, show a slightly hypo-eutectic behavior according to their solidification model.

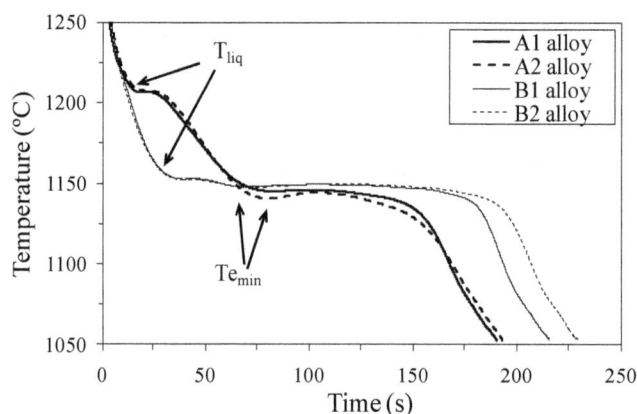

Figure 2. Cooling curves recorded from inoculated and non-inoculated samples.

Table 2. Thermal analysis (TA) parameters obtained from the cooling curves plotted in Figure 2.

Alloy	Inoculation	T_{liq} (°C)	Te_{min} (°C)	Rc (°C)	T_{sol} (°C)	C_{AT} (%)
A1	yes	1206.4	1145.3	0.9	1102.4	3.20
A2	no	1207.5	1140.9	3.6	1097.5	
B1	yes	1152.8	1148.0	1.5	1107.6	3.52
B2	no	1152.4	1147.7	2.1	1102.1	

Bulk eutectic reaction, where the massive austenite-graphite precipitation and growth take place, starts at lower temperatures for the A alloys, assessed by lower minimum eutectic temperature (Te_{min}). Inoculation effects only seem to be observed on the bulk eutectic areas as the graphite nodule precipitation must be accelerated from the beginning of this period and consequently lower recalescence ($Rc = Te_{max} - Te_{min}$) values are detected. Additionally inoculation effects become less relevant for the A cast irons due to the nucleation promoter effect of high carbon and silicon contents in these alloys. These results are in good agreement with the ones previously reported by the authors for ductile irons [7] and for gray cast irons [13].

Figure 3 shows the different time steps when quenching experiments were primarily carried out for each of the four alloys studied in this work. Note that quenching effectiveness strongly depends on the fast cooling applied to each sample so that deviations between the steps illustrated in Figure 3 and by the corresponding fs values, metallographically obtained, are possible. On the other hand it is necessary here to draw to attention that the A22, A24, B21 and B22 samples were discharged due to the appearance of pearlite in the metallic matrix after quenching which is associated to lack of achievement of sufficient cooling rate during the experiment. Although the quenching process is quite useful, unfortunately its methodology is not similarly efficient in all fields of a given sample. Moreover low quenching efficiencies are expected to be more important in those fields that are less exposed to quenching media, i.e., #1 to #6 or the internal ones (see Figure 1). As a consequence it becomes very convenient here to analyze the quenching method effectiveness by means of fs evolution with all fields present in each sample.

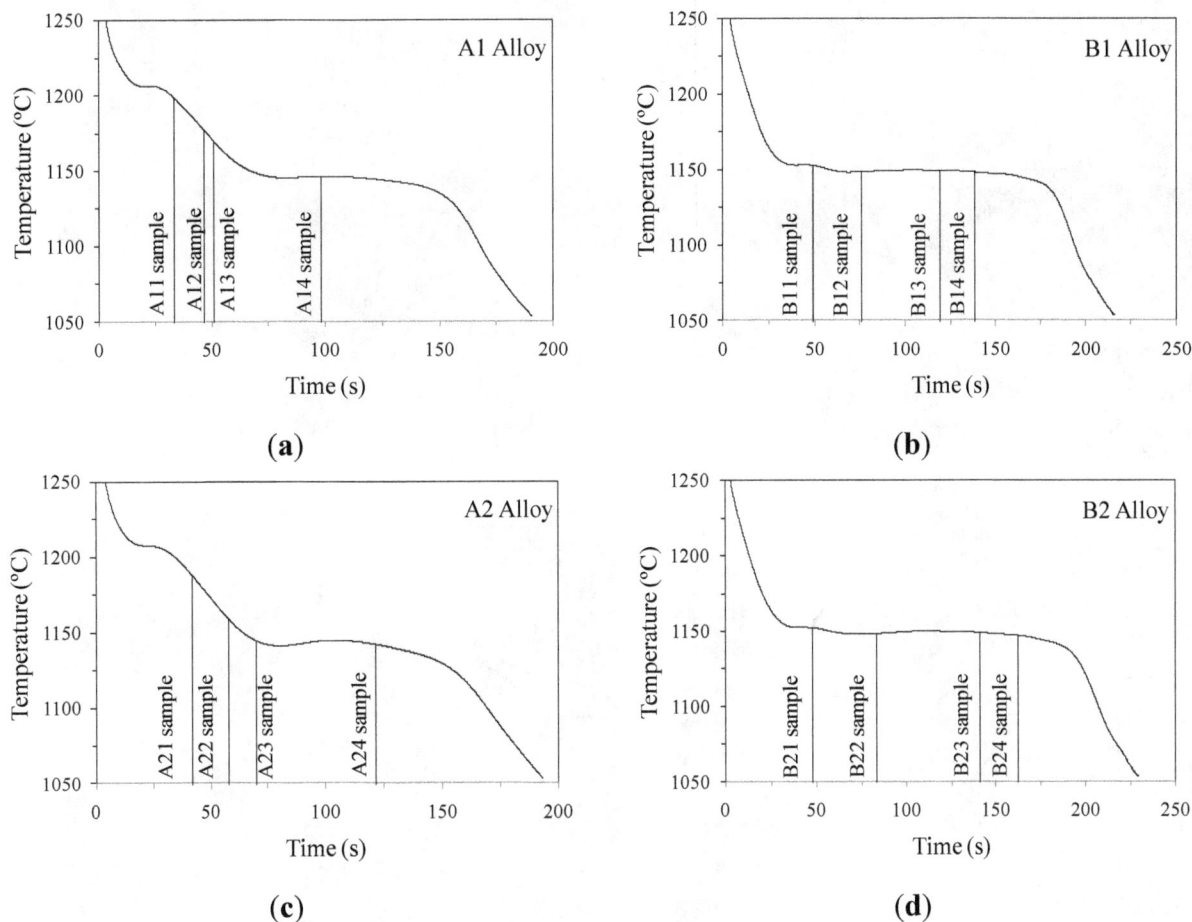

Figure 3. Quenching experiments made for each cast iron alloy. (**a**) A1 Alloy; (**b**) B1 Alloy. (**c**) A2 Alloy; (**d**) B2 Alloy.

3.2. Solid Fraction and Graphite Area Fraction Evolution during Solidification

Quenching experiments performed in such a casting size show some difficulties associated to heat release from the sample. At the same time, the top part is in direct contact with the quenching bath, which assures sufficient cooling rate to succeed in freezing the structure, the bottom part has more difficulty to release the heat and so higher fs are expected in those areas. For this reason, in the same sample, different fs and graphite precipitation can arise from the quenching process. As the main aim of this work was to correlate graphite amount evolution and fs, it was considered convenient to analyze the 12 different fields described in Figure 1. Solid fraction evolutions of both inoculated A and B alloys in field No. 11, *i.e.*, the most sensible to quenching, are shown in Figures 4 and 5 respectively.

Metallographic analyses on inoculated A samples show dendrites growing from the liquid while graphite nodules precipitate close to the austenite-liquid interface. This observation agrees with the hypo-eutectic solidification model showed in Figure 2. In the case of inoculated B alloy, solidification starts with the graphite nodules surrounded by small austenite shells together with some isolated dendrites that also exhibit some nodules in contact with them. These last characteristics are usually found in ductile cast irons with a near eutectic solidification [7].

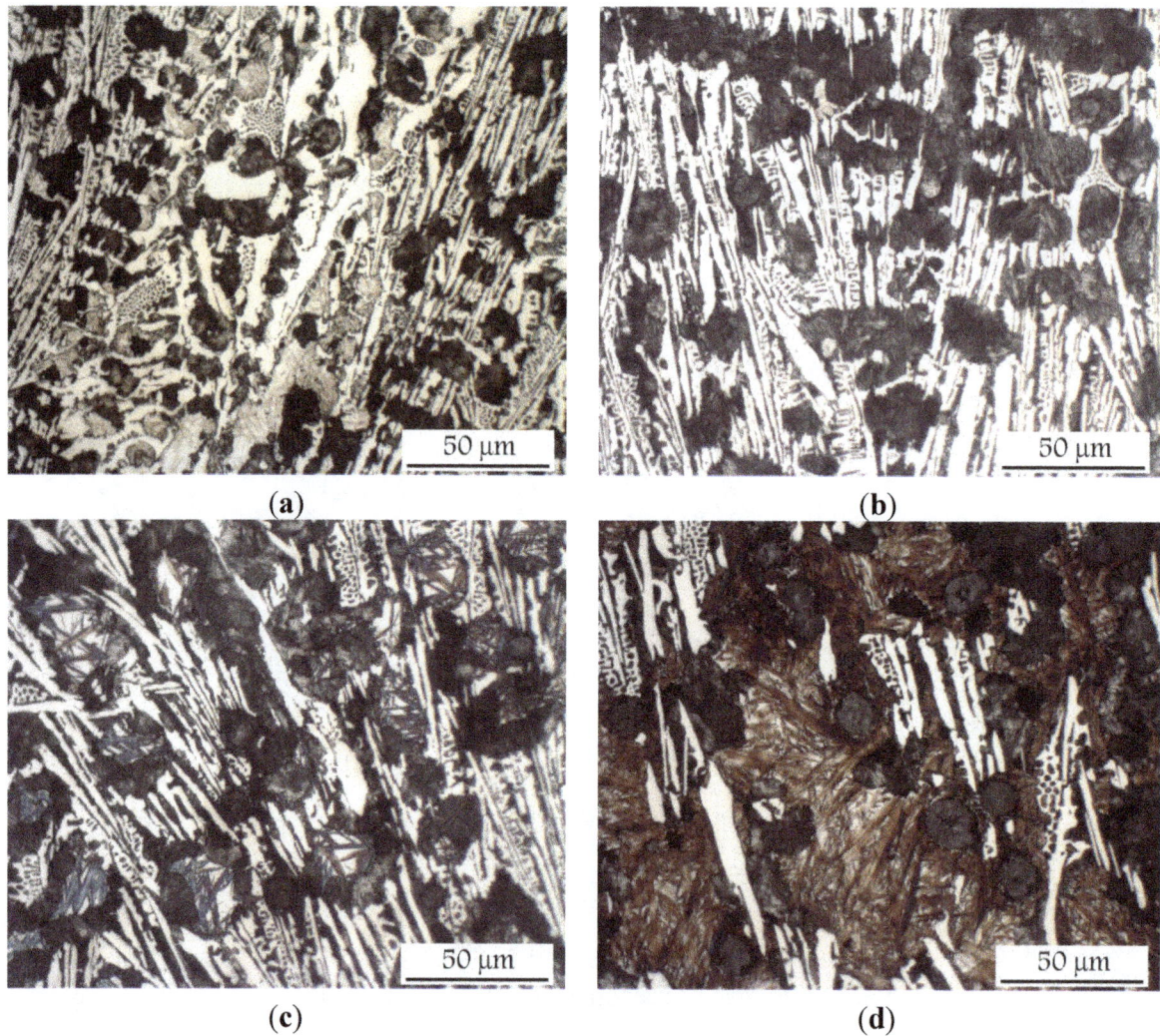

Figure 4. Solid fraction evolution of A1 samples. (**a**) A11 Sample; (**b**) A12 Sample; (**c**) A13 Sample; (**d**) A14 Sample.

When comparing A and B alloys higher f_S and graphite area fraction (f_{GA}) values are observed in the later one than in the first one. This result is more evident in case of the more sensitive fields to quenching process, *i.e.* #10 to #12. This behaviour, besides to higher carbon content, must be mainly related to the higher ability to precipitate graphite directly from the liquid in the B alloy which might increase additionally the corresponding austenite precipitation in contact with graphite. Although the non inoculated alloys have shown a similar behaviour to the inoculated ones regarding the effect of CE, inoculation seems to affect more the number of precipitating sites than to f_S evolution in the sample fields (Figure 6). It can be observed in this figure that inoculation promotes high nodule counts at the very beginning of the eutectic reaction, which confirms the differences observed in their respective cooling curves.

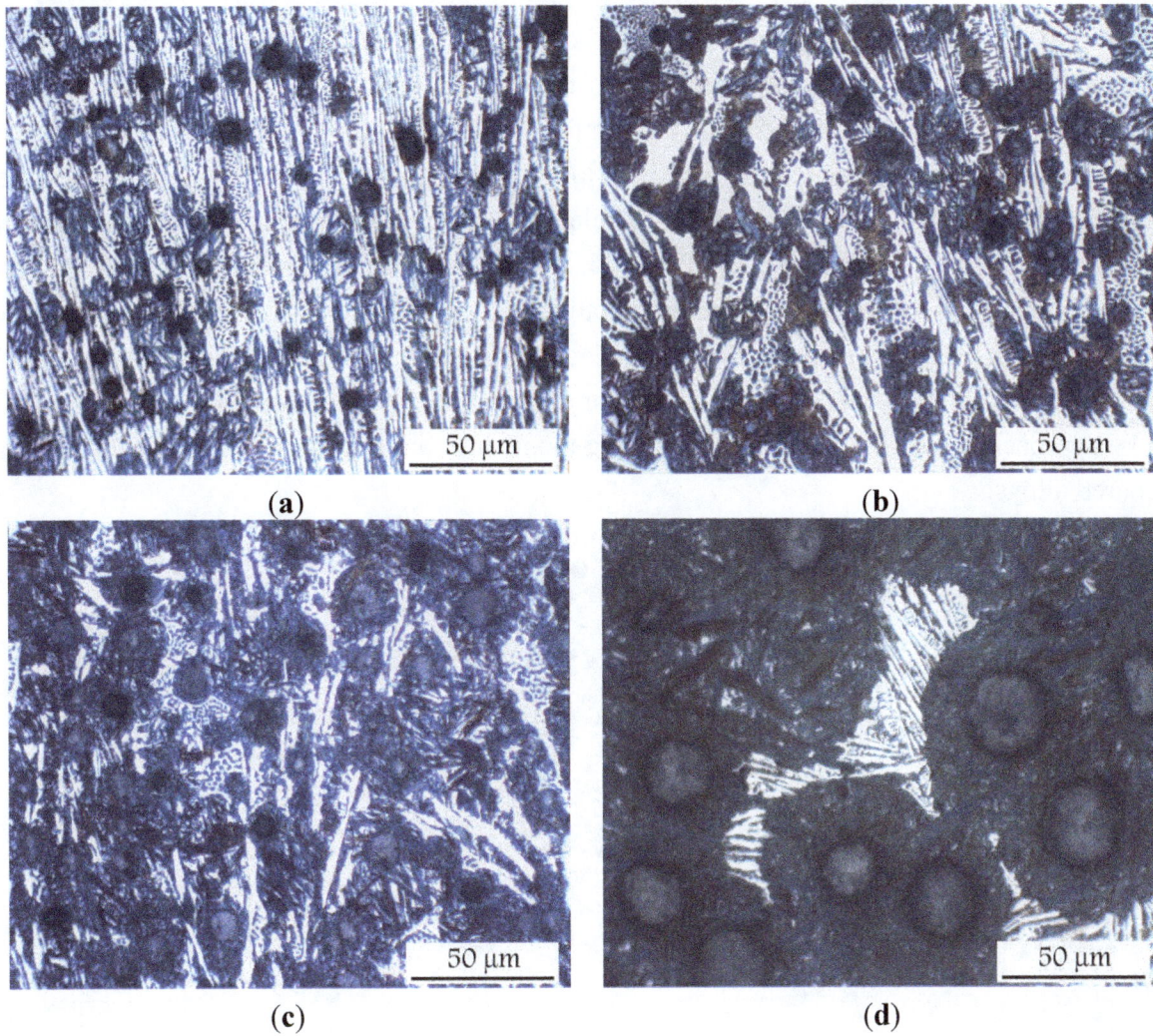

Figure 5. Solid fraction evolution of B1 samples. (**a**) B11 Sample; (**b**) B12 Sample; (**c**) B13 Sample; (**d**) B14 Sample.

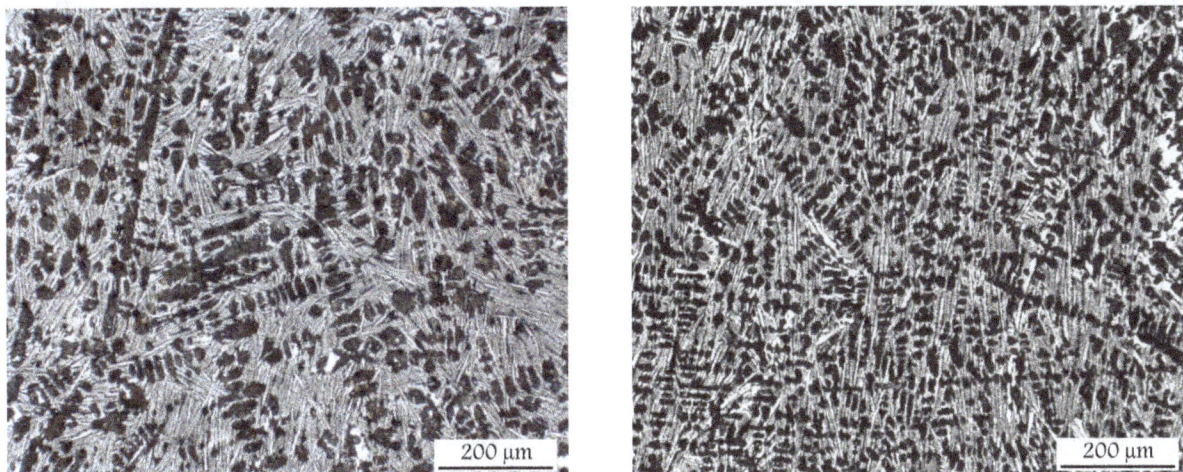

Figure 6. Coarser structure of A alloy in non-inoculated sample (**left**); compared to the inoculated one (**right**) for similar solid fraction (0.53 and 0.49 respectively).

Surprisingly f_{GA} values found during solidification of the non-inoculated samples are even higher than those detected on the inoculated ones. This observation can be a consequence of the different conception between the number of graphite nodules and their size, *i.e.*, a specific f_{GA} value can be obtained from both a big number of small nodules and/or from a low amount of big ones. Figure 7 shows an example of such a possibility for two different samples analyzed in the present work. However the graphite area fraction values at the end of solidification become slightly lower in the non-inoculated samples than in the inoculated ones. This result is related to lower nodule count, which enlarges the distances that carbon has to go through by diffusion. Thus a less overall final precipitated graphite amount and a more carbon saturated austenite are obtained for the same solidification time. This effect agrees with the common observation where non-inoculated samples show a higher tendency to shrinkage defect and carbide appearance than inoculated ones for a given composition. Further discussion on this issue will be approached in Chapter 3.3 when studying the nodule count evolution during solidification of the prepared alloys.

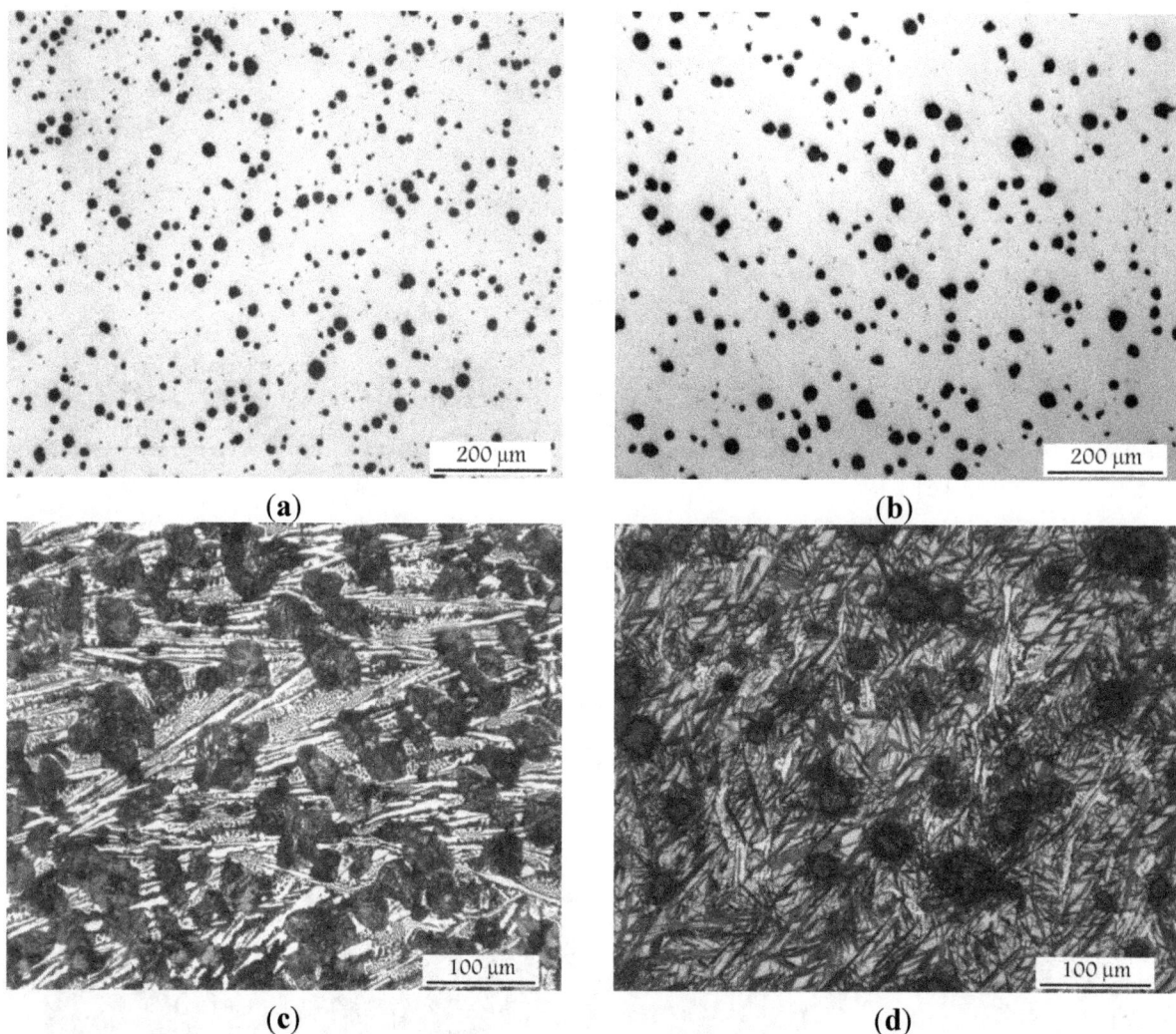

Figure 7. Different fields for different samples with the same f_{GA} value (6.1%). (**a**) Field #5 of A11 sample. $N = 527$ mm^{-2}; (**b**) Field #2 of A23 sample. $N = 218$ mm^{-2}; (**c**) Field #5 of A11 sample. $f_S = 0.44$; (**d**) Field #2 of A23 sample. $f_S = 0.90$.

Figure 8 illustrates the correlation between f_S and f_{GA} for the two inoculated ductile iron alloys. As expected [6] this last parameter increases when solidification progresses (represented here by the means of f_S). Scattering of data in the two graphs situated in the top level of Figure 8 is high likely due to the different quenching degrees present in each field. Therefore only the most sensitive fields, *i.e.*, #10 to #12, were selected for plotting the same correlation (bottom level of Figure 8). This fact is effective for the A alloy samples but it is only partially useful for the B alloy ones probably due to the high graphite nucleation ability present in the latter that stresses the quenching variability. This effect is also supported by the low scattering found in the A2 samples (all metallographic fields considered) when compared to the A1 samples.

Again composition effect on f_{GA} is observed here as B alloy samples lead to the highest values of this parameter (right side of Figure 8). Additionally this increment of the graphite area fraction appears during the whole solidification period and interestingly the primary graphite areas start to form at early f_S values for the A alloy case. Thus early graphite nucleation should be expected when increasing CE in ductile iron alloys which must be found when analyzing the nodule count evolution during solidification (see Chapter 3.3). Once again f_{GA} evolution during solidification of the non-inoculated samples can be estimated as comparable to the inoculated ones but with lower scattering.

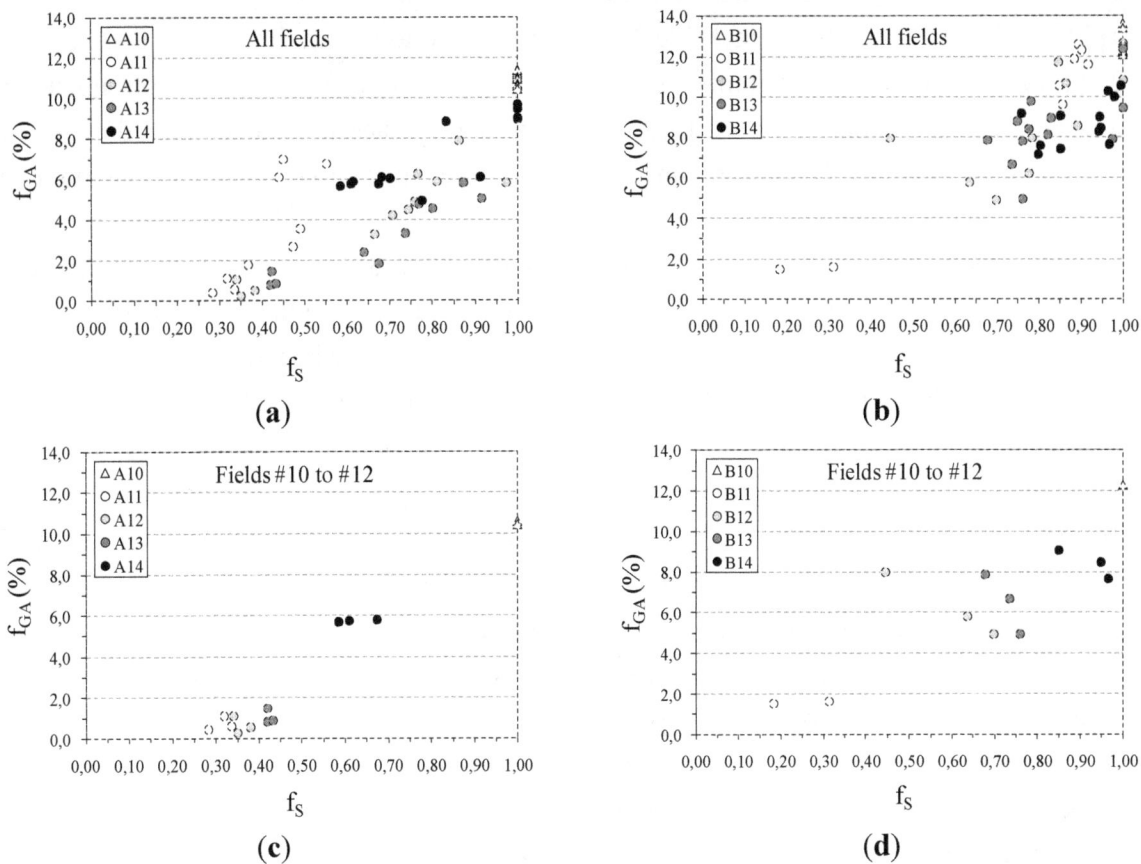

Figure 8. f_S *vs.* f_{GA} correlations for the two inoculated cast irons. (**a**) all fields for A1 series; (**b**) all fields for B1 series; (**c**) Fields #10 to #12 for A1 series; (**d**) Fields #10 to #12 for B1 series.

3.3. Nodule Count Evolution during Solidification

Basically, it seems to be worthwhile here to determine comparatively the constituents present in the most sensitive fields when quenching (#10 to #12) where f_S values are the lowest. Table 3 shows the f_S, f_{GA}, and N average values distributed by size. B21 is not considered due to pearlite appearance as mentioned above.

Table 3. Solid fraction and nodule count average values around the liquidus arrest in fields #10 to #12.

Sample	f_S	f_{GA} (%)	$N < 0.015$ mm (mm^{-2})
A11	0.31	0.7	146
A21	0.33	0.8	163
B11	0.25	1.6	344

As expected in this early step of solidification both intermediate- and big-sized graphite nodules were not found. Thus the f_{GA} values included in Table 3 only correspond to the small range of nodules. The comparative analysis of data shows the composition effect on the liquidus reaction, *i.e.*, both f_{GA} and N are much higher for the B11 sample than for the A11 one (both alloys are inoculated) even though f_S is higher in the last case than in the former one. Thus graphite nucleation is strongly favored by increasing CE.

However a comparison between A11 and A21 samples leads to the idea that the inoculation effect is not the expected one at the beginning of the solidification. Data obtained from the most sensitive part of these two samples indicate that quite similar f_{GA} and N values are obtained when adding inoculant to the A alloy. This result is different to the effect found in a previous work [7] where a near eutectic ductile iron alloy increased the nodule count (from 40–67 mm^{-2}) at the beginning of the liquidus arrest ($f_S = 0.06$) when adding inoculant. A possible reason for this lack of agreement could be the quite lower Te_{min} value of such non inoculated alloy (1133.3 °C) in contrast to alloy A2 (1140.9 °C) as this parameter has been successfully related to the graphite nucleation potential of solidifying melts [2]. Unfortunately the B21 sample was discarded in this work so this item will now be a subject for further investigations.

Quenching variability on the different fields of samples was studied in the previous chapters so it is interesting to approach a similar analysis here. Apart from the graphite nucleation potential exhibited by melts, nodule count is affected by cooling conditions. Thus N parameter can be increased by accelerating the cooling process. However it will become very low if the cooling rate is too fast as in quenching experiments. In spite of admitting that some nucleation and growth may occur due to the quenching method, it must be taken into account that the time for this to happen is limited, so the final graphite amount is the sum of both effects. In any case, the obtained results should be considered as comparable in order to determine the effect of the chemical composition and inoculation process.

It was observed in this work that the distribution of nodule count strongly depends on nodule size. A comparison between the three different nodule sizes found in this work is illustrated in Figure 9 where the two inoculated ductile iron alloys only are included. Thus the central fields (#4 to #9) contain the highest small nodule count during solidification while the bottom part of the samples (fields #1 to #3) shows the highest amount of nodules 0.015–0.030 mm and 0.030–0.060 mm in size. Note also that N

values at the end of solidification are quite low for the smallest size (50–60 mm^{-2}) while they almost achieve the maximum value of the two other big sizes.

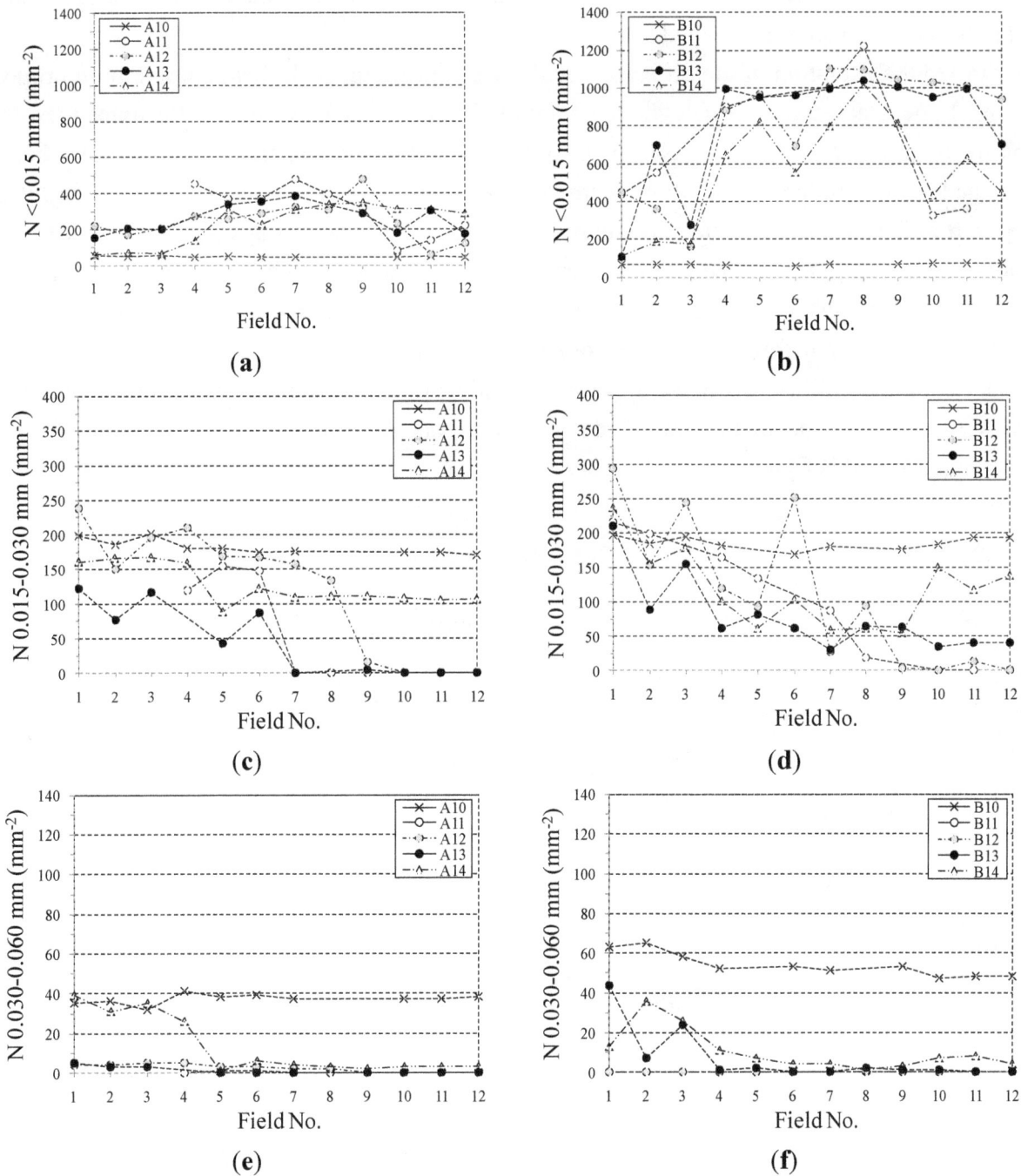

Figure 9. Nodule count distribution in all fields for the inoculated samples. (**a**) Nodule count for the smallest nodule size (<0.015 mm) of A1 series; (**b**) Nodule count for the smallest nodule size (<0.015 mm) of B1 series; (**c**) Nodule count for the nodule size 0.015–0.030 mm of A1 series; (**d**) Nodule count for the nodule size 0.015–0.030 mm of B1 series; (**e**) Nodule count for the nodule size 0.030–0.060 mm of A1 series; (**f**) Nodule count for the nodule size 0.030–0.060 mm of B1 series.

When avoiding inoculation on samples included in Figure 9, nodule count values during solidification become significantly lower than in the inoculated samples. However the affected fields and the shape of such distributions are quite similar to the inoculated cases. After completing the solidification period the non-inoculated samples also show a very low nodule count for the smallest nodule size (similar to the one detected for the inoculated samples) but N is significantly lower for the two other sizes when compared to the corresponding inoculated samples.

In order to see the evolution of the different nodule sizes during the solidification in a more proper way, it seems to be necessary to include the f_S parameter in this chapter. Figure 10 illustrates these N evolutions for both the A1 alloy and the B1 one (only the smallest nodule size distribution has been included in the last alloy for comparison). As it has been previously stated the smallest graphite nodules apparently start to form from the liquid alloy in the very early solidification steps ($f_S = 0.05–0.20$) as it was previously reported [6,8,12]. Then a maximum N value is reached in the 0.50–0.70 solid fraction range for the A1 samples while this range becomes 0.50–0.80 for the B1 ones. Finally N values decrease at the end of liquid-solid transformation. On the other hand the two big nodule sizes show an increasing evolution during the solidification process (see right side plots in Figure 10) until achieving the maximum N value at the end of this transformation. Note that N scales are different in two of the four graphs included in this figure.

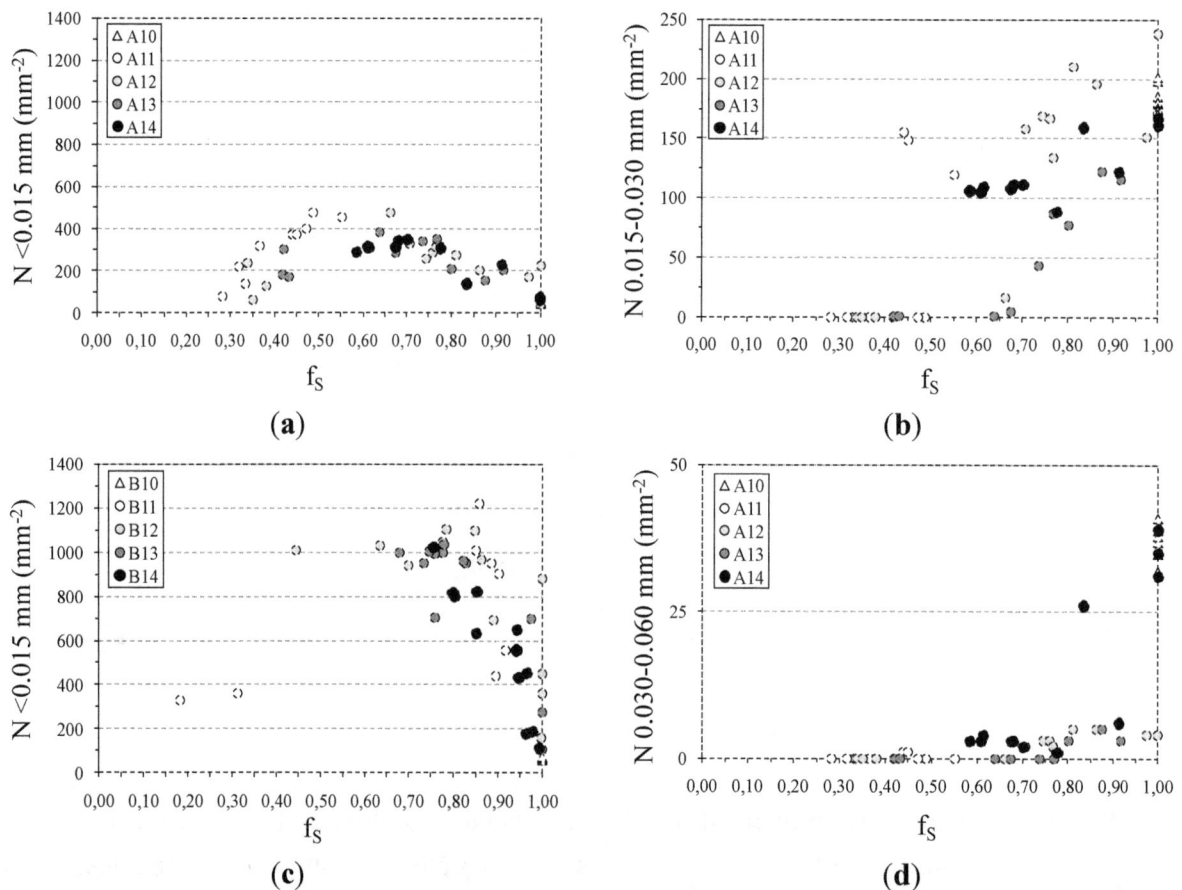

Figure 10. Nodule count evolution during solidification process (all fields are included). (**a**) Smallest nodule count evolution *vs.* fs for A1 series; (**b**) 0.030–0.060 mm nodule count evolution *vs.* fs for A1 series; (**c**) Smallest nodule count evolution *vs.* fs for B1 series; (**d**) 0.030–0.060 mm nodule count evolution *vs.* fs for A1 series.

When increasing CE in the alloy it is observed that all N evolution shapes are comparable to the A alloy case though the nodule count values strongly increase (see the two graphs on the left side of Figure 10). Note here, that the smallest nodule count also decreases (50–70 mm^{-2}) at the end of solidification, even for the B alloy samples.

Regarding the inoculation effect on N, no relevant changes on these evolution shapes due to the lack of inoculation have been detected in this study. On the other hand it was confirmed that comparatively minor increments on nodule count values are obtained due to inoculant addition over the whole range of solidification, *i.e.*, the composition effect becomes the most powerful one. However the weak but clear increase of N due to inoculation observed here can be more potent in those alloys that exhibit a low graphite nucleation potential before adding an inoculant, as already detected in a previous work [7]. The most important increment of N due to inoculation affects the graphite nodules 0.030–0.060 mm in size, *i.e.*, the biggest ones. These nodules are also called "primary nodules" because they nucleate at the very beginning of the solidification. Therefore it is possible to assume that inoculation promotes the early precipitation of graphite particles. This result should support the unusually high graphite nucleation potential of A2 alloy as being the main cause of the apparent negligible inoculation effect on A21 sample with respect to the A11 one (see Table 3).

Correlations between N and f_{GA} are similar to those obtained when using the solid fraction as the parameter for analyzing the solidification progression. However scattering of data is high for the first case (especially when analyzing hyper-eutectic samples). This fact should be related to the uncertain balance among f_{GA}, nodule count and the size of nodules as was mentioned in Chapter 3.2.

Figure 11 shows the evolution of total nodule count values obtained from the quenched samples and from the non-quenched one for all ductile irons prepared in this work. The downward evolutions found in the two B alloys are a consequence of the observed strong contribution coming from the small-sized graphite nodule evolution during solidification. On the contrary A alloys show much more stable tendencies though they can also be considered as descending ones.

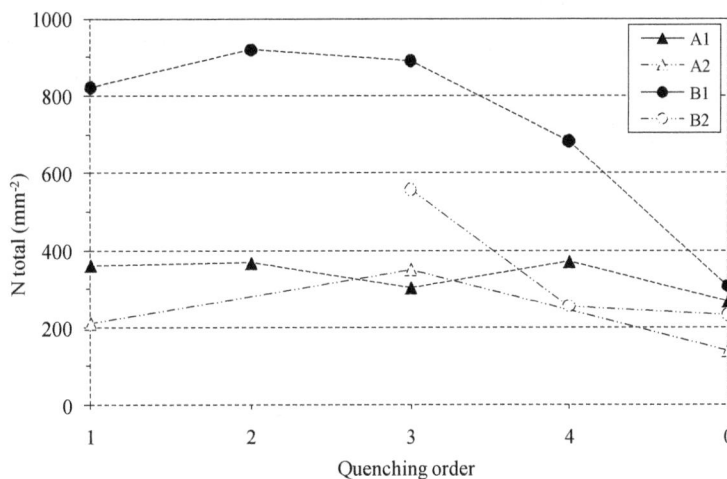

Figure 11. Total nodule count evolution with all quenched samples.

It has been already stated up to now that an increase of the graphite area fraction at the end of solidification is obtained when increasing the CE content of the alloy (see Figure 8). However this result has to be combined with two other observed phenomena: an important reduction of the smallest nodule count and an increase of the big-sized nodules. Two different explanations were considered in the present work to tackle this difficulty. One of them pre-supposes that the smallest-sized graphite nodules mainly found in the central areas of samples with maximum nodule count values, detected in $f_S = 0.50$–0.80, are formed due to the quenching process. Thus these observations must not be found when samples solidify under normal cooling conditions. It is assumed that some undercooling can be expected before freezing the resulting matrix structure due to quenching. Thus the alloy in contact with the quenching media should freeze almost immediately so very low extra undercooling would be expected on it. Thus the former solidified layer makes the heat release from the inner zones more difficult in such a case, slowing down the very high cooling rate expected from quenching, and slightly increasing undercooling before the quenching effect reaches these internal zones. Such phenomena would lead to obtaining sufficient time for small graphite particles to be precipitated. This effect would be more noticeable on increasing the distance from the quenching media and would explain the f_S and nodule count distributions showed in the top level of Figure 9. According to this explanation, the quenching methodology used in the present study would change the observed solidification development although, at the same time, it would mean that the conditions to form these phases already existed at this particular time.

Another way to explain the small-sized nodule count evolution during solidification is to assume that the quenching process is fully effective although some cooling variability is present in the whole of the samples. Thus a true precipitation of small graphite nodules would occur in the way shown in Figures 9 and 10. This fact means that some of the nodules do not achieve the required size that would allow further growth on the nodules, so they become less stable and disappear. According to this assumption, coarse graphite nodules would grow by carbon diffusion, also coming from these small graphite nodules which would be disappearing during the last period of the solidification process. Furthermore, it implies that not every nodule is formed at the very beginning of the solidification but rather nucleation development occurs during all the transitions. The high smallest nodule count values found in the center of the fields of many samples, with a solid fraction in the range 0.70–1.00, could be explained by means of this second postulation.

From a general point of view high CE contents would lead to formation of a huge amount of small-sized graphite nodules. Some of them nucleate and grow from the liquid by carbon atom buildup, meanwhile others grow probably by accepting carbon atoms which come from other nodules by diffusion. Thus these last nodules would progressively disappear and the total nodule count would decrease during solidification until achieving the final value once f_S becomes 1.00. Inoculation would show a similar effect to the one described for CE.

Unfortunately neither of these two possible explanations was verified in the present work and therefore will be approached in further investigations. However quenching experiments must be useful for comparing the different results obtained when varying processing parameters as CE and inoculation. Thus the downward trends found in all four alloys (Figure 11) contrast with some others previously reported in the literature for ductile irons [7]. In such a work FGE1i denotes a hyper-eutectic and inoculated alloy that showed a similar trend to the one found for B1 alloy (Figure 11). In the same study the nodule count evolutions for FGE1ni (hyper-eutectic and non-inoculated alloy), FGE2i, and FGE2ni (both near eutectic alloys which are inoculated and non-inoculated respectively) are, however, upward while the opposite trends are found for the B2, A1, and A2 alloys.

These different behaviors mean that another important aspect is acting on N evolution in addition to CE and inoculation. Once again the graphite nucleation potential of melts should play an important role in this way. In fact the Te_{min} values for FGE2i (1144.8 °C) and for FGE2ni (1133.3 °C) alloys are lower than the ones obtained for A1 and A2 (see Table 2). Thus a high nucleation potential would also increase the obtained number of graphite nodules at the beginning of solidification and tendencies like the ones plotted in Figure 11 become downward for a given CE and inoculation condition. The authors already correlated this aspect with N and the shrinkage tendency of ductile irons [2] although only final nodule count values were considered in this previous work.

4. Conclusions

The following conclusions can be drawn from the present work:

1. Small nodules rapidly form at the beginning of solidification. As the solidification advances medium-sized graphite nodules together with the maximum amount of the smallest nodules are achieved. When f_S is higher than 0.8, medium- and big-sized nodules become predominant while the amount of the smallest ones decreases until reaching a minimum value at the end of the solidification.
2. CE and inoculation promote the increase of graphite size during the solidification process. In the present study, CE was shown to be a more influential parameter than inoculation. Notwithstanding, the biggest graphite nodules are observed in low CE alloys.
3. Inoculation increases the final precipitated graphite content.
4. As CE increases, nodule count shows a tendency to decrease with f_S while low CE maintains a fairly similar nodule count during the whole solidification process in the present study. These tendencies were not found in a previous work already published by the authors. A comparative analysis of the thermal records in the two works shows that this disparity may be related to the graphite precipitation ability of the melt, *i.e.*, the amount of seeds where graphite can nucleate.

Acknowledgments

The authors acknowledge Basque Government for the funding received in its "Ikertu" program. The authors would like to thank Technology Quality Control, S.L.U. for all the collaborating efforts made in the experimental tests.

Author Contributions

All the authors contributed equally to the present work. A.N. designed and performed experiments and collected data; R.S. designed experiments and gave technical support and conceptual advice; J.S. designed experiments, analyzed data and wrote the paper; P.L. designed experiments, analyzed data and wrote the paper; All authors discussed the results and implications and commented on the manuscript at all stages.

Conflicts of Interest

The authors declare no conflict of interest.

References

1. Stefanescu, D.M. Solidification and modeling of cast iron—A short history of the defining moments. *Mater. Sci. Eng. A* **2005**, *413–414*, 322–333.

2. Larrañaga, P.; Gutiérrez, J.M.; Loizaga, A.; Sertucha, J.; Suárez, R. A computer-aided system for melt quality and shrinkage propensity evaluation based on the solidification process of ductile iron. *AFS Trans.* **2008**, *116*, 547–561.

3. Oldfield, W. A quantitative approach to casting solidification: Freezing of cast iron. *Trans. ASM* **1966**, *59*, 945–961.

4. Lora, R.; Diószegi, A.; Elmquist, L. Solidification study of gray cast iron in a resistance furnace. *Key Eng. Mater.* **2011**, *457*, 108–113.

5. Guo, X.; Stefanescu, D.M. Solid phase transformation in ductile iron. *AFS Trans.* **1997**, *105*, 533–543.

6. Mampaey, F. Quantitative description of solidification morphology of lamellar and spheroidal graphite cast iron. *AFS Trans.* **1999**, *107*, 425–432.

7. Larrañaga, P.; Sertucha, J.; Suárez, R. Análisis del proceso de solidificación en fundiciones grafíticas esferoidales. *Revista de Metalurgia* **2006**, *42*, 244–255.

8. Boeri, R.E. The Solidification of Ductile Cast Iron. Doctoral Thesis, Faculty of Graduate Studies, Department of Metals and Materials Engineering, The University of British Columbia, Vancouver, BC, Canada, November 1989.

9. Liu, Y.; Yang, S. Growth mode and modification of graphite in cast iron melt. *Acta Metall. Sinica* **1992**, *5*, 263–267.

10. Alonso, G.; Stefanescu, D.M.; Larrañaga, P.; Suárez, R. Understanding compact graphite iron solidification through interrupted solidification experiments. In Proceedings of the 10th International Symposium on the Science and Processing of Cast Iron (SPCI-10), Mar de Plata, Argentina, 10–13 November 2014.

11. Fidos, H. A study of the graphite morphology in nodular cast iron. *FWP J.* **1977**, *17*, 39–54.

12. Lacaze, J.; Castro, M.; Aichoun, N.; Lesoult, G. Influence de la vitesse de refroidissement sur la microstructure et la cinétique de solidification de fontes G.S.: Expérience et simulation numérique de solidification dirigée. *Mémoires et Etudes Scientifiques Revue de Métallurgie* **1989**, *2*, 85–97.

13. Larrañaga, P.; Sertucha, J. Estudio térmico y estructural del proceso de solidificación de fundiciones de hierro con grafito laminar. *Revista de Metalurgia* **2010**, *46*, 370–380.

12

Ultrafine-Grained Austenitic Stainless Steels X4CrNi18-12 and X8CrMnNi19-6-3 Produced by Accumulative Roll Bonding

Mathis Ruppert *, Lisa Patricia Freund, Thomas Wenzl, Heinz Werner Höppel and Mathias Göken

Department of Materials Science and Engineering, Institute I: General Materials Properties, Friedrich-Alexander Universität Erlangen-Nürnberg, Martensstr. 5, 91058 Erlangen, Germany; E-Mails: lisa.freund@fau.de (L.P.F.); thomas.wenzl@fau.de (T.W.); hwe.hoeppel@fau.de (H.W.H.); mathias.goeken@ww.uni-erlangen.de (M.G.)

* Author to whom correspondence should be addressed; E-Mail: mathis.ruppert@fau.de

Academic Editor: Hugo F. Lopez

Abstract: Austenitic stainless steels X4CrNi18-12 and X8CrMnNi19-6-3 were processed by accumulative roll bonding (ARB). Both materials show an extremely high yield strength of 1.25 GPa accompanied by a satisfactory elongation to failure of up to 14% and a positive strain rate sensitivity after two ARB cycles. The strain-hardening rate of the austenitic steels reveals a stabilization of the stress-strain behavior during tensile testing. Especially for X8CrMnNi19-6-3, which has an elevated manganese content of 6.7 wt.%, necking is prevented up to comparatively high plastic strains. Microstructural investigations showed that the microstructure is separated into ultrafine-grained channel like areas and relatively larger grains where pronounced nano-twinning and martensite formation is observed.

Keywords: accumulative roll bonding (ARB); austenitic steel; ultrafine-grained microstructure; strength; nano-twinning; strain rate sensitivity

1. Introduction

Accumulative roll bonding (ARB) [1] as a process of severe plastic deformation is one of the most effective methods for the production of bulk ultrafine-grained (UFG) materials with a median grain size smaller than 1 μm. The microstructural evolution during ARB is well described in literature, see for example [2–4] for details. The mechanical properties of those materials are frequently claimed to be favorable compared to their conventionally grained (CG) counterparts, as a good combination of high strength and satisfactory ductility can be achieved [5–7]. The enhanced ductility is often brought into connection with the enhanced strain rate sensitivity [3,6,8,9] of those materials. Moreover, strain rate sensitivity is strongly related to an increased fraction of high angle boundaries [4], as they can act as sources and sinks for dislocations [3]. Although there are numerous publications on ARB available, only a couple of them are dealing with ARB-processing of steel-sheets. Among those, the majority concerns interstitial free bcc steels (IF-steels) with a rather low content of alloying elements, see for example [1,10–12]. However, there is very little literature available about ARB of austenitic steels, which will be focused on in the following. Kitahara *et al.* [13] investigated the martensite transformation of an ultrafine-grained Fe-28.5at.-%Ni-alloy with single phase metastable austenite at room temperature. They performed the accumulative roll bonding process with sheets that were pre-heated at 500 °C for 600 s up to 5 cycles. However, each cycle was divided into two passes with a thickness reduction <50%. They achieved an ultrafine-grained microstructure with a mean grain size of 230 nm and a yield strength that was increased by a factor of 4.9 compared to the initial material. Moreover, they showed that the martensite transformation starting temperature decreases with the number of ARB cycles. Another study about ARB of austenitic steel was published by Jafarian *et al.* [14]. They investigated the microstructure and texture development in a Fe-24Ni-0.3C (wt.%) austenitic steel up to 6 ARB-cycles and subsequent annealing. The processing was performed at 600 °C. The texture was found to change from a copper orientation after 1 cycle towards a strong brass component after 6 cycles of ARB. Shen *et al.* [15] performed an accumulative cold rolling process with a thickness reduction of 17% each pass, using sheets of a commercial austenitic stainless steel 304SS, which was pre-heated up to 400 °C prior to each pass. Due to various subsequent heat-treatments, they could achieve materials with different grain sizes. The tensile samples reached a yield strength of up to 1.8 GPa, yet rather small elongation to failure of about 6%. Furthermore twinning and slip of partial dislocations were found to dominate plastic deformation in the ultrafine-grained state. Li *et al.* [16] processed sheets of an austenitic 36%Ni (mass-%) steel up to 6 cycles, with a pre-ARB heat-treatment at 500 °C. They found a rather small grain size of 150 nm and a high misorientation concerning the high angle boundaries.

As no information is available in literature, this work is focused on the strain rate sensitivity of ARB-processed ultrafine-grained austenitic stainless steels. Moreover, the effect of ultrafine grains and pronounced nano-twinning on the mechanical properties of commercially available and technically relevant alloys was addressed.

2. Experimental Section

Austenitic steels X4CrNi18-12 (1.4303) and X8CrMnNi19-6-3 (1.4376) were used as initial sheet materials for accumulative roll bonding. Due to the high Ni-content, the austenitic phase of X4CrNi18-12 is stabilized, which leads to a higher cold forming capability. This is desirable, as the material is severely strained during ARB-processing. The austenitic steel X8CrMnNi19-6-3 is a metastable one, which contains 6.7 wt.% of manganese. This induces twinning by plastic deformation, which might lead to satisfactory ductility. The material was delivered by Thyssen Krupp Nirosta GmbH and the chemical composition can be found in Table 1.

Table 1. Chemical compositions of the processed austenitic steels X4CrNi18-12 and X8CrMnNi19-6-3.

Alloy	Composition in wt.%								
	C	Si	Mn	Cr	Ti	P	S	Mo	Ni
X4CrNi18-12	0–0.06	-	0–2.0	17.0–19.0	-	-	-	-	11.0–13.0
X8CrMnNi19-6-3	0.025	0.46	6.76	17.43	0.001	0.029	0.0008	0.23	4.03

The sheets of X4CrNi18-12 and X8CrMnNi19-6-3 had an initial geometry of 25 × 150 × 1 mm (width × length × thickness) and were processed up to three and two cycles of ARB, respectively. Henceforth, the number of performed ARB cycles is denoted by N0-N3. Hereby one cycle of ARB equals a v. Mises equivalent strain of 0.8. During each cycle, the sheets were degreased in acetone and wire brushed with a rotating steel brush. Afterwards, the sheets were pre-heated in an electrical furnace for 5 min at 300 °C and finally roll bonded with a thickness reduction of 50%. After each cycle, edge cracking was cut off and the sheets were prepared accordingly to the scheme described above before the next cycle. In order to determine the mechanical properties of the processed sheets, both Vickers hardness measurements and tensile testing were performed. Therefore, a hardness measurement unit Leco V-100A and an Instron 4505 universal testing machine (Hegewald & Peschke MPT GmbH, Nossen, Germany) for uniaxial tensile testing were utilized. The hardness measurements were conducted at the sheet plane, the rolling plane and the transversal plane. Tensile testing was conducted in rolling direction at room temperature and at strain rates of $10^{-3}s^{-1}$, $10^{-4}s^{-1}$ and $10^{-5}s^{-1}$ in order to determine the strain rate sensitivity. Moreover, microstructural characterization was done using a Zeiss Cross Beam 1540 EsB (Carl Zeiss AG, Oberkochen, Germany) scanning electron microscope in backscattered electron contrast at an acceleration voltage of 11 kV and a working distance of 7–8 mm, as well as a Philips CM 200 transmission electron microscope operated at 200 kV (FEI, Hillsboro, OR, USA).

3. Results and Discussion

In Figure 1, the results of the hardness measurements in sheet plane, rolling plane and transversal plane are shown for X4CrNi18-12 and for X8CrMnNi19-6-3. Generally an increase of the hardness with the number of ARB cycles can be observed. The largest increase is found for the sheet plane, which is due to the high friction between the sheets and the rolls. This leads to a large shear strain at the surface regions of the sheet [17] and, therefore, to a 15% higher hardness compared to the other planes. Moreover, the hardness of both alloys increases severely during the first cycle of ARB, but only slightly during subsequent cycles. While the hardness of X4CrNi18-12 increases rather constantly between one and three cycles, the hardness of X8CrMnNi19-6-3 already appears to saturate after two ARB cycles.

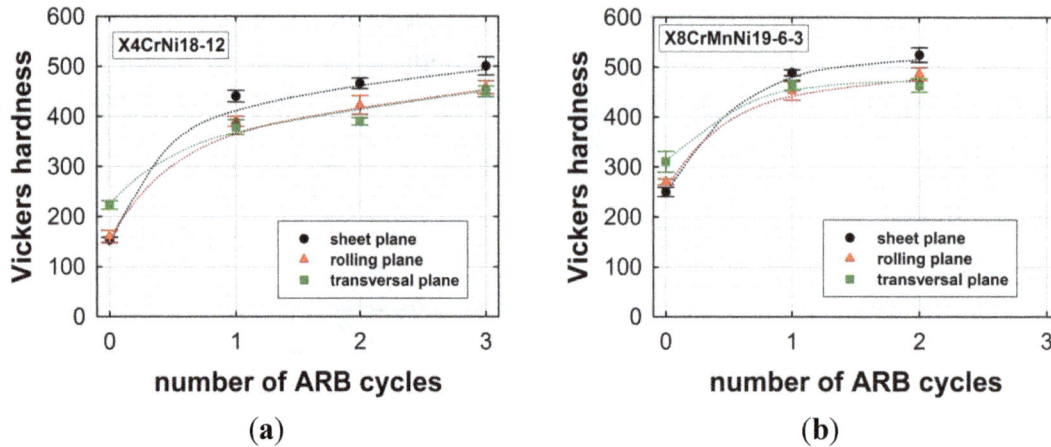

Figure 1. Vickers hardness measurements at the sheet, rolling and transversal planes of the samples after different cycles of ARB for (**a**) X4CrNi18-12 and (**b**) X8CrMnNi19-6-3.

The mechanical performance of both austenitic steels during tensile testing is shown in Figure 2. The ultimate tensile strength (UTS) and the yield strength (YS) increase significantly during the first ARB cycle and the general trend is similar to the hardness measurements. That is to say, both steels reach a YS of around 1.25 GPa after N2 and N3, respectively. The highest increase in strength is found after the first cycle. Therefore, X4CrNi18-12 shows an increase of the UTS by a factor of 1.9 and an increase of the YS by a factor of 3.8. During subsequent ARB-cycles, the strength is further increased, although the relative increase is smaller. The total increase in yield strength compared to the initial material equals a factor of 4.7. The uniform elongation is reduced from 50.4% to 1.2%, while the elongation to failure is reduced from 58% to 7.7%. Similar behavior is found for X8CrMnNi19-6-3, which also shows a strong increase in strength and a strong decrease in ductility after the first ARB cycle. Thereby, the YS increases by a factor of 2.4. The reduction in ductility can be attributed to a decreased hardening rate after severe plastic deformation. Nevertheless, both materials show an excellent combination of strength and ductility. The ARB-processed X8CrMnNi19-6-3 especially performs very well and reaches an UTS between 1.3 and 1.5 GPa, while it maintains an elongation to failure between 10 and 14%. Generally, the shape of the stress-strain curves changes after the first ARB cycle. While the N0 samples deform mainly uniformly during the tensile testing, the ARB-processed ones start necking at pretty low strains but show long post-necking deformation. This transition of tensile deformation behaviors in ultrafine-grained materials was discussed in detail for aluminum by Yu *et al.* [18]. Referring to this publication, tensile stress-strain curves can be categorized into four different characteristic types, in dependence of the grain size and the testing temperature. The N0-curves obtained in the present study a can be assigned to Type IV, which means that the curve shows continuous strain-hardening. This is typically observed in coarse-grained materials with a grain size larger than 4 μm. The stress-strain curves of the ARB processed

X4CrNi18-12 can be assigned to Type II, which means that the curves exhibit a distinct yielding peak followed by strain-softening. Type II behavior is observed for grain sizes between 0.4 µm and 1 µm. According to Yu *et al.* [18], the yielding peak is brought into connection with a lack of mobile dislocations, due to the large dislocation sink area provided by grain boundaries. This so-called yield-drop was also found for ultrafine-grained aluminum AA1100 and IF-steel [19], UFG Cu [20], UFG Ti [21] and also for cold-rolled high-manganese austenitic steel [22], which shows twinning induced plasticity. Concerning X8CrMnNi19-6-3, also a distinct yield point followed by strain softening can be observed. In Figure 2d, the true strain hardening rate is determined between the uniform elongation and the elongation to failure and plotted over the true plastic strain for both austenitic steels after N2. The curves were obtained by calculating the true stress strain curves (Figure 2c) from the engineering data and by determining the derivative of those curves. It has to be considered, that the actual cross-section of the tensile samples during tensile testing was not measured. However, the strain hardening curves can be compared qualitatively. It can be observed, that both materials initially show the same behavior up to 5% of plastic strain. That is to say, the strain hardening rate is reduced between the yield point and 3% of true plastic strain. Up to 5% the hardening rate for both alloys is increasing again. Afterwards, the strain-hardening rate of X4CrNi18-12 drastically decreases, while for X8CrMnNi19-6-3 the strain hardening rate increases until it suddenly drops as soon as the sample breaks. This is most likely due to failure because of ARB-related bonding defects. On the one hand, the stabilization of the stress-strain curves could be influenced by an enhanced strain rate sensitivity, which is typically observed for ultrafine-grained fcc metals. Thereby, thermally activated annihilation of dislocations is assumed to play a decisive roll and might lead to increased post-necking strains. On the other hand, both steels show a three-stage work hardening behavior, which is typical for materials with pronounced twinning activity and which is also found for high manganese TWIP (twinning induced plasticity) steels. It appears that the second stage is more distinct for X8CrMnNi19-6-3 compared to X4CrNi18-12. This might be due to the elevated manganese content, which leads to a higher twinning activity during tensile deformation that could stabilize the deformation behavior. The decreasing hardening rate after 5% of plastic strain, which is observed for X4CrNi18-12, might be attributed to a saturation in the amount of twinned grains. This saturation is not reached for X8CrMnNi19-6-3 during plastic straining. However, also martensitic transformation might contribute to the stabilization of the stress-strain curves. To gain more insight, the strain rate sensitivity and the twinning behavior of the materials were investigated.

The strain rate sensitivity (SRS) of the austenitic steels was determined from tensile testing experiments at strain rates of $10^{-3}s^{-1}$, $10^{-4}s^{-1}$ and $10^{-5}s^{-1}$. SRS has to be determined under constant microstructural conditions. Thus, the determination of the SRS from tensile tests is rather crucial, as a microstructural stable condition is hardly achieved. In order to minimize this problem, all stress-strain curves were analyzed at maximum stress, which appears to be a good compromise between the evolution of the microstructure and the limitations of the onset of necking. Therefore, true stress–true strain diagrams were plotted and the maximum stress was evaluated from the different curves. Afterwards, the determined values where plotted over the corresponding strain rate according to [23,24]. Figure 3 reveals that in the initial N0 condition X4CrNi18-12 shows a small positive SRS of around 0.007 at room temperature, while the SRS of X8CrMnNi19-6-3 is around zero. After the first and the second ARB cycle, the SRS of X4CrNi18-12 is slightly decreased to around zero.

However, it increases distinctly during the third ARB cycle to 0.017. A similar trend can be found for X8CrMnNi19-6-3, where the SRS remains around zero after the first cycle, but increases after the second cycle to 0.021. The SRS after two cycles in the case of X8CrMnNi19-6-3 and after three cycles in the case of X4CrNi18-12 are in the range of SRS found for other ultrafine-grained fcc materials in literature [6,9,25]. Pronounced strain rate sensitivity is frequently brought into connection with an increased fraction of high angle boundaries, which is typically found in ultrafine-grained fcc metals, see for example [4]. Those high angle grain boundaries can act as sources and sinks for dislocations [3], which are able to stabilize the stress-strain behavior and eventually lead to a higher elongation to failure [6,25]. Consequently, Figure 2 reveals a higher elongation to failure for X8CrMnNi19-6-3 compared to X4CrNi18-12, as the strain rate sensitivity was determined to be higher. However, the SRS usually increases with the number of ARB cycles. Nevertheless, the SRS of X4CrNi18-12 is decreased after the first ARB cycle, which is a rather untypical behavior. In order to clarify more about this point, the microstructures of both alloys were investigated in detail by means of SEM and TEM.

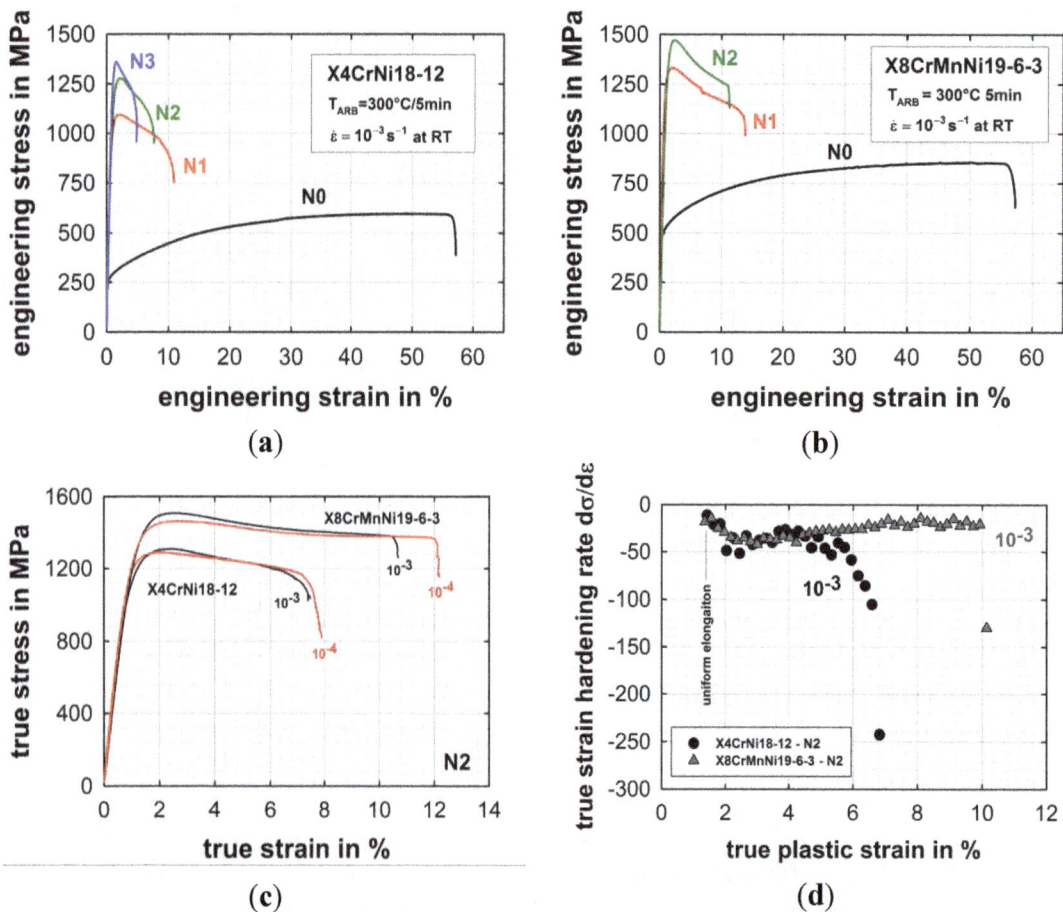

Figure 2. Engineering stress-strain curves at RT and a strain rate of $10^{-3}s^{-1}$ for (**a**) X4CrNi18-12 and (**b**) X8CrMnNi19-6-3. (**c**) Comparative plot of true stress-strain curves of X4CrNi18-12 and X8CrMnNi19-6-3 at strain rates of $10^{-3}s^{-1}$ and $10^{-4}s^{-1}$. Stress-strain curves for $10^{-5}s^{-1}$ are omitted for reasons of clarity. (**d**) True strain hardening rate for X4CrNi18-12 and X8CrMnNi19-6-3 during tensile testing at a strain rate of $10^{-3}s^{-1}$ after N2 cycles of ARB between the uniform elongation and the elongation to failure.

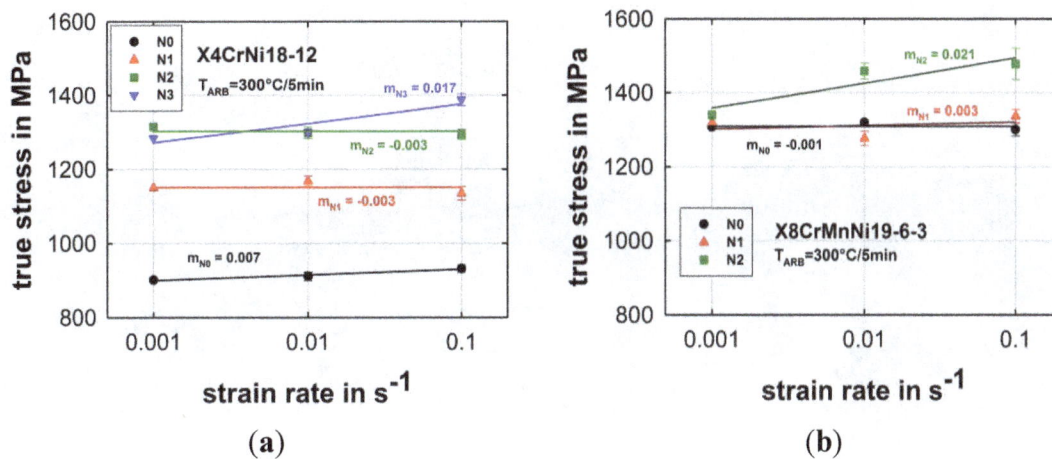

Figure 3. True maximum stress in MPa *vs.* strain rate for **(a)** X4CrNi18-12 and **(b)** X8CrMnNi19-6-3. The slope of the curves equals the strain rate sensitivity m_{Nx}.

Concerning the stabilized austenitic steel X4CrNi18-12 in the initial N0 condition, which can be seen in Figure 4a, the grain size of the equiaxial austenite grains is found to be in the range between 15–30 μm. Moreover, various annealing twins can be observed. The coarse grain structure leads to a high strain hardening capability and therefore to a high uniform elongation. Figure 4b reveals that after one ARB-cycle ($\varepsilon = 0.8$) the grains appear highly deformed and pronounced mechanical twinning and martensite formation are clearly visible. Generally, the microstructure shows various types of microstructural features. On the one hand, large grains with a grain size clearly above 1 μm. These grains are filled up by nano-twins and martensite needles. On the other hand, shear band like regions were found, indicating highly localized plastic deformation with very small grain sizes in the UFG regime. With increasing number of ARB cycles, the fraction of the UFG area is extended and the grain size is further reduced far below 1 μm, see Figure 4c. Moreover, the heterogeneity of the microstructure related to the above described features decreases. Nevertheless, some of the ultrafine grains still contain nano-sized twins in the range of 5–20 nm, which could be observed in the TEM, see Figure 4d. According to Mangonon *et al.* [26], martensitic transformations in Fe-Cr-Ni alloys proceed in the sequence γ (fcc)→ε (hcp)→α' (tetragonal-bcc). Thereby, α' is preferably built at intersections of two ε bands and near regions where ε bands are adjacent to twins or grain boundaries. Furthermore, α' first appears to be needle-shaped and becomes more lath-shaped during subsequent growth. Hereby, the growth of α' leads to a reduction of the ε phase. Shen *et al.* [27] found that both ε and twins act as an intermediate phase during the transformation form γ-austenite to α'-martensite. Above a certain strain level, they observed that the twin density is decreasing, while the martensite density is further increasing. Huang *et al.* [28] performed detailed TEM-investigations on the nucleation mechanism of deformation-induced martensite in an austenitic steel under ECAP-deformation. They observed similar behavior as described above; however they could also find α' nucleating at the intersection of two deformation twins and micro shear-bands. Accordingly, the diffraction patterns of those zones become rather complex (compare inset of Figure 5d).

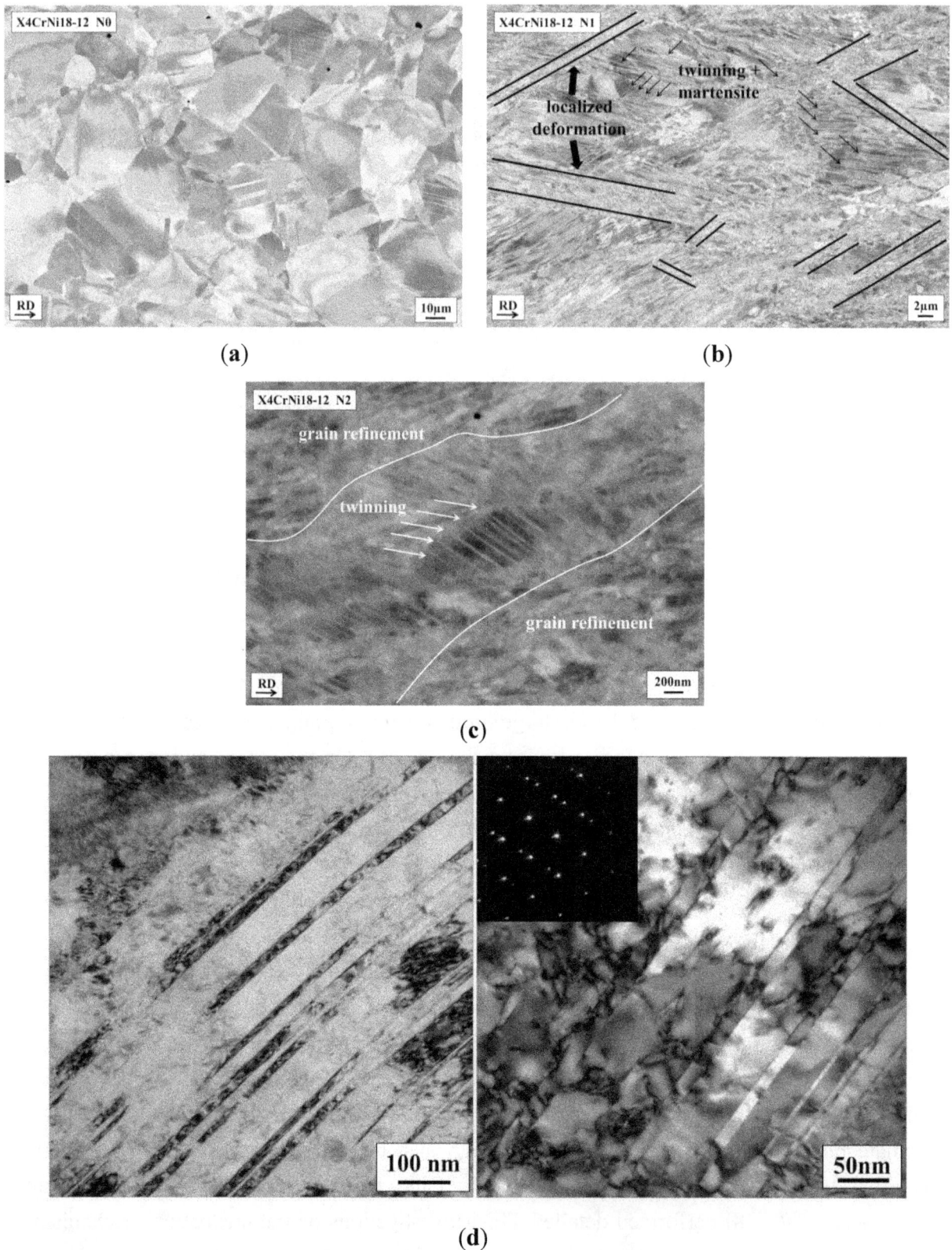

Figure 4. SEM micrographs of ARB-processed austenitic steel X4CrNi18-12 after (**a**) N0, (**b**) N1 and (**c**) N2. The arrows indicate areas where twins but also martensite were found. In (**d**) TEM captions of a highly twinned region with the corresponding diffraction pattern is shown after N2 cycles of ARB.

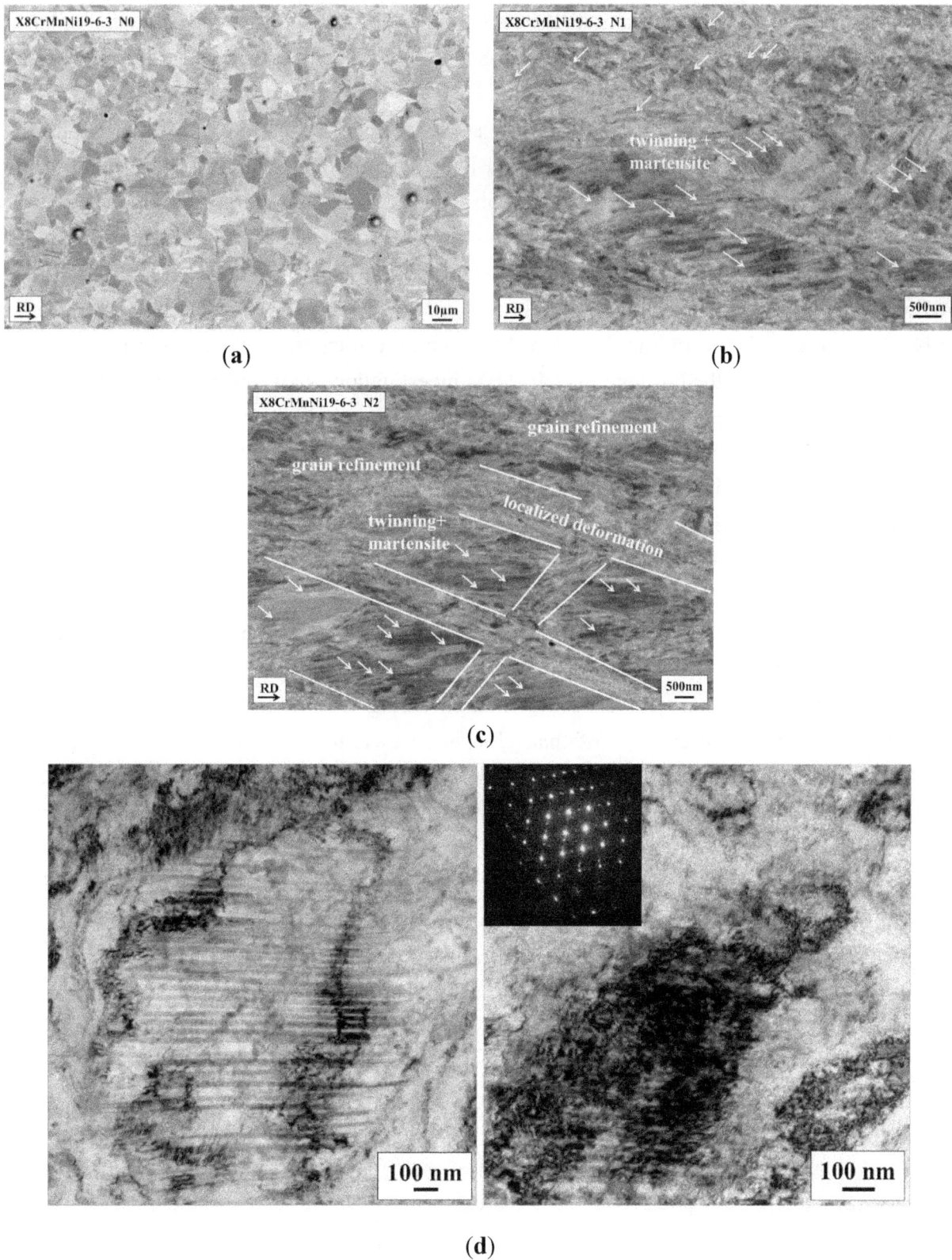

Figure 5. SEM micrographs of ARB-processed austenitic steel X8CrMnNi19-6-3 after (**a**) N0, (**b**) N1 and (**c**) N2 cycles. The arrows indicate areas where twins but also martensite was found. (**d**) TEM bright-field caption after N2 ARB-cycles and the corresponding diffraction pattern of this area.

Generally, the microstructural evolution of the austenitic steel X8CrMnNi19-6-3 is quite similar to that of X4CrNi18-12. That is to say, X8CrMnNi19-6-3 also shows equiaxed coarse grains in the initial N0 condition. The grain size is about 10 µm, which is slightly smaller than that of X4CrNi18-12 and there is also a smaller amount of annealing twins, see Figure 5a. After the first ARB cycle, the microstructure already appears highly deformed, and both twinning and martensite are found in the whole microstructure, which can be seen in Figure 5b. Furthermore, some larger grains in the range between 0.5 and 1.5 µm are found, which are rather randomly surrounded by very small grains. After the second ARB cycle, oval-shaped areas with a diameter of 2–5 µm are forming, see Figure 5c. Those areas are divided by zones of localized plastic deformation and very small grain sizes, which appear to be shear bands oriented 45° to rolling direction. Moreover, the fraction of ultrafine-grained areas is clearly increased compared to the N1 condition. In TEM investigations γ-austenite, martensitic areas as well as twinning were found. Accordingly, in [28], the observed diffraction patterns were quite complex. In Figure 5d the diffraction pattern of such an area is representatively shown. Although indications for twinning and martensite could be found, the heavily deformed structures did not allow clear identification and separation of the different phases.

After the first ARB cycle, the microstructure of both alloys is dominated by larger grains containing mechanical twins and martensitic needles. As the fraction of ultrafine grains is rather small, twinning and martensitic transformation can be regarded as the dominating deformation mechanisms. Thus, the highly deformed microstructure contributes to the pronounced increase in strength. Twinning is also assumed to significantly contribute to plastic deformation during tensile testing, which stabilizes the stress-strain behavior. As a consequence of that, elongation to failure becomes relatively high, see Figure 2. Because of pronounced twinning after one ARB cycle and reduced interaction of dislocations with grain boundaries the strain rate sensitivity of both austenitic steels is approaching zero after the first ARB cycle, as shown in Figure 3. This appears to be in contrast to observations by Lu et al. [29], who report a positive effect of twins on the SRS in UFG Cu. Nevertheless, both the grain size and the twin thickness are in a much smaller regime in that study. Moreover. martensite formation found in this work might affect the SRS. As the fraction of ultrafine-grained regions is increasing again during subsequent ARB cycles and the twinned areas are reducing, the dislocation grain boundary interaction is enhanced and the strain rate sensitivity is also increasing again. This leads to a positive strain rate sensitivity, which is usually observed in many fcc metals. Mechanisms based on the thermally activated interaction of dislocations with grain boundaries, which might provide a good explanation for the observed increase in SRS, are given by Blum et al. [3] in terms of thermally activated annihilation of dislocations at grain boundaries or by Kato et al. [30] in terms of thermally activated dislocation depinning at grain boundaries. Moreover, grain boundary sliding cannot be completely neglected and might also contribute to the enhanced SRS to a certain extent. However, when ductility is regarded, the situation becomes more complex: On the one hand, with increasing number of ARB cycles, decreasing elongation to failure is observed as the twinning capability is further decreased and the fraction of martensitic areas becomes more pronounced. On the other hand, the SRS, which is known to positively affect ductility, see Figure 2c, is increased for both alloys. Consequently, the elongation to failure is still rather satisfactory. Therefore, it is assumed that the increased strain rate sensitivity, martensitic transformation as well as twinning appear to contribute to the good ductility during tensile testing.

4. Conclusions

In the present study, commercially available austenitic steels X4CrNi18-12 and X8CrMnNi19-6-3 were processed by accumulative roll bonding at elevated temperatures. For both alloys, pronounced twinning and needles like martensite were found after the first ARB cycle. An increase in yield strength by a factor of 3.9 and 2.4, respectively, was found. During subsequent ARB cycles, the fraction of ultrafine-grained microstructure was clearly increased and a very high yield strength of 1.25 GPa was achieved. For X8CrMnNi19-6-3, the high strength was combined with a satisfactory ductility of more than 10%. The ultrafine-grained regions appeared in channels, dividing coarser areas where pronounced twinning with twin size of 5–20 nm was found. Typical for ultrafine-grained fcc metals, positive strain rate sensitivity between 0.017 and 0.021 was revealed for both alloys, as soon as the fraction of ultrafine-grained regimes was increased. Strain rate sensitivity, twinning and martensitic transformation appear to contribute to the satisfactory ductility.

Acknowledgments

The authors would like to thank the German Research Council (DFG) for their financial support and the Cluster of Excellence "Engineering of Advanced Materials", which is funded within the framework of its "Excellence Initiative".

Author Contributions

Mathis Ruppert and Thomas Wenzl performed the mechanical characterization and the SEM investigations. Lisa Patricia Freund prepared the TEM images. Heinz Werner Höppel and Mathias Göken supervised the work and discussed the results with the other authors.

Conflicts of Interest

The authors declare no conflict of interest.

References

1. Saito, Y.; Tsuji, N.; Utsunomiya, H.; Sakai, T.; Hong, R.G. Ultra-fine grained bulk aluminum produced by accumulative roll-bonding (ARB) process. *Scr. Mater.* **1998**, *39*, 1221–1227.

2. Huang, X.; Tsuji, N.; Hansen, N.; Minamino, Y. Microstructural evolution during accumulative roll-bonding of commercial pure aluminum. *Mater. Sci. Eng. A* **2003**, *340*, 265–271.

3. Blum, W.; Zeng, X.H. A simple dislocation model of deformation resistance of ultrafine-grained materials explaining Hall-Petch strengthening and enhanced strain rate sensitivity. *Acta Mater.* **2009**, *57*, 1966–1974.

4. Hughes, D.A.; Hansen, N. High angle boundaries formed by grain subdivision mechanisms. *Acta Mater.* **1997**, *45*, 3871–3886.

5. Saito, Y.; Utsunomiya, H.; Tsuji, N.; Sakai, T. Novel ultra-high straining process for bulk materials—Development of the accumulative roll bonding (ARB) process. *Acta Mater.* **1999**, *47*, 579–583.

6. Höppel, H.W.; May, J.; Göken, M. Enhanced strength and ductility in ultrafine-grained aluminium produced by accumulative roll bonding. *Adv. Eng. Mater.* **2004**, *6*, 781–784.

7. Valiev, R.Z.; Alexandrov, I.V.; Zhu, Y.T.; Lowe, T.C. Paradox of strength and ductility in metals processed by severe plastic deformation. *J. Mater. Res.* **2002**, *17*, 5–8.

8. Wei, Q. Strain rate effects in the ultrafine grain and nanocrystalline regimes-influence on some responses. *J. Mater. Sci.* **2007**, *42*, 1709–1727.

9. May, J.; Höppel, H.W.; Göken, M. Strain rate sensitivity of ultrafine-grained aluminium processed by severe plastic deformation. *Scr. Mater.* **2005**, *53*, 189–194.

10. Tsuji, N.; Ueji, Y.; Minamino, Y. Nanoscale crystallographic analysis of ultrafine grained IF steel fabricated by ARB process. *Scr. Mater.* **2002**, *47*, 69–76.

11. Kamikawa, N.; Sakai, T.; Tsuji, N. Effect of redundant shear strain on microstructure and texture evolution during accumulative roll-bonding in ultralow carbon IF steel. *Acta Mater.* **2007**, *55*, 5873–5888.

12. Lee, S.-H.; Saito, Y.; Park, K.-T.; Shin, D.H. Microstructures and mechanical properties of ultra low carbon IF steel processed by accumulative roll bonding process. *Mater. Trans.* **2002**, *43*, 2320–2325.

13. Kitahara, H.; Tsuji, N.; Minamino, Y. Martensite transformation from ultrafine grained austenite in Fe-28.5 at. % Ni. *Mater. Sci. Eng. A* **2006**, *438–440*, 233–236.

14. Jafarian, H.; Eivani, A. Texture development and microstructure eolution in metastable austenitic steel processed by accumulative roll bonding and subsequent annealing. *J. Mater. Sci.* **2014**, *49*, 6570–6578.

15. Shen, Y.F.; Zhao, X.M.; Sun, X.; Wang, Y.D.; Zuo, L. Ultrahigh strength of ultrafine grained austenitic stainless steel induced by accumulative rolling and annealing. *Scr. Mater.* **2014**, doi:10.1016/j.scriptamat.2014.05.001.

16. Li, B.; Tsuji, N.; Minamino, Y. Microstructural evolution in 36%Ni austenitic steel during the ARB process. *Mater. Sci. Forum* **2006**, *512*, 73–78.

17. Lee, S.-H.; Saito, Y.; Tsuji, N.; Utsunomiya, H.; Sakai, T. Role of shear strain in ultragrain refinement by accumulative roll-bonding (ARB) process. *Scr. Mater.* **2002**, *46*, 281–285.

18. Yu, C.Y.; Kao, P.W.; Chang, C.P. Transition of tensile deformation behaviors in ultrafine-grained aluminum. *Acta Mater.* **2005**, *53*, 4019–4028.

19. Tsuji, N.; Ito, Y.; Saito, Y.; Minamino, Y. Strength and ductility of ultrafine grained aluminum and iron produced by ARB and annealing. *Scr. Mater.* **2002**, *47*, 893–899.

20. An, X.H.; Wu, S.D.; Zhang, Z.F.; Figueiredo, R.B.; Gao, N.; Langdon, T.G. Enhanced strength-ductility synergy in nanostructured Cu and Cu-Al alloys processed by high-pressure torsion and subsequent annealing. *Scr. Mater.* **2012**, *66*, 227–230.

21. Li, Z.; Fu, L.; Fu, B.; Shan, A. Yield point elongation in fine-grained titanium. *Mater. Lett.* **2013**, *96*, 1–4.

22. Saha, R.; Ueji, R.; Tsuji, N. Fully recrystallized nanostructure fabricated without severe plastic deformation in high-Mn austenitic steel. *Scr. Mater.* **2013**, *68*, 813–816.

23. Hart, E.W. Theory of the tensile test. *Acta Metall.* **1967**, *15*, 351–355.

24. Ghosh, A.K. On the measurement of strain-rate sensitivity for deformation mechanism in conventional and ultra-fine grain alloys. *Mater. Sci. Eng. A* **2007**, *463*, 36–40.

25. Höppel, H.W.; May, J.; Eisenlohr, P.; Göken, M. Strain-rate sensitivity of ultrafine-grained materials. *Zeitschrift für Metallkunde* **2005**, *96*, 566–571.

26. Mangonon, P.; Thomas, G. The martensite phases in 304 stainless steel. *Metall. Trans.* **1970**, *1*, 1577–1586.

27. Shen, Y.F.; Li, X.X.; Sun, X.; Wang, Y.D.; Zuo, L. Twinning and martensite in a 304 austenitic stainless steel. *Mater. Sci. Eng. A* **2012**, *552*, 514–522.

28. Huang, C.X.; Yang, G.; Gao, Y.L.; Wu, S.D.; Li, S.X. Investigation on the nucleation mechanism of deformation-induced martensite in an austenitic stainless steel under severe plastic deformation. *J. Mater. Res.* **2007**, *22*, 724–729.

29. Lu, K.; Lu, L.; Suresh, S. Strengthening materials by engineering coherent internal boundaries at the nanoscale. *Science* **2009**, *324*, 349–352.

30. Kato, M. Thermally activated dislocation depinning at a grain boundary in nanocrystalline and ultrafine-grained materials. *Mater. Sci. Eng. A* **2009**, *516*, 276–282.

Permissions

All chapters in this book were first published in Metals, by MDPI; hereby published with permission under the Creative Commons Attribution License or equivalent. Every chapter published in this book has been scrutinized by our experts. Their significance has been extensively debated. The topics covered herein carry significant findings which will fuel the growth of the discipline. They may even be implemented as practical applications or may be referred to as a beginning point for another development.

The contributors of this book come from diverse backgrounds, making this book a truly international effort. This book will bring forth new frontiers with its revolutionizing research information and detailed analysis of the nascent developments around the world.

We would like to thank all the contributing authors for lending their expertise to make the book truly unique. They have played a crucial role in the development of this book. Without their invaluable contributions this book wouldn't have been possible. They have made vital efforts to compile up to date information on the varied aspects of this subject to make this book a valuable addition to the collection of many professionals and students.

This book was conceptualized with the vision of imparting up-to-date information and advanced data in this field. To ensure the same, a matchless editorial board was set up. Every individual on the board went through rigorous rounds of assessment to prove their worth. After which they invested a large part of their time researching and compiling the most relevant data for our readers.

The editorial board has been involved in producing this book since its inception. They have spent rigorous hours researching and exploring the diverse topics which have resulted in the successful publishing of this book. They have passed on their knowledge of decades through this book. To expedite this challenging task, the publisher supported the team at every step. A small team of assistant editors was also appointed to further simplify the editing procedure and attain best results for the readers.

Apart from the editorial board, the designing team has also invested a significant amount of their time in understanding the subject and creating the most relevant covers. They scrutinized every image to scout for the most suitable representation of the subject and create an appropriate cover for the book.

The publishing team has been an ardent support to the editorial, designing and production team. Their endless efforts to recruit the best for this project, has resulted in the accomplishment of this book. They are a veteran in the field of academics and their pool of knowledge is as vast as their experience in printing. Their expertise and guidance has proved useful at every step. Their uncompromising quality standards have made this book an exceptional effort. Their encouragement from time to time has been an inspiration for everyone.

The publisher and the editorial board hope that this book will prove to be a valuable piece of knowledge for researchers, students, practitioners and scholars across the globe.

List of Contributors

Aleksander Lisiecki
Welding Department, Silesian University of Technology, Konarskiego 18A, 44-100 Gliwice, Poland

Chul Kyu Jin
Graduate School of Mechanical and Precision Engineering, Pusan National University, San 30 Chang Jun-dong, Geum Jung-Gu, Busan 609-735, Korea

Chang Hyun Jang
Graduate School of Mechanical and Precision Engineering, Pusan National University, San 30 Chang Jun-dong, Geum Jung-Gu, Busan 609-735, Korea

Chung Gil Kang
School of Mechanical Engineering, Pusan National University, San 30 Chang Jun-dong, Geum Jung-Gu, Busan 609-735, Korea

Edward Fraś
Foundry Institute, AGH University of Science and Technology, Reymonta 23, Cracow 30-059, Poland

Hugo F. Lopez
Department of Materials Science and Engineering, University of Wisconsin Milwaukee, 3200 N. Cramer Street, Milwaukee, WI 53211, USA

Magdalena Kawalec
Foundry Institute, AGH University of Science and Technology, Reymonta 23, Cracow 30-059, Poland

Marcin Gorny
Foundry Institute, AGH University of Science and Technology, Reymonta 23, Cracow 30-059, Poland

Triratna Shrestha
Department of Chemical and Materials Engineering, University of Idaho, Moscow, ID 83844-3024, USA

Sultan F. Alsagabi
Department of Chemical and Materials Engineering, University of Idaho, Moscow, ID 83844-3024, USA

Indrajit Charit
Department of Chemical and Materials Engineering, University of Idaho, Moscow, ID 83844-3024, USA

Gabriel P. Potirniche
Department of Mechanical Engineering, University of Idaho, Moscow, ID 83844-0902, USA

Michael V. Glazoff
Energy Systems Integration, Idaho National Laboratory, Idaho Falls, ID 83415-3710, USA

Alexandre Deltell
Department of Physics, University of Girona, Campus Montilivi s/n, 17071 Girona, Spain

Lluisa Escoda
Department of Physics, University of Girona, Campus Montilivi s/n, 17071 Girona, Spain

Joan Saurina
Department of Physics, University of Girona, Campus Montilivi s/n, 17071 Girona, Spain

Joan Josep Suñol
Department of Physics, University of Girona, Campus Montilivi s/n, 17071 Girona, Spain

Seul-Kee Kim
Department of Naval Architecture and Ocean Engineering, Pusan National University, Busan 609-735, Korea

Chi-Seung Lee
Department of Naval Architecture and Ocean Engineering, Pusan National University, Busan 609-735, Korea

Jeong-Hyeon Kim
Department of Naval Architecture and Ocean Engineering, Pusan National University, Busan 609-735, Korea

Myung-Hyun Kim
Department of Naval Architecture and Ocean Engineering, Pusan National University, Busan 609-735, Korea

Byeong-Jae Noh
Hyundai Heavy Industries, Ulsan 682-792, Korea

Toshyuki Matsumoto
Research Institute of ClassNK, Tokyo 102-8567, Japan

Jae-Myung Lee
Department of Naval Architecture and Ocean Engineering, Pusan National University, Busan 609-735, Korea

Gang Chen
Department of Chemical and Materials Engineering, the University of Auckland, Private Bag 92019, Auckland State Key Laboratory of Porous Metal Materials, Northwest Institute for Nonferrous Metal Research, Xi'an 1142, New Zealand 710016, Shaanxi, China

Klaus-Dieter Liss
Australian Nuclear Science and Technology Organisation, New Illawarra Road, Lucas Heights, NSW 2234, Australia Quantum Beam Science Directorate, Japan Atomic Energy Agency, 2-4 Shirakata-Shirane Tokai-mura, Naka-gun, Ibaraki-ken 319-1195, Japan

Peng Cao
Department of Chemical and Materials Engineering, the University of Auckland, Private Bag 92019, Auckland 1142, New Zealand

Riadh Slimi
Unité de Technologies Chimique et Biologiques pour la Santé (UTCBS), UMR 8258 CNRS, U 1022 Inserm, Ecole Nationale Supérieure de Chimie de Paris (Chimie ParisTech)/PSL Research University, 11 rue Pierre et Marie Curie, 75005 Paris, France

Christian Girard
Unité de Technologies Chimique et Biologiques pour la Santé (UTCBS), UMR 8258 CNRS, U 1022 Inserm, Ecole Nationale Supérieure de Chimie de Paris (Chimie ParisTech)/PSL Research University, 11 rue Pierre et Marie Curie, 75005 Paris, France

Tae Won Kim
Sekjin E&T Co., Ltd., Sinpyeong-dong 642-19, Saha-Gu, Busan 604-030, Korea

Chul Kyu Jin
Graduate School of Mechanical and Precision Engineering, Pusan National University, San 30 Chang Jun-dong, Geum Jung-Gu, Busan 609-735, Korea

Ill Kab Jeong
Sekjin E&T Co., Ltd., Sinpyeong-dong 642-19, Saha-Gu, Busan 604-030, Korea

Sang Sub Lim
Sekjin E&T Co., Ltd., Sinpyeong-dong 642-19, Saha-Gu, Busan 604-030, Korea

Jea Chul Mun
Graduate School of Mechanical and Precision Engineering, Pusan National University, San 30 Chang Jun-dong, Geum Jung-Gu, Busan 609-735, Korea

Chung Gil Kang
School of Mechanical Engineering, Pusan National University, San 30 Chang Jun-dong, Geum Jung-Gu, Busan 609-735, Korea

Hyung Yoon Seo
Department of Computer Science and Engineering, Pusan National University, San 30 Chang Jun-dong, Geum Jung-Gu, Busan 609-735, Korea

Jong Deok Kim
Department of Computer Science and Engineering, Pusan National University, San 30 Chang Jun-dong, Geum Jung-Gu, Busan 609-735, Korea

Alex Escobar
Departamento de Ingeniería Metalúrgica, Universidad de Santiago de Chile (USACH), Av. Bernardo O'Higgins 3363, Santiago, Chile

Diego Celentano
Departamento de Ingeniería Mecánica y Metalúrgica, Centro de Investigación en Nanotecnología y Materiales Avanzados (CIEN-UC), Pontificia Universidad Católica de Chile (PUC), Av. Vicuña Mackenna 4860, Santiago, Chile

Marcela Cruchaga
Departamento de Ingeniería Mecánica, Universidad de Santiago de Chile (USACH), Av. Bernardo O'Higgins 3363, Santiago, Chile

Bernd Schulz
Departamento de Ingeniería Metalúrgica, Universidad de Santiago de Chile (USACH), Av. Bernardo O'Higgins 3363, Santiago, Chile

Amaia Natxiondo
Veigalan Estudio 2010, Durango E-48200, Spain

Ramón Suárez
Veigalan Estudio 2010, Durango E-48200, Spain

Jon Sertucha
Engineering, R&D and Metallurgical Processes Department, IK4-Azterlan, Durango E-48200, Spain

Pello Larrañaga
Engineering, R&D and Metallurgical Processes Department, IK4-Azterlan, Durango E-48200, Spain

Mathis Ruppert
Department of Materials Science and Engineering, Institute I: General Materials Properties, Friedrich-Alexander Universität Erlangen-Nürnberg, Martensstr. 5, 91058 Erlangen, Germany

Lisa Patricia Freund
Department of Materials Science and Engineering, Institute I: General Materials Properties, Friedrich-Alexander Universität Erlangen-Nürnberg, Martensstr. 5, 91058 Erlangen, Germany

Thomas Wenzl
Department of Materials Science and Engineering, Institute I: General Materials Properties, Friedrich-Alexander Universität Erlangen-Nürnberg, Martensstr. 5, 91058 Erlangen, Germany

Heinz Werner Höppel
Department of Materials Science and Engineering, Institute I: General Materials Properties, Friedrich-Alexander Universität Erlangen-Nürnberg, Martensstr. 5, 91058 Erlangen, Germany

Mathias Göken
Department of Materials Science and Engineering, Institute I: General Materials Properties, Friedrich-Alexander Universität Erlangen-Nürnberg, Martensstr. 5, 91058 Erlangen, Germany